ELECTRICAL AND ELECTRONICS DRAWING

FOURTH EDITION

Charles J. Baer

Professor of Engineering
The University of Kansas

John R. Ottaway

Engineer
Western Electric Company, Inc.

Gregg Division
McGraw-Hill Book Company

New York	Düsseldorf	Panama
Atlanta	Johannesburg	Paris
Dallas	London	São Paulo
St. Louis	Madrid	Singapore
San Francisco	Mexico	Sydney
Auckland	Montreal	Tokyo
Bogotá	New Delhi	Toronto

Library of Congress Cataloging in Publication Data

Baer, Charles J
   Electrical and electronics drawing.

   Bibliography: p.
   Includes index.
   1. Electric drafting. 2. Electronic drafting.
I. Ottaway, John R., joint author. II. Title.
TK431.B3   1980   604'.2'6213   79–15837
ISBN 0–07–003010–3

Sponsor: Gordon Rockmaker
Editing supervisor: Alice V. Manning
Design supervisors: Caryl V. Spinka and Nancy Axelrod
Art supervisor: George T. Resch
Production supervisors: Kathleen Morrissey and Priscilla Taguer

Cover designer: Tracy Glasner
Technical Studio: Fine Line, Inc.

Contents

Preface   v

1   Techniques and lettering   1

2   Pictorial drawing   29

3   Device symbols   44

4   Production drawings   72

5   Flow diagrams and logic diagrams   141

6   The schematic (elementary) diagram   164

7   Microelectronics and microprocessors   209

8   Industrial controls   252

9   Drawings for the electric power field   300

10   Electrical drawing for architectural plans   330

11   Graphical representation of data   358

12   Computer-aided drafting and design   398

Appendix A   Glossary of electronics and electrical terms   415
Electrical device reference designations   420
Abbreviations for drawings and technical publications   422

Appendix B   The frequency spectrum   425
Width of copper foil conductors for printed circuits   425
Resistor color code   426
Color code for chassis wiring   427
Circuit-identification color code for industrial control wiring   427
Transformer color codes   427

Control-device designations   428
Approximate radii for aluminum alloys for 90° cold bend   429
Thickness of wire and metal-sheet gages (inches)   430
Metric conversion table   430
Minimum radius of conduit (inches)   431
Decimal equivalents of fractions   431
Wire numbers and sizes   432
Small drills—metric   433
Twist drill sizes   434
Standard unified thread series   436
Metric screw threads   438

Appendix C   Symbols for electrical and electronic devices   440
The relationship of basic logic symbology between various standards   449
Symbols for electrical and electronic devices   453
Electrical symbols for architectural drawings   457

Bibliography   **466**

Index   **468**

Preface

This book was designed to serve both as a text and a reference for all fields of electrical and electronics drawing.

Organizing the subject matter in logical blocks paralleling the typical curriculum and including abundant problem materials at the end of the chapters provides the student with a comprehensive textbook for classroom and individual study. The comprehensive coverage of practical applications also makes the book especially appropriate as a professional reference for drafters, designers, engineers, and production personnel in the main areas of electricity and electronics.

This fourth edition uses ANSI/IEEE Y32E, which contains symbols for electric wiring and logic diagrams, as well as for devices, circuit paths, and many symbols from the international standard, IEC No. 117.

Coverage of microprocessors, standards for industrial electronics, and a chapter on computer graphics make this book completely modern and attuned to contemporary practices. On the other hand, we have retained all the strong features of the previous edition. The sequence of chapters, for example, is the same as in the previous edition to allow for continuity of use from one edition to another. And, as in the previous edition, practically all examples are real ones, not hypothetical.

The book treats electronics, automation, microelectronics, electric power, and wiring for power and lighting in industrial, commercial, and residential building. It also includes large portions of many important standards (Appendixes B and C), as well as a glossary of terms (Appendix A) and a bibliography. The standards have been established by agencies such as ANSI (American National Standards Institute), a private agency founded by industry many years ago, and the IEEE (Institute for Electrical and Electronics Engineers, Inc.), a professional engineering society. Other standards have been formed as needed by JIC (Joint Industrial Council), NMTBA (National Machine Tool Builders' Association), NEMA (National Electrical Manufacturers Association), and agencies of the U.S. government. The reader may obtain lists of the available standards by writing the organizations listed below.

ASME   American Society of Mechanical Engineers, 345 East 47th Street, New York, NY 10017

American National Standards Institute, 1430 Broadway, New York, NY 10018

EIA   Electronic Industries Association, 2001 Eye Street, N.W., Washington, DC 20006

IEEE   Institute of Electrical and Electronics Engineers, Inc., 345 East 47th Street, New York, NY 10017

JIC   Joint Industrial Council, 7901 Westpark Drive, McLean, VA 23101

NEMA   National Electrical Manufacturers Association, 2101 L Street, N.W., Suite 300, Washington, DC 20037

NMTBA   National Machine Tool Builders' Association, 7901 Westpark Drive, McLean, VA 23101

Superintendent of Documents, U.S. Government Printing Office, Washington, DC 20402

The authors wish to make known their gratitude to the officials and engineers of the many companies and firms who were interested and thoughtful enough to assemble drawings and photographs for use in this book. We have endeavored to print these drawings as closely as possible to the way they were originally drawn.

CHARLES J. BAER
JOHN R. OTTAWAY

Chapter 1
Techniques
and lettering

Because drawing for the electrical and electronics industry includes so many types of drawings, it utilizes all the various graphical techniques in one place or another, at one time or another. In the beginning stages of some projects, rough freehand sketches may be the most appropriate way to show ideas graphically. On the other hand, the final stages of some projects may require precision drafting that necessitates the use of sophisticated and expensive equipment.

This book will not attempt to instruct the reader in the uses of these devices and methods. The Bibliography lists some excellent textbooks on engineering graphics (engineering drawing). A reader who is not reasonably proficient in the use of conventional drawing instruments (triangles, T square, irregular curve, compass, and lettering guide) may wish to consult one of these texts. Such books are often found in school libraries and sometimes in city libraries. These and other books, at both the high school and college levels, are probably adequately written to enable people to teach themselves how to use the more commonly known instruments well enough to work 95 percent of the problems in this book. The chances are, however, that these people will not be able to do as fine work or as rapid work as students who have had one or more courses in mechanical drawing or engineering drawing.

As for the use of some of the more sophisticated drawing equipment, even the texts mentioned above do not attempt to do much. Very little, if any, mention is made of such devices or methods as photodrawing, coordinatographs, automatic drafting machines, and negative scribing, for example. The only ways in which the reader can become familiar with such devices are (1) to work with the equipment or watch experienced persons using the equipment, and (2) to read about them in technical journals such as the *Journal of Engineering Design Graphics, Machine Design,* or *Electronics.*

1·1 Line work

Most electrical drawings require the drawing of many parallel lines, often horizontal and vertical. In certain cases these lines are drawn in black ink. In many cases, including work preliminary to ink drawing, the lines are drawn

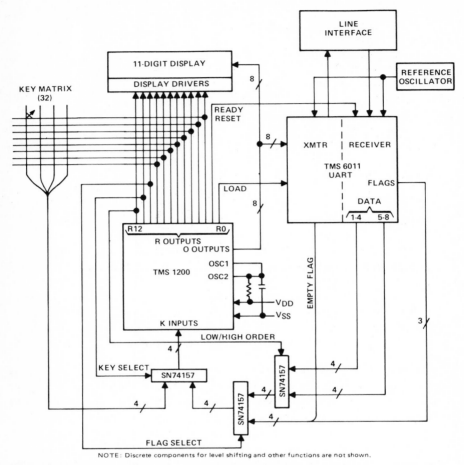

Figure 1·1 Block diagram of a single-board computer used as a data terminal.

in pencil, with one weight of line predominating. Figure 1·2 shows the different weights used in making different types of lines. Sometimes it is difficult to get much difference in weights of lines made with a pencil, but this figure should give the reader an idea of what we strive for in the matter of lines.

There are quite a number of ways in which one can draw parallel lines, depending upon the accuracy required and the equipment available. Such methods are:

1. With a plotting device such as the coordinatograph
2. With a drafting machine
3. With a parallel straightedge
4. With special section-lining devices
5. With a T square (and triangles)

——————— Medium ———————	*General purpose. Outlines.* *Circuit path. Symbols*
— — — — Medium — — — —	*Shielding. Mechanical connection.* *Hidden line. Future circuits.*
— — — Medium — — —	*Bracket - connecting line.*
——————— Thin ———————	*Dimension line. Leader. Bracket.*
——— Thin — — ——— — —	*Alternate position. Adjacent part.* *Mechanical grouping boundary.*
━━━━━━ Thick ━━━━━━	*For emphasis.*

Figure 1·2 Different lines and their uses in mechanical and electrical drawings.

6. With two triangles
7. With tape appliqués
8. Freehand

In general, the cost of equipment and the accuracy obtained decrease as we go from 1 to 8 in the above list. Figures 1·3 to 1·5 show some of the methods listed. For small drawings, a second triangle can be substituted for the T square, although this is not as convenient as using the latter.

Figure 1·3 Two styles of drafting machines. *(Keuffel & Esser Co.)*

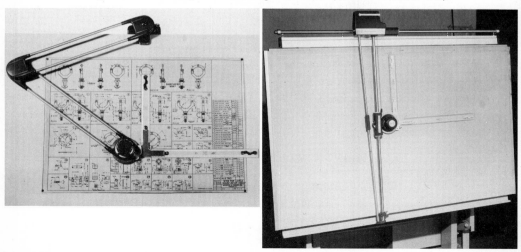

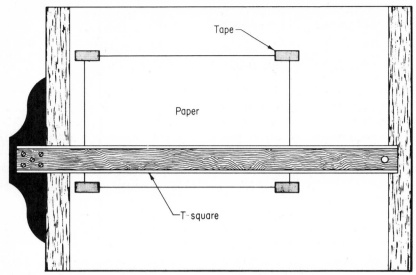

Figure 1·4 Positioning of a small-size sheet of drawing paper on a drawing board.

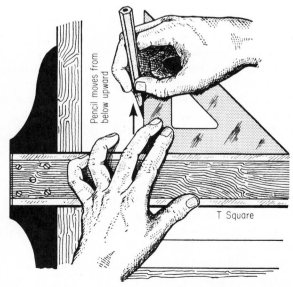

Figure 1·5 Using a T square and triangle to draw vertical lines. (*From Frank Zozzora*, Engineering Drawing, *2d ed., McGraw-Hill Book Company, New York, 1958. Used by permission.*)

This text will not discuss the making of inked drawings, a topic well treated by most of the standard drawing texts.

1·2 Circles, symbols, and other shapes

Circles are generally made with the compass or with templates. Because circles in most electrical drawings are smaller than 1 in., templates are used more frequently than the compass for circle construction. Templates are flexible, thin (0.020- to 0.060-in.) pieces of plastic with generally accurate shapes punched in them. Many models are designed for the purpose of making geometric shapes such as circles, rectangles, triangles, etc.; while others, for use in electrical and electronic drawings, are for making symbols, component outlines, and wiring diagrams.

Over the years, designers and drafters accumulate a large arsenal of templates to do their drawings. Templates are a more convenient, faster, and more accurate method of drawing shapes and electrical symbols compared with conventional methods. For example, drawing a transformer symbol as in Fig. 3·11 would take 2 min and 10 s by conventional methods, whereas drawing it using a template would take only 15 s. (The times mentioned are those expended by one of the authors.)

Care should be used in selecting templates in terms of the quality, accuracy, symmetry, arrangement of the shapes, and the shapes and symbols required. Many template manufacturers will make special templates upon request. Figure 1·6 shows several templates that could be used in doing drawings. Figure 1·6a is a template having a variety of geometric shapes. Figure 1·6b is a logic-symbol template that is used for making logic diagrams. Figure 1·6c is an electronic-components template used for making printed-circuit-board layouts and wiring diagrams. Figure 1·6d is an electromechanical-control-symbol template used in making control schematics. These are just a few examples of the many varieties of templates on the market today.

Several concepts should be followed in using a template.

1. In order to draw any geometric shape at the precise location desired, construct centerlines (along the axes, at least as long as the longest diameter of an equivalent circle) on the drawing before the template itself is positioned for the drawing of the shape. Many times, for small quick modifications to existing drawings, experienced designers use one line and an "eyeballed" centerpoint; however, this is not recommended for new drawings, large modifications, or beginners.

2. For proper alignment, the template must be held firmly against a T square, drafting machine arm, or some suitable (horizontal) straightedge.

3. The pencil must be held *vertically* as the figure is being drawn. This is the only way to be certain of the correct alignment and shape of the figure. It may also be necessary to use a slightly duller pencil point than normal in order to achieve the desired line thickness of the figure.

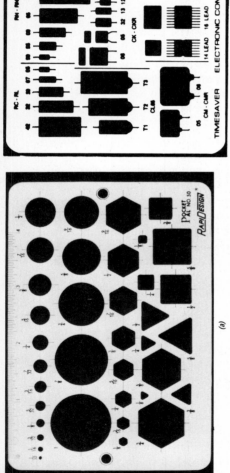

(a)

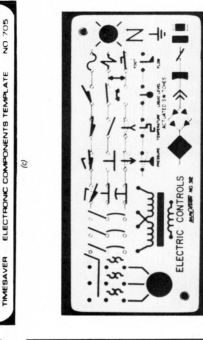

(c)

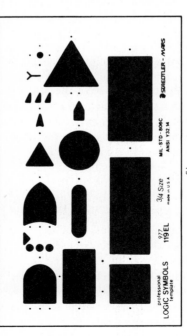

(d)

(b)

Figure 1·6 Typical drafting templates. *(a)* Template having geometric shapes of different sizes. *(Berol USA, division of Berol Corp.) (b)* A logic-symbol template. *(Staedtler-Mars.) (c)* Electronics-component outline template. *(Timesaver Templates, Inc.) (d)* Electromechanical control-device symbol template. *(Berol USA.)*

4. When inking, the template should be slightly raised off the drawing medium by a template guide and a sheet of Mylar or tape should be fastened to the back of the template to prevent the ink from *feathering*. Figure 1·7 shows the general procedure on how to use a template.

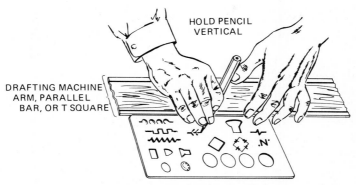

Figure 1·7 Using a template.

Templates are very convenient for making shapes and electrical symbols; however, a designer should also know how to construct symbols with the basic drafting tools because the necessary templates might not be available. These procedures are covered in Chap. 3.

1·3 The importance of lettering

Electrical drawings utilize all the conventions and shortcuts of engineering drawings in order to present a graphical concept of devices or systems. However, the line work, symbols, and other patterns of a drawing are not sufficient to give the complete picture. Considerable lettering is required on most electrical and electronics drawings. Figure 1·1 is a typical drawing for showing the amount and kind of lettering required on an electrical drawing. Some drawings require even more in proportion to the amount of line work than this drawing. In many cases, as much time is used in doing the lettering on the drawing as is used in doing the line work. In drawings which are to be used for the manufacture, construction, or installation of equipment, lettering is usually done freehand. In drawings that are to be used over and over again or in those which for certain reasons are to be printed in books, manuals, or journals, lettering is usually drawn with mechanical lettering devices.

In every case, lettering must be good. Poor lettering not only ruins the appearance of a drawing that is otherwise good but improves the chances of costly mistakes being made in the reading of the drawing. It so happens that the lettering in Fig. 1·1 was done mechanically, mainly because it was a part of a printed instruction manual. But before this final drawing was made, another one using freehand lettering had been drawn. The majority of electrical drawings include freehand lettering.

1·4 Types of engineering lettering

Practically all the alphabets used in technical drawings are single-stroke Commercial Gothic. Four types of these single-stroke letters in use today are:

1. Vertical uppercase (capitals)
2. Inclined uppercase
3. Inclined lowercase ("small" letters)
4. Vertical lowercase

The alphabets above are arranged in what the authors believe is the order in which each type is used in United States industry, with type 1 being used the most, type 2 the next, etc. One who wishes to become expert in the making of electrical or electronics drawings should be proficient in doing freehand lettering of types 1, 2, and 3.

The person who has had no lettering instruction and who wishes to become proficient at freehand lettering should practice one type of alphabet at a time. A logical sequence of alphabets would be (1) vertical uppercase, (2) inclined uppercase, and (3) inclined lowercase. With each alphabet the student should develop a set of numbers that appear to match or fit with the letters. As a rule, numbers are usually made narrower than the letters of the companion alphabet. Thus, the number zero should be narrower than the letter O of the same alphabet.

1·5 Standard alphabets

Lettering for drawings is standard throughout the United States. Three standardized alphabets are shown in Fig. 1·8. There may be slight differences between alphabets authorized by one agency and another, and also between alphabets shown in drawing textbooks, but these differences are minor. The shapes of the letters are, in general, the same.

Figure 1·9 shows vertical uppercase letters placed in blocks that are six small squares high. Borrowed from a famous textbook, this illustration shows the comparative widths of all letters and the suggested order of strokes for right-handed persons. This order does not have to be followed, but it is used by many drafters. Left-handers will probably have to develop their own order of stroking, using whatever is most comfortable and effective.

1·6 Guidelines

Practically all freehand lettering must make use of guidelines. These are very light, thin lines that locate the tops and bottoms of capital letters, numbers,

TYPE 1

ABCDEFGHIJKLMNOPQRST
UVWXYZ &
1234567890 $\frac{1}{2}$ $\frac{3}{4}$ $\frac{5}{8}$

TYPE 4

ABCDEFGHIJKLMNOPQRSTUVWXYZ &
1234567890 $\frac{1}{2}$ $\frac{3}{4}$ $\frac{5}{8}$ $\frac{7}{16}$
FOR BILLS OF MATERIAL, DIMENSIONS
& GENERAL NOTES

TYPE 6

abcdefghijklmnopqrstuvwxyz
Type 6 may be used in place of
Type 4, for Bills of Material and
Notes on Body of Drawing.

Figure 1·8 Typical American National Standard alphabets.

etc. They should be just dark enough to see, because erasing them is not practicable after letters are penciled in; therefore they should be as nearly invisible as possible.

There are a number of ways to draw guidelines. The most convenient way is to use a device especially made for that purpose. Two such devices are shown in Fig. 1·10. In this illustration, guidelines for capital letters have been constructed with both the Braddock and the Ames lettering devices. In the lower figure, lines have also been partially drawn for the bodies of lowercase letters but were stopped near the letter C and number 3. Guidelines may also be made with T squares and triangles in conjunction with scales.

Different combinations of letters and figures require different arrangements of guidelines. If only uppercase letters are to be used, just the top and bottom guidelines are necessary (see Fig. 1·11a). For lowercase letters, the intermediate, or "waist," line must be used with the other two lines, as shown in Fig. 1·11b. A "drop" line may be placed below the baseline to facilitate making the descenders. This is not done very often, probably because the normal spacing provided by lettering devices does not allow enough space for drop lines. Occasionally it is desirable to have a line exactly centered between the upper and lower guidelines. This would be helpful, for example, when many fractions

Figure 1·9 Proportions and suggested strokings for letters. (*From Thomas E. French and Charles J. Vierck*, Fundamentals of Engineering Drawing, *McGraw-Hill Book Company, New York, 1960. Used by permission.*)

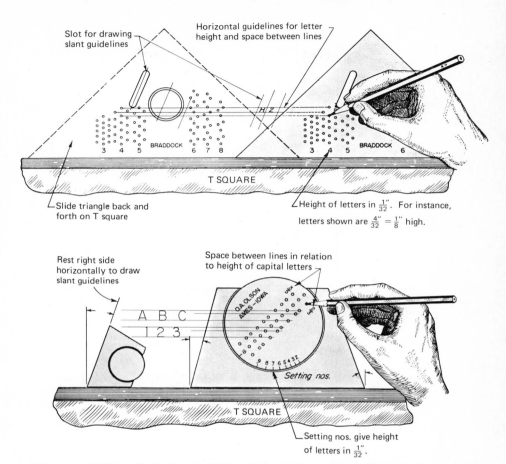

Figure 1·10 Use of lettering guideline devices. (*From Frank Zozzora,* Engineering Drawing, *2d ed., McGraw-Hill Book Company, New York, 1958. Used by permission.*)

are part of a note (see Fig. 1·11c). Because fraction heights are usually made twice those of whole numbers and capital letters, additional lines may be added above and below, as shown at the right end of Fig. 1·11c. This would produce a series of lines all equally spaced. This type of spacing can be obtained with the Ames lettering instrument for nine different sizes of letters and with some models of the Braddock-Rowe lettering triangle for one size of letter.

Figure 1·12 shows another type of freehand lettering guide that is easy and simple to use. Lettering guides of this type have a number (usually four) of guide slots for various heights of letters—ranging from $\frac{3}{32}$ to $\frac{3}{8}$ in. Either inclined or vertical freehand lettering may be done using this type of guide with the assurance of uniform-height letters due to the upper and lower edges

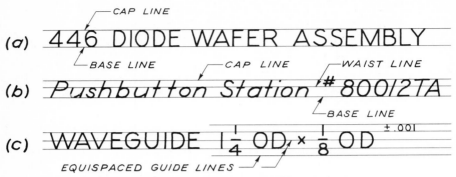

(a) 446 DIODE WAFER ASSEMBLY

(b) Pushbutton Station #80012TA

(c) WAVEGUIDE 1¼ OD × ⅛ OD

Figure 1·11 Use of guidelines for different situations.

Figure 1·12 Freehand lettering guide. *(Timesaver Templates, Inc.)*

of the guide slots. A drafter can letter with great speed using this type of guide; however, care should be exercised to avoid flattening rounded letters against the edges of the guide.

Another way to align the tops and bottoms of letters is to use guidelines already drawn on a piece of paper. This paper is positioned under the sheet of tracing paper or transparent plastic material on which the drawing is to be made. The guidelines show through the drawing medium and are followed just as if they were drawn on the medium itself.

1·7 Parts lists and tables

Figure 1·13 shows a part of a typical parts list, or bill of materials. In practice, the heavy horizontal dividing lines are usually spaced $\frac{1}{4}$, $\frac{5}{16}$, or $\frac{3}{8}$ in. apart, and the letters are made not more than one-half as high as this vertical spacing. One convenient way to organize such a list is to use equally spaced guidelines, as shown in the lower part of Fig. 1·13. For instance, if the lines are spaced $\frac{1}{16}$ in. apart (using a lettering guide set for No. 4 letters, as in Fig. 1·13), letters and numbers will be $\frac{1}{8}$ in. high and centered between the horizontal dividing lines, which will be $\frac{1}{4}$ in. apart. Abbreviations are frequently used, examples being CRS for cold-rolled steel and AL for aluminum. Capital letters are almost always used. Parts lists and other tabular arrangements often accompany, or are part of, electrical drawings. Their exact arrangements (formats) are not standardized.

ITEM	REFERENCE	DESCRIPTION	MAT'L
1	310-19506	CHASSIS	
2	35A-19472	BRACKET, MOUNTING	
3	DD-6040A	CAPACITOR, 1 MF 25 WVDC	
4	50C-19503	CABLE ASSEMBLY	
5	30103744	LOCK SHAFT	CRS
6	60104321	SCREW #4 - 40 x $\frac{1}{4}$	STEEL

Figure 1·13 Layout for a parts list with suggested vertical spacing.

1·8 Mechanical lettering devices

A number of devices for making engineering letters mechanically have been successfully tried and used in the drawing room. They fall into two classes: (1) the stencil, or incised-letter, type, with which the drafter puts the letters on the drawing medium, and (2) the special typewriter, with which a typist puts the letters on the drawing after the drafter or engineer has written or otherwise indicated what material is to be typed. Examples of the first type are shown in Figs. 1·14 and 1·15. These can be used for penciled or inked letters. The fountain-pen type of lettering pen (the Rapidograph, for example) can be used with the stencil type such as that shown in Fig. 1·15.

Figure 1·14 A mechanical lettering device. Leroy lettering template and scriber. *(Leroy is a registered trademark of Keuffel & Esser Co., by whose courtesy the above photograph is shown.)*

13

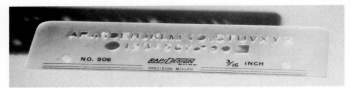

Figure 1·15 Another mechanical lettering device. *(This Rapidesign lettering template is sold by Gramercy Guild Group, Inc.)*

Although mechanical lettering devices have been on the market for more than 30 years, they have not replaced freehand lettering on many engineering or electrical drawings. For this reason, the drafting or engineering student cannot expect to get along in the profession without being able to do neat, proficient freehand lettering.

1·9 Product and manufacturers' directories, catalogs, and other sources of design information

One drafting technique a designer has to master sooner or later is how to find and understand information about an electrical or electronic component. This information is required for schematics, layouts, bills of materials, and other electrical and electronics drawings. Usually, the technique is gained through experiences of finding out "where to look," "where to go," or "whom to talk to."

Generally, a designer who does not know what specific product to use and/or what manufacturer makes the product or provides a service will consult one of the product and manufacturers' directories available today. These directories vary from large multivolume sets to single volumes and from a very general to a very specialized scope of products. They are published by technical magazines, specialized publishers, and large publishing companies. Essentially all product and manufacturers' directories have a product index and a manufacturers' index, and some have a trade-name index and/or manufacturers' catalogs. The product index is arranged alphabetically by product classification, and, sometimes, subclassification [e.g., ICs (integrated circuits), transistor array; or interrupters, fault]. Under these product classifications, there is an alphabetical listing of manufacturers that make the product, giving their addresses, phone numbers, and, sometimes, teletype (TWX) numbers and branch offices. Some product indexes give more complete listings than others. The manufacturers' index is an alphabetical list of the manufacturers, and for each manufacturer gives the manufacturer's address, phone number, and, sometimes, TWX number, gross sales, management personnel, number of employees, branch offices, and distributors. A trade-name index is an alphabetical listing of trade names of products and their respective manufacturers. Some product and manufacturers' directories are listed below.

For general use:

1. *Thomas Register*
   Thomas Publishing Co.
   One Penn Plaza
   New York, NY 10001

2. *U.S. Industrial Directory*
   Cahners Publishing Co.
   1200 Summer St.
   Stamford, CT 06905

3. *Sweets Construction and Renovation Catalog File and Plant Engineering Extension*
   Sweets Division
   McGraw-Hill, Inc.
   1221 Avenue of the Americas
   New York, NY 10020

4. *Plant Engineering Directory and Specifications Catalog*
   Technical Publishing Co.
   1301 S. Grove Ave.
   Barrington, IL 60010

For electrical equipment:

1. *Design, Professional Product Bulletin Directory*
   Consulting Engineer
   Technical Publishing Co., Div. of Dun & Bradstreet
   1301 S. Grove Ave.
   Barrington, IL 60010

2. *Electrical Products Yearbook*
   Electrical Construction & Maintenance
   McGraw-Hill, Inc.
   1221 Avenue of the Americas
   New York, NY 10020

3. *Annual Specifiers & Buyers Guide*
   Transmission and Distribution
   Cleworth Publishing Co., Inc.
   One River Rd.
   Cos Cob, CT 06807

4. *Index of Advertisers and Products*
   Contractors Electrical Equipment
   172 S. Broadway
   White Plains, NY 10605

For electrical/electronic equipment:

1. *Electrical/Electronics Reference Edition*
   Design News
   Cahners Publishing Co.
   270 St. Paul St.
   Denver, CO 80206

For electronic equipment:

1. *Electronics Designs Gold Book—Master Catalog & Directory of Suppliers*
   to Electronics Manufacturers
   Hayden Publishing Co., Inc.
   50 Essex St.
   Rochelle Park, NJ 07662

2. *Electronics Buyers Guide*
   McGraw-Hill, Inc.
   1221 Avenue of the Americas
   New York, NY 10020

3. *Electronic Engineers Master Catalog*
   United Technical Publications
   645 Stewart Ave.
   Garden City, NY 11530

4. *Electronics Industry Telephone Directory*
   Harris Publishing Co.
   33140 Aurora Rd.
   Cleveland, OH 44139

5. *Electronics Buyers Handbook and Directory*
   CMP Publications, Inc.
   280 Community Dr.
   Great Neck, NY 11021

6. *D.A.T.A. Book Electronics Information Series*
   D.A.T.A., Inc.
   45 U.S. Highway
   Pine Brook, NJ 07058

7. *IC Master*
   United Technical Publications
   645 Stewart Ave.
   Garden City, NY 11530

For control and instrumentation:

1. *Control Equipment Master*
   Chilton Publishing Co.
   Chilton Way
   Radnor, PA 19089

Other sources of product and manufacturer information are technical and trade magazines and journals, salespeople, distributors, other engineers and designers, and purchasing departments. The authors have even used stockbrokers and Dun & Bradstreet Reports to find the address of a known manufacturer which could not be found elsewhere.

After a designer has selected a product, a manufacturer's or distributor's catalog is obtained within the office from the catalog file or library or from another engineer, or it is obtained from a salesperson. Usually, engineering firms have large and fairly complete catalog files, such as the one shown in Fig. 1·16. Although catalogs vary considerably, designers should observe the following concepts when looking at an unfamiliar catalog.

1. Check the issue date or number to see if it is the most up-to-date version.
2. Scan the table of contents, index, and the catalog for an overview or "big picture" of the contents of the catalog.
3. Examine the conditions of sale, discount schedules, and other general information, usually found in the front of the catalog.
4. Then find the product description in the catalog, using the index.
5. Once the product description is located in the catalog, look at the pictures of the product, read all the information about the product, including the

Figure 1·16 Typical engineering catalog file. This one belongs to an electrical contractor. *(Boese-Hilburn Co.)*

notes and footnotes, which are sometimes found at the front or end of the product class section. Observe all the operating parameters and specifications, physical dimensions, available modifications, and costs associated with the product.

Once the product is selected, record the relevant information from the catalog onto the drawings, paying particular attention to getting all necessary ordering information on the bill of materials—such as a general product description, manufacturer, all manufacturer's model and type numbers, operating parameters, general industry class numbers.

SUMMARY

The techniques of making electrical drawings are the same as those used in making other engineering drawings. With proper study, practice, and equipment, nearly any person can make most types of electrical drawings. Some types, however, require expensive precision equipment which is not described in most instructional books or courses.

A good knowledge of the shapes of letters belonging to several standard alphabets is required of the person who is engaged in drawing electrical or electronics drawings. One can learn to letter well by practicing fundamental strokes and imitating good samples of lettering. In the foreseeable future there is little likelihood that mechanical lettering or computers will replace freehand lettering in most electrical drawings.

The knowledge of how to use product and manufacturers' directories and catalogs is extremely important, because one must know how to find and use information on products to be illustrated on the electrical drawings.

QUESTIONS

1·1 Name or list several ways to draw circles.

1·2 In a note that includes capital letters and fractions, how high should the fractions be with respect to the letters?

1·3 In general, how many types of lettering are used in United States industry?

1·4 Name five different ways of drawing parallel horizontal lines.

1·5 Why is it necessary to have different line widths and different types of lines in electronics drawings?

1·6 For what purposes are templates used in the drawing room?

1·7 In addition to achieving uniform height of capital letters, in what other respects do we strive to achieve uniformity?

1·8 What are some advantages of uppercase lettering? Of lowercase?

1·9 Are the letters of an alphabet generally narrower or wider than the numbers which accompany the same alphabet?

1·10 What is a drop line?

1·11 Why do some agencies and publishers require mechanical lettering?

1·12 Show, by sketching, an effective way of making guidelines for a parts list.

1·13 How does the shape of a curve differ between inclined and vertical upper-case letters? (Sketch or describe.)

1·14 Name several product and manufacturers' directories.

1·15 In a product and manufacturers' directory, what information is contained in a product index? a manufacturers' index? a trade-name index?

1·16 Describe how to use a catalog.

PROBLEMS

Lettering Exercises

1·1 In $\frac{1}{8}$-in. uppercase letters, letter the following terms or sentences:
  a. Identification notch
  b. Pin spacing 0.100 TP
  c. Plastic dual-in-line
  d. High-level V_{IH}
  e. 40-pin DIP package
  f. Glass-filled nylon

1·2 With $\frac{1}{8}$-in. uppercase letters, letter the following:
  a. Random-access memory
  b. $V_{cc} = 4.75$ V
  c. P-11C425 waveguide screw
  d. KS-19087 L-13 connector detail
  e. All resistors are $\frac{1}{4}$ watt $\pm$ 5%

1·3 With No. 3 or 4 inclined lowercase letters, letter the following notes:
  a. 12.47-KV switchgear location
  b. Wheel-pulse input pickup: master warning panel
  c. $C1$, $C2$, $C3$ are for connections of power-factor-correcting capacitors
  d. Tinned leads .040 $\pm$ .003 dia.

1·4 In a space about 2 × 4 in., make a set of general notes as follows: Resistance value in ohms. Capacitance values less than one in MF, one and above in PF, unless otherwise noted. Direction of arrows at controls indicates clockwise rotation. Voltages should hold within $\pm$20% with 117 V ac supply.

1·5 With uppercase letters $\frac{3}{32}$ or $\frac{1}{8}$ in. high, letter the following specific notes:
  a. Knockout for $1\frac{1}{2}$-in. conduit may enter top or bottom of cabinet.
  b. $\frac{9}{16}$ DIA (4 holes) for wall mtg.
  c. $5\frac{1}{2}$ × 6 cutouts for sec. cable conn.
  d. Tie-bolt holes $\frac{1}{2}$ max.

e. All pins 0.093 ± 0.003 DIA

f. 187 ± .003 DIA, 4 pins

1·6 Make a parts list with the following headings: Item No., Part No., Description, No. Required, Remarks. Place the following items in the appropriate boxes, 1 through 6.

1	21–4	100-ohm resistor	2	
2	31–14	.001 UF capacitor	4	Disc
3	40–66	Antenna coil	1	
4	431–10	Terminal strip	4	3-lug
5	511–4	Transistor	1	2N155
6	19–27	Control with switch	1	1 Megohm

1·7 Letter a lighting-fixture schedule using the following headings: Mark, Manufacturer, Cat. No., Mounting, Watts, Lamps, Finish. Place the following items in the appropriate boxes, A through E.

A   Skylite, 50, Recessed, 300W, Silver Bowl, Std.

B   Lightcraft, 60R, Ceiling, 150W, Par 40, Satin Alum.

C   Prescolite, 1313–6630, Recessed, 300W, 1F, Std.

D   Prescolite, 1015–6615, Recessed, 150W, 1F, Std.

E   Prescolite, 90-L, Recessed, 150W, 1F, Std.

1·8 Make a schedule, using No. 4 or 5 letters, for a printed wiring board. Use the block format which is used with parts lists.

Hole No.	X-coord.	Y-coord.	Position Tolerance	Diameter
11	.486	2.120	.028	.304/.314
12	1.988	2.100	.028	.304/.314
13	1.520	2.120	.028	.304/.314
14	.838	1.712	.028	.063/.068*
15	1.738	1.590	.014	.090/.097

* .063/.068 hole to be plated through

1·9 Using inclined lowercase letters, letter the following equipment descriptions, one below the other, as they would appear beside a vertical one-line diagram:

a. 1/0 ACSR

b. 3.30KV Line-type Lightning Arresters

     c. 23KV 200 AMP S&C Hook-operated Disconnect

     d. 3000KVA SC-3917KVA FAC West. Transformer 22000Δ — 4160Y / 2400 volts

     e. 3-CT's 1000/5

     f. Watthour Meter W/Demand

     g. 5KV 600A Air Circuit Breaker

     h. 400-AMP 5KV Enclosed Disconnect @ 3–4/0 cu.

1·10 Using No. 3 or 4 vertical uppercase letters, letter the following statements:

     a. Unless otherwise specified, resistance values are in ohms and capacitance values are in picofarads.

     b. DC coil resistance values under one ohm are not shown on the schematic diagram.

     c. Arrows on controls indicate clockwise rotation.

     d. Control viewed from shaft end

     e. Saturable-coil line reactors

1·11 Using No. 3, or 0.10-in-high, vertical uppercase letters, letter the following graph terms:

     a. Discriminator output, volts DC

     b. Base current, milliamperes

     c. 1024-bit shift register

     d. Frequency response curve of amplifier

     e. Duration of fades in seconds

     f. E_k in electrons per ion

     g. Read-only memories

1·12 Using a mechanical lettering device or a lettering template, letter the terms shown in Prob. 1·9 or 1·10 in ink.

1·13 Using $\frac{5}{32}$- or $\frac{6}{32}$-in. uppercase letters, letter the following titles, centering them from left to right on the sheet:

     a. Series 104 Read-only memory

     b. Schematic diagram of Model 130 television receiver

     c. One-line diagram of Littlefield substation

1·14 Letter the titles shown in Prob. 1·13, using a mechanical lettering device or template.

1·15 Make a title plate for the drawings you have made in this course. Details concerning the size of the plate and the exact information to be lettered on the plate will be supplied by the instructor.

1·16 Lay out and letter a block title for one or more drawings of this course. The information which should be included in the title block will be supplied by the instructor.

Mechanical Drawing Exercises

1·17 Make a drawing of the diffused resistor crossover shown in Fig. 1·17. Drawn to twice the proportions shown, it will fit on half a sheet of $8\frac{1}{2}$ × 11 paper.

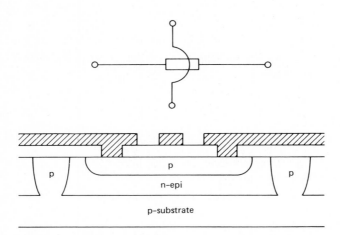

Figure 1·17 (Prob. 1·17) Schematic view of a diffused resistor crossover.

1·18 Make a drawing of the photodiode shown in Fig. 1·18. Use $8\frac{1}{2}$ × 11 paper. Include lettering. (May be combined with Prob. 1·17, 1·25, or 1·26.)

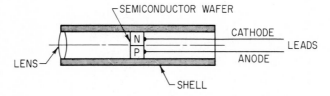

Figure 1·18 (Prob. 1·18) Diagram of section of a junction-type photodiode. The length is four times the diameter.

1·19 Make a drawing of the third-rail detail shown in Fig. 1·19. Some of the dimensions are supplied. Make your drawing to similar proportions so that it will fit nicely on $8\frac{1}{2}$ × 11 paper.

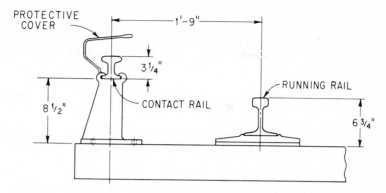

Figure 1·19 (Prob. 1·19) Third-rail detail for mass-transit system.

1·20 Make a drawing of the end view of a satellite oscillator assembly to the dimensions shown in Fig. 1·20. Include lettering. Use $8\frac{1}{2} \times 11$ paper (may be combined with the next problem on a larger sheet).

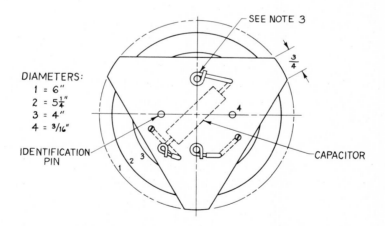

Figure 1·20 (Prob. 1·20) End view of a satellite oscillator assembly. Some dimensions: capacitor $\frac{1}{4} \times 20$ in.; distance between identification pins, 2 in.; between centers of large loops, 2 in.

1·21 Make a drawing of the front view of the oscillator assembly shown in Fig. 1·21. Use $8\frac{1}{2} \times 11$ paper. Suggested scale: full size.

1·22 Make a two-view (top and front or top and side) drawing of the waveguide cover flange shown in Fig. 1·22. Dimensions may be converted to common

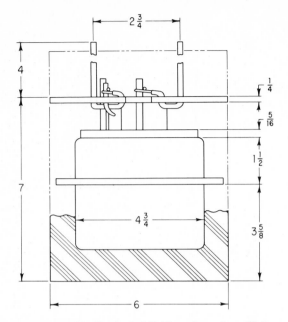

Figure 1·21 (Prob. 1·21) Elevation of oscillator assembly.

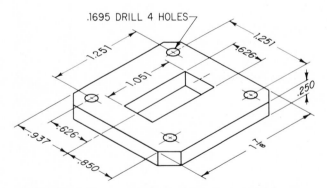

Figure 1·22 (Prob. 1·22) Isometric drawing of cover flange for waveguide. Material: brass.

fractions (see conversion table in Appendix B) or to two-place decimals. Suggested scale: twice actual size. (This will fit on $8\frac{1}{2} \times 11$ paper.) This may be dimensioned, if your instructor so indicates.

1·23 Draw two views of the connector shown in Fig. 1·23. Draw to twice the actual size (2:1). Show the dimensions that are given in the illustration. For drawing purposes, convert dimensions to common fractions (see Appendix B) or two-place decimals. Use $8\frac{1}{2} \times 11$ paper.

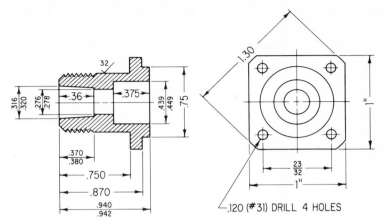

Figure 1·23 (Prob. 1·23) Two related views of a connector. The threads are $\frac{5}{8}$-24 ($\frac{5}{8}$-in. diameter, 24 threads per inch).

1·24 Make outline drawings of the semiconductors shown, respectively, in Figs.
1·25 1·24 and 1·25. Use an enlarged scale so that one or two of these objects (two views of each, as shown) will fit on an $8\frac{1}{2} \times 11$ drawing sheet. Many of the dimensions are supplied as *limits;* that is, the upper and lower limits of each dimension are shown. Select the average of the two limits for your drawing dimension. Optional additions to drawing the outline are:

a. Showing the dimensions on the drawing

b. Making a specification drawing which includes:

    (1) The outlines

    (2) Manufacturer's name and model number

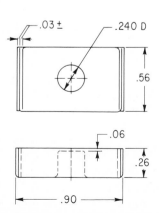

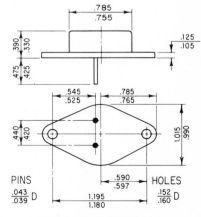

Figure 1·24 TO-5 transistor. (See pictorial view, Fig. 2·18.)

Figure 1·25 Diode.

(3) A list of specifications such as contact resistance, dielectric strength, ampere rating, life expectancy, and other information

(4) The dimensions

(5) Component symbol

The above-listed information will have to be obtained from the instructor or from a manufacturer's transistor manual.

1·26 Make a two-view drawing of the microprocessor socket shown in Fig. 1·26.

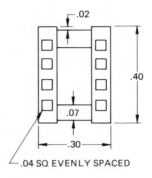

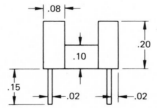

Figure 1·26 (Prob. 1·26) DIP socket.

1·27 Make an outline drawing (two or three views) of the integrated-circuit package shown in Fig. 1·27. Use an enlarged scale and show the dimensions. Then underneath your drawing, letter the following title: 16-PIN CERAMIC DUAL-IN-LINE PACKAGE; and add the note: Each pin $\mathrm{C\!\!\!L}$ is located within .010 of its true longitudinal position.

1·28 Make two or three views (front, top, side) of the subchassis shown in Fig. 1·28. May be shown as a development. Show the dimensions. Holes are to be drilled. Their diameters are: No. 1,—$1\frac{1}{2}$; No. 2,—$\frac{1}{4}$; No. 3,—$\frac{5}{8}$; No. 4,—$\frac{1}{4}$; and No. 5,—$\frac{1}{4}$. Material: SAE 1020 steel. Add notes: Break all sharp edges; remove all burrs; degrease per 51606; bright dip per 51606; dry per 51606. Hole sizes are in inches; other dimensions are in mm.

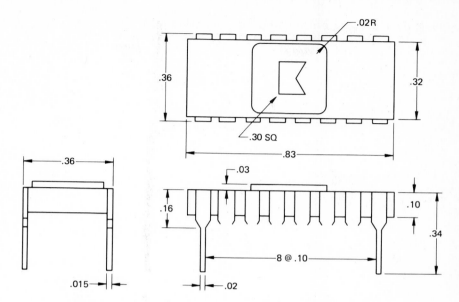

Figure 1·27 (Prob. 1·27) Sixteen-pin DIP ceramic package.

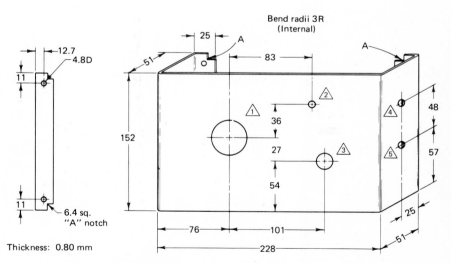

Figure 1·28 (Prob. 1·28) Pictorial view of subchassis.

1·29 Make a drawing of the alarm top shown in Fig. 1·29.

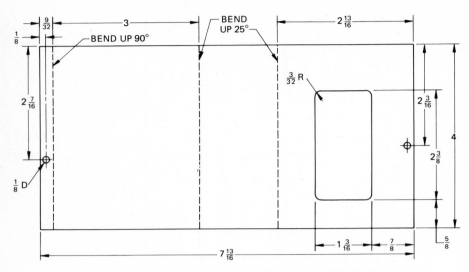

Figure 1·29 (Prob. 1·29) Top of burglar-alarm chassis.

Chapter 2
Pictorial drawing

As the electronics and electrical fields continue their phenomenal expansion, more and more devices that require pictorial representation, either in place of or as adjuncts to standard orthographic projection, are encountered. Also, electrical drawings are being read by more and more technical personnel who have not had training in the reading and making of engineering drawings. These are two factors which have been responsible for the increase in the amount of pictorial drawing being done in the electrical industry.

Good examples of pictorial drawing used in the electrical field are those used by our armed forces maintenance personnel, drawings used by assembly-line workers for assembling electronic equipment, and those used by companies who manufacture do-it-yourself kits. Figure 2·1 is a typical example of the latter.

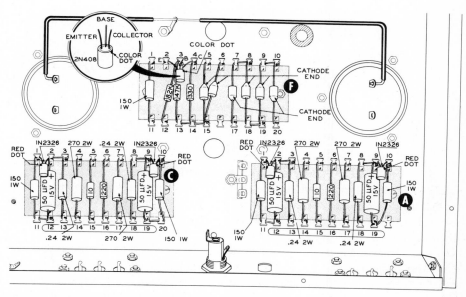

Figure 2·1 Pictorial drawing used for the assembly of electronics equipment. *(Heath Company.)*

The types of pictorial drawings most often found in the electrical field are:

1. Isometric drawing
2. Oblique drawing
3. Dimetric drawing
4. Perspective drawing

Brief treatment of the first three categories of pictorial drawing will follow. References will be given for the more complicated subject of perspective drawings.

2·1 Isometric drawing[1]

To draw an object at such an angle or position that all edges will be foreshortened equally is conveniently possible. For example, all lines of the cube shown in Fig. 2·2 have been drawn to the same length; that is, they have been foreshortened equally. This drawing, called an isometric drawing or isometric projection, was obtained by using three axes, as shown in Fig. 2·2b. Rectangular-shaped objects lend themselves readily to this type of drawing because all lines, or edges, are drawn along or parallel to the three isometric axes and can be scaled directly.

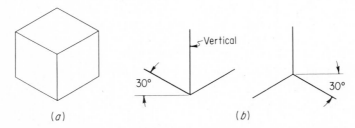

(a)   (b)

Figure 2·2 An isometric drawing of a cube and the isometric axes. *(a)* Cube. *(b)* Two arrangements of isometric axes.

A simple and direct way to make an isometric drawing of an object composed of rectangular-shaped surfaces is to draw an outline of a cube and then to measure off distances from edges of this cube to the corners and edges of the object. Such step-by-step procedure is shown in Fig. 2·3b; the completed drawing appears in Fig. 2·3c.

Isometric drawings may be made with instruments (the 30–60 triangle is a natural choice), or they may be drawn freehand. If an instrument drawing is to be made, any suitable scale may be used, and only the one scale selected will be used in making measurements along all three axes. Isometric sketches

---

[1] Liberal translation of the word *isometric* from the Greek gives "iso" meaning "the same" and "metric" meaning "measurement."

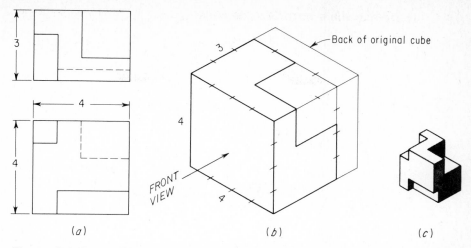

Figure 2·3 Preparation of an isometric drawing begun as a 4-in. cube. *(a)* Front and top views (given) of the object. *(b)* Starting to draw the notch in the rear, upper right part. *(c)* Completed isometric drawing.

are sometimes shaded to give additional clarity. Hidden lines are not usually shown. Dimensions may be added if necessary.

2·2 Examples of isometric drawing

Figure 2·4 shows isometric drawings of a flatpack integrated-circuit package and a full-wave bridge. For these particular objects, one pictorial drawing may be more descriptive and informative than two or three orthographic (principal) views. An isometric drawing of a console for use in the automatic assembly of circuit boards appears in Fig. 2·5.

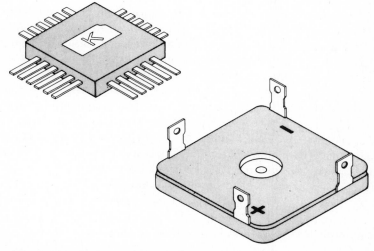

Figure 2·4 Isometric drawings of a flatpack IC and a full-wave bridge.

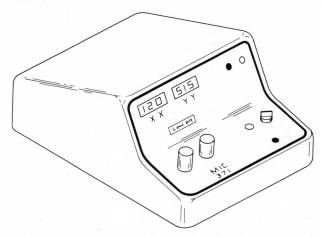

Figure 2·5 Isometric drawing of a console for an automatic circuit-board assembler.

2·3 Oblique drawing

Another type of pictorial representation which is easily made is oblique drawing, or oblique projection. In this type of drawing, one face of the object is drawn in its true shape and the other visible faces are shown by parallel lines, or projectors, drawn at the same angle (usually 30 or 45°) with the horizontal. Figure 2·6 shows two objects drawn in this manner.

The projectors of the cube (Fig. 2·6a) are drawn to the right at a 30° angle and are shortened to preserve the appearance of the cube. Such foreshortening of lines along the oblique axis is called *cabinet drawing,* and more often than not the projectors are drawn to one-half the size of the lines they represent. The rectifier (Fig. 2·6b) is suitable for oblique drawing for two reasons: (1) The face drawn perpendicular to the line of sight (in the plane of the paper) contains circular parts which can be drawn with a compass or circle template, and (2) the other axis is comparatively short, thus making it unnecessary to shorten the projectors. The projectors of the rectifier were drawn 45° to the left and are not foreshortened. If the rectifier were drawn in the isometric manner, they would have been drawn at a 30° angle and all circles and arcs would have to be drawn as ellipses or parts of ellipses. Like isometric drawings, oblique drawings may be made freehand or with instruments.

2·4 Examples of oblique drawings

Figure 2·7 shows an oblique drawing of a transistor with radiator. Figure 2·8 is an oblique drawing of the same console which appeared in Fig. 2·5.

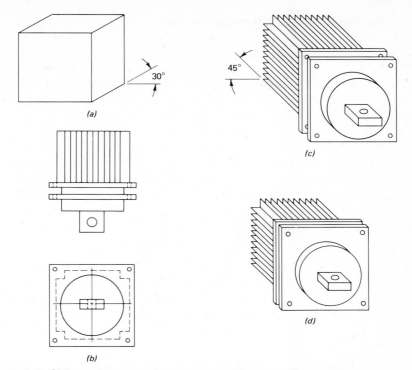

Figure 2·6 Oblique drawings of a cube and silicon rectifier. *(a)* Cube. *(b)* Front and top views. *(c)* Oblique drawing of rectifier. *(d)* Modified oblique (cabinet) drawing.

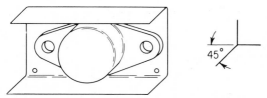

Figure 2·7 Oblique drawing of a transistor with a heat-radiation attachment.

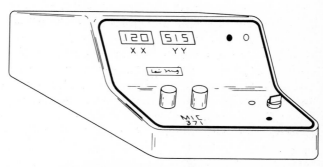

Figure 2·8 Oblique drawing of a console. See Fig. 2·5 for comparison purposes.

2·5 Dimetric projection

Another form of pictorial representation, similar to isometric drawing, is dimetric drawing, or dimetric projection. In this type of drawing, the reference cube is viewed from such an angle that two edges are equally foreshortened. The third axis (the lines marked 3 in Fig. 2·9a) is shortened a different amount, however, and therefore cannot be measured with the same scale that is used for the other two axes. Examples of dimetric drawing are shown in Fig. 2·9. Dimetric drawing is not as popular as isometric and oblique drawing for two reasons: (1) Difficulty is encountered in drawing circular parts as represented dimetrically, and (2) it is necessary to use two different scales when laying out a dimetric drawing. However, a dimetric drawing will produce a less distorted, more pleasing effect than will either of the other two types of drawing.

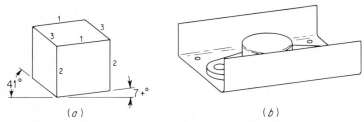

Figure 2·9 Dimetric drawing. *(a)* Cube. *(b)* TO-66 transistor can with radiator.

2·6 Perspective drawing

The three methods of pictorial representation described thus far in this chapter produce only approximate representations of objects as they appear to the eye. Each type produces varying degrees of distortion of any device or system so drawn. However, because of the ease and quickness of their execution, they are the types of pictorial drawing most often used in the electrical industry.

On certain occasions, the exact pictorial representation of an object as it actually appears to the eye may be necessary or desirable. To do this requires

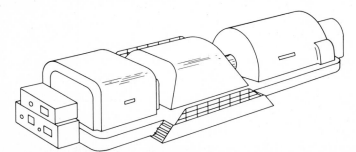

Figure 2·10 Perspective drawing of a steam turbine generator.

that the principles of perspective drawing be observed. An example of such a drawing is shown in Fig. 2·10.

Perspective drawing is a time-consuming process which requires more explanation than can conveniently be given in this book. Most texts on engineering drawing (see Bibliography) cover this subject.

2·7 Pictorial sections

Occasionally, the interior detail of a device can be appropriately shown by means of a pictorial section. All types of pictorial views—isometric, oblique, and perspective—can be used to show unusual interior details with full sections, half sections, or broken-out sections.

Figure 2·11 is a section view showing the construction of a typical silicon point-contact diode. It is a full isometric section, in which the main axis is horizontally oriented. This is in contrast to the other isometric drawings in this chapter in which this axis has been drawn in a vertical position.

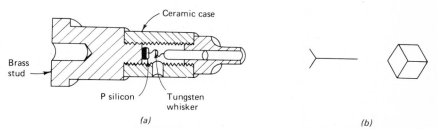

(a)

(b)

Figure 2·11 *(a)* An isometric section of a diode. *(b)* Showing how the isometric axes have been oriented.

The same figure also illustrates the fact that standard section-lining (cross-hatching) symbols are not always adequate for electrical and electronics drawings. There is no standard section symbol for silicon. The silicon slice in this case was not section-lined. Each material has been labeled—a necessary, or at least highly desirable, feature when the American National Standard (ANSI) section symbols are not followed.

2·8 Hidden lines and centerlines

Hidden lines and centerlines are not usually drawn in pictorial drawings. However, hidden lines may be shown if special hidden details are desired to be shown, and centerlines may be required if an object is to be dimensioned.

2·9 References

More detailed explanation of the construction required for oblique, isometric, and dimetric views may be found in the reference books listed in the Bibliography at the end of the book.

SUMMARY

The use of pictorial drawings has increased greatly in the last decade. This is true for all technical areas and in particular for the electronics field. Most of the well-known types of pictorial drawing have been used. Probably the two types that are the easiest to make are isometric and oblique drawings. The choice of which to use could well depend upon the shape of the object to be represented pictorially. Isometric, dimetric, and oblique drawings do not quite show the true shape of any object so drawn; thus some distortion results. If this distortion is objectionable, then one must use perspective drawing, which is considerably more difficult and therefore more expensive. Pictorial sections are often desirable. Hidden lines, section lines, and centerlines as a rule are not shown. However, they may be placed on a pictorial drawing if the occasion so demands.

QUESTIONS

2·1 Why are pictorial drawings becoming more numerous in the electrical and electronics fields?

2·2 Refer to Fig. 1·2. Which of these types of lines are used in pictorial drawings?

2·3 What are the advantages of an oblique drawing as compared with an isometric drawing?

2·4 Is a cabinet drawing a refinement of an isometric drawing or of an oblique drawing?

2·5 Name a situation in which there would be an advantage in making an oblique drawing instead of an isometric drawing.

2·6 Name a situation in which making an isometric drawing would have an advantage over making an oblique drawing.

2·7 Why are hidden lines not usually drawn on a pictorial drawing?

2·8 Name two requirements that make a dimetric drawing more difficult to execute than an oblique drawing.

2·9 Can dimension lines be added to an isometric drawing of an object?

2·10 Some pictorial drawings contain all three of the following: centerlines, dimension lines, and hidden lines. Is this statement correct?

2·11 What are the angles most often used in drawing "projectors" in oblique drawings?

PROBLEMS

In many of these problems, one or more types of pictorial drawing can be used to depict the object or system satisfactorily. In only a few cases, therefore, has a specific type of pictorial view been required. Many of the objects can be drawn pictorially as freehand sketches or as mechanical drawings. Your instructor may, in addition, require you to put dimensions on the pictorial drawing.

2·1   Make a pictorial drawing of the bracket shown in Fig. 2·12. Diameters

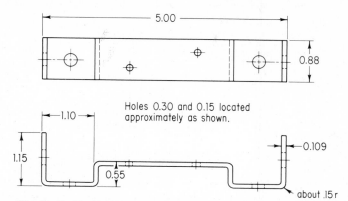

Figure 2·12 (Prob. 2·1) Front and top views of a bracket.

of the holes are 0.15 and 0.30 in. Position the holes as closely as you can to the way they appear in Fig. 2·12. Use $8\frac{1}{2} \times 11$ paper.

2·2   Make a pictorial drawing of the U-shaped chassis shown in Fig. 2·13. The 1-in. flanges (with holes) are bent at 90° to the $4\frac{3}{4}$-in. sides. Reorient so that these flanges are horizontal. Use $8\frac{1}{2} \times 11$ paper.

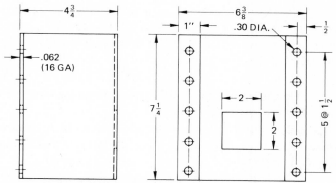

Figure 2·13 (Prob. 2·2) A U-shaped chassis.

2·3   Make a pictorial drawing of the microcircuit waveguide accessory shown in Fig. 2·14. Use the rough scales along the left edge for correct proportions. Use $8\frac{1}{2} \times 11$ paper.

2·4   Make a pictorial drawing of the plastic microprocessor package shown in Fig. 2·15. Estimate those dimensions that are not shown. Use $8\frac{1}{2} \times 11$ paper.

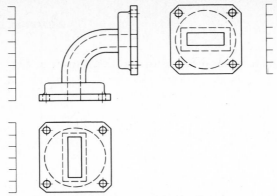

Figure 2·14 (Prob. 2·3) Waveguide bend.

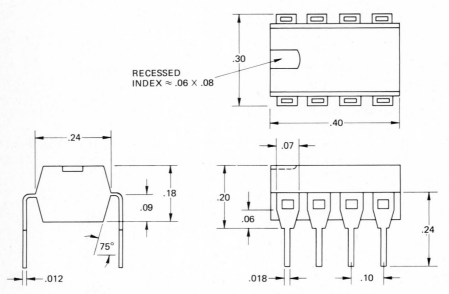

RECESSED
INDEX ≈ .06 × .08

Figure 2·15 (Prob. 2·4) Eight-pin plastic package.

2·5 Draw a pictorial section view of the phototransistor shown in Fig. 2·16. Estimate those sizes or distances which are not dimensioned in the book. Use $8\frac{1}{2} \times 11$ paper.

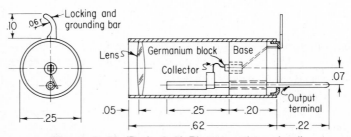

Figure 2·16 (Prob. 2·5) Phototransistor details.

2·6   Draw a pictorial view of one or both objects shown in Fig. 2·17. These will have to be drawn many times their actual size. Use $8\frac{1}{2} \times 11$ paper.

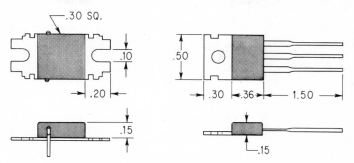

Figure 2·17 (Prob. 2·6) Plastic TO-3 (left) and Triac (right).

2·7   Draw a pictorial view of the transistor shown in Fig. 2·18. This will have to be enlarged if a good accurate drawing is to be made. Make an isometric or oblique pictorial, or both, as assigned by your instructor. Use $8\frac{1}{2} \times 11$ paper.

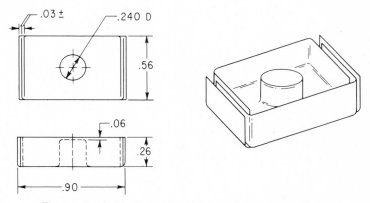

Figure 2·18 (Prob. 2·7) TO-5 transistor with radiator.

2·8   Make a pictorial drawing of the silicon diode rectifier of Fig. 2·19. Use an enlarged scale. Use $8\frac{1}{2} \times 11$ paper.

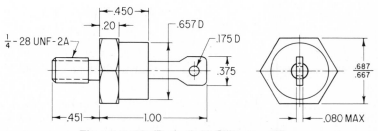

Figure 2·19 (Prob. 2·8) Silicon rectifier.

2·9   Make a pictorial drawing of the bracket shown in Fig. 2·20. It is made of 16-gage (0.062-in.) steel. Thickness may be exaggerated.

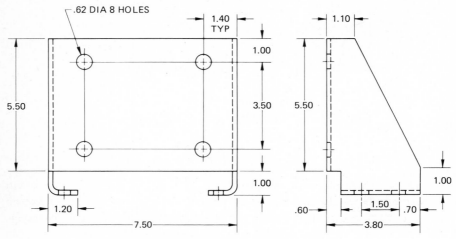

Figure 2·20 (Prob. 2·9) Mounting bracket.

2·10 Make a pictorial view of the console shown in Fig. 2·21. This may be an isometric or oblique drawing. Use $8\frac{1}{2} \times 11$ paper.

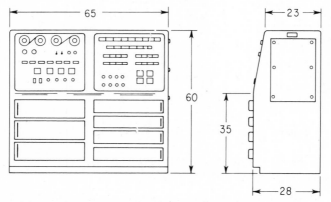

Figure 2·21 (Prob. 2·10) Front and side views of Saturn IC checkout console.

2·11 Make a complete pictorial diagrammatic drawing of the industrial television arrangement shown in Fig. 2·22. Label all items. Use 11×17 or 12×18 paper.

2·12 Construct a complete pictorial drawing of the automatic paging system shown in Fig. 2·23. The front panel of each of the six blocks is shown below. The rest of each unit can be shown as the plant PA system, upper

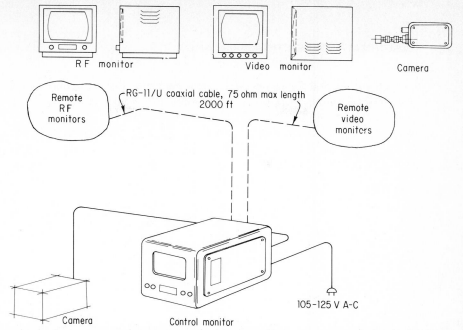

RF monitor Video monitor Camera

RG-11/U coaxial cable, 75 ohm max length 2000 ft

Remote RF monitors

Remote video monitors

Camera Control monitor

105-125 V A-C

Figure 2·22 (Prob. 2·11) A partial pictorial drawing of a closed-circuit television system and orthographic views of some of its components.

Figure 2·23 (Prob. 2·12) A partially completed pictorial diagram of an automatic paging system.

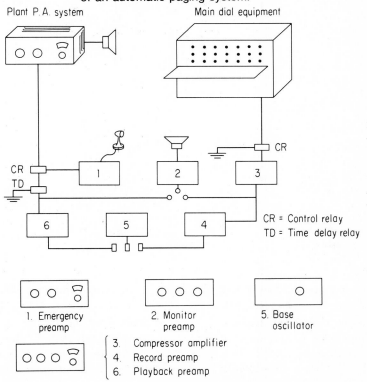

Plant P.A. system Main dial equipment

CR
TD
1 2 3

CR = Control relay
TD = Time delay relay

6 5 4

1. Emergency preamp
2. Monitor preamp
5. Base oscillator

3. Compressor amplifier
4. Record preamp
6. Playback preamp

41

left, appears. If oblique projection is used, the emergency preamp (block 1) will appear almost the same as the plant PA, upper left. Label each unit and the relays. Use 11 × 17 or 12 × 18 paper.

2·13 Draw the sonar system in Fig. 2·24. Label each unit and show the schematic location as it is shown in the upper center. Label all parts. Use 11 × 17 or 12 × 18 paper.

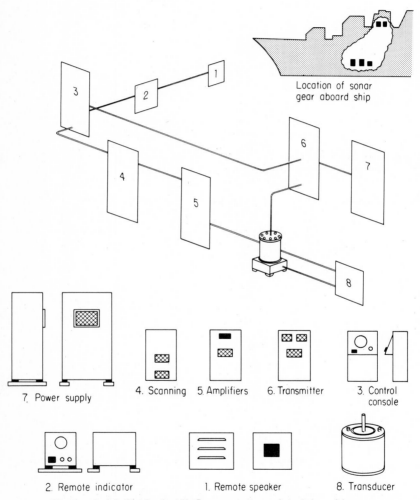

Figure 2·24 (Prob. 2·13) Sonar-system pictorial problem.

2·14 Make a pictorial drawing of the automatic drafting machine shown in Fig. 2·25. Key dimensions have been shown. Use 11 × 17 or 12 × 18 paper.

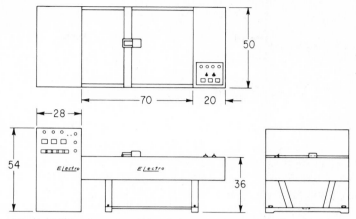

Figure 2·25 (Prob. 2·14) Outline drawing of an automatic drafting machine.

2·15 Make a pictorial drawing of the 16-pin ceramic package shown in Fig. 2·26. Use $8\frac{1}{2} \times 11$ paper.

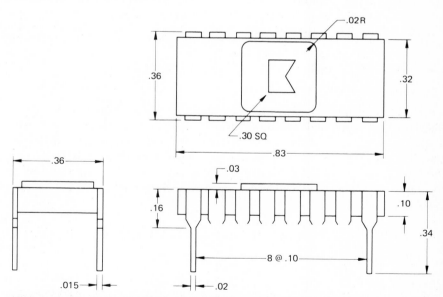

Figure 2·26 (Prob. 2·15) Three views of a 16-pin ceramic package for microprocessor.

Chapter 3
Device symbols

A very large portion of drawing in the electrical and electronics fields is of a diagrammatic nature. This diagrammatic drawing makes great use of symbols. Originally, these symbols were drawn to look something like the parts they were to represent. However, the parts and the symbols have changed considerably through the years, until now there is not much physical resemblance between symbols and the respective parts. In some cases, though, a symbol may give a suggestion of the shape of the part it represents. Examples are the inductor, resistor, headset, antenna, and knife switch.

In order to facilitate making and reading electrical drawings, representatives of industry and government have come together to make up lists of standard symbols to be used in drawings. Two important standards are the following:

76-ANSI/IEEE[1] Y32E, "Electrical and Electronics Graphic Symbols and Reference Designations." (This standard includes the most recent editions of ANSI Y32·2, IEEE 315, and CSA Z99.)

IEC (International Electrotechnical Commission) 117, "Graphical Symbols."

Some of the commonly used symbols will be shown and explained in this chapter. Many more symbols will be shown in Appendix C. To improve coordination with IEC Pub. No. 117, IEC-approved versions of certain symbols have been added to the ANSI/IEEE standard as alternatives. We have included some of these, such as the capacitor symbol. Another set of symbols, approved by the Joint Industry Committee, will be explained later in the book and is also included in Appendix C.

Symbols of Common Nonelectronic Devices
3·1 The battery

One of the simplest symbols is that which represents a single-cell battery, shown in Fig. 3·1a. (The horizontal line represents the path of the signal, or current, and is not a part of the battery symbol itself.) The longer of the two lines always represents the positive terminal. The short line is about half the length of the long one. Multicell batteries can be shown with four or more lines (four is enough), as shown in Fig. 3·1b. Polarity symbols have been shown but theoreti-

---

[1] The Institute of Electrical and Electronics Engineers, Inc.

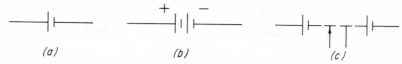

Figure 3·1 Symbols for batteries. *(a)* Single cell. *(b)* Multicell with polarity marks added. *(c)* Multicell with two taps.

cally are not necessary. However, these are sometimes used for emphasis or when it is believed the reader may be unsure. The third battery symbol shows two taps—one fixed and one (with arrow) adjustable.

3·2 The capacitor

Another often-used symbol that is easy to draw is the capacitor—sometimes called a condenser; its main function is to store electrical charge. The straight line can be made about the length of, or larger than, the long line of the battery symbol. The entire symbol is sometimes drawn twice the size of the single-cell battery symbol. (More about sizes will be found in Sec. 3·11.) The curved line represents the outside electrode in fixed-paper and ceramic-dielectric capacitors, the moving element in variable and adjustable types, and the low-potential element in feedthrough capacitors. Showing the polarity sign usually indicates an electrolytic capacitor. Good drawing practice should be observed in the drawing of the shielded symbol. Corners should be full and complete, and the dashed lines should not touch or intersect the signal path. In the case of Fig. 3·2e, the capacitance of one part increases as the capacitance of the other part decreases. But in the split-stator capacitor, the capacitances of both parts increase simultaneously.

A symbol that was not recently included in the standards now appears in 1976 ANSI/IEEE Y32E. It is the parallel straight-line symbol listed as preferred by the IEC. Shown in Fig. 3·3a, it is almost identical to the contact symbol, which has a wider gap between the lines. See Fig 3·7f.

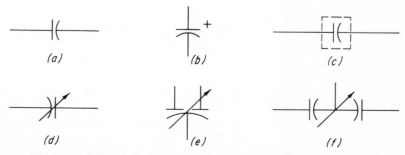

Figure 3·2 Symbols for capacitors. *(a)* General. *(b)* Polarized. *(c)* Shielded. *(d)* Adjustable or variable. *(e)* Adjustable or variable differential. *(f)* Split-stator.

Figure 3·3 IEC-approved symbols for capacitors. *(a)* General. *(b)* Polarized. *(c)* Split-stator. The distance between the plates shall be between 20 and 33% of the length of the plates.

3·3 Chassis, ground, circuit return

It is usually necessary to connect parts of a circuit to a chassis, ground, frame, etc. If the conducting connection is to a chassis or frame that may have substantially higher potential than ground or the surrounding structure, the chassis symbol (Fig. 3·4*a*) should be used. If the conducting connection is to earth, a body of water, or a structure which serves the same function (such as a land, sea, or air frame), the symbol shown in Fig. 3·4*b* should be used. The common connection symbol should be used for common-return connections at the same potential level. This triangle symbol is used when a common conductor such as a ground bus or battery bus is used. Then, in the lower part of the diagram, a key is used in which the symbol is shown and the words **GRD BUS** or **BAT BUS**—24 V or other appropriate terms are indicated. Sometimes a letter system is used if more than one type of common circuit return is shown on the same diagram. The triangle may be omitted and proper identification made where the asterisk appears on the right-hand symbol of Fig. 3·4*c,* although the triangle is more meaningful to the authors than the end of a line without a symbol. (See Figs. 6·18 and 6·30.)

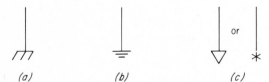

Figure 3·4 Symbols for circuit return. *(a)* Chassis connection. *(b)* Ground (frame) connection. *(c)* Common connections.

3·4 Connections and crossovers

Two systems are approved for showing connections (junctions) or crossovers. One is the dot system, as shown in Fig. 3·5*a* to *c*. The other is the no-dot system, used in Fig. 3·5*d* to *f.* The no-dot system has some weaknesses, but it has equality in the standards. However, many users (a majority in the authors' estimation) prefer the dot system for clarity.

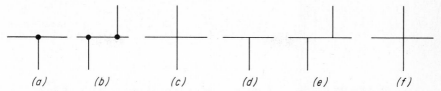

Figure 3·5 Connections and crossovers. *(a)* and *(b)* Connections for dot system. *(c)* Crossover for dot system. *(d)* and *(e)* Connections when using no-dot system. *(f)* Crossover when using no-dot system.

3·5 The inductor

The inductor, or induction coil, is used in a great many ways. In different situations it may be a transformer winding, a reactor, a radio-frequency coil, or a retardation coil. Some recent standards have shown the two symbols in Fig. 3·6a as approved. ANSI/IEEE Y32E stresses the more rudimentary symbol shown at the left. Many United States companies, however, still use the more complicated helical symbol. Because induction increases as the frequency increases, no cores (or air cores) are usually necessary at high frequencies. But at low frequencies, magnetic cores are often used to increase the induction. Ceramic cores are often composed of magnetic materials called *ferrites*. A new alternative symbol appears in the latest standard. It is two large circles about 40 percent overlapped aligned with the signal path.

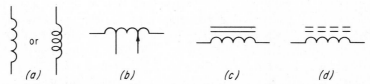

Figure 3·6 Inductor symbols. *(a)* General symbols. *(b)* With fixed and variable taps. *(c)* With magnetic core. *(d)* With ceramic-type core.

3·6 The relay coil

Three symbols are approved for the relay coil, also known as the solenoid. These are shown in Fig. 3·7a to c. The asterisk in Fig. 3·7c indicates that a letter or value, such as the relay number, should be in the circle. The semicircular dot in Fig. 3·7d indicates the inner end of the winding. Figure 3·7e and f shows combinations of relay coils with switches; the whole combinations are called *contactors*. The switches shown here are often called *contacts* and are turned off or on by action of the relay coils. In Fig. 3·7f, the left-hand contact is normally open and the right-hand contact (with inclined line drawn at 60° to the horizontal) is normally closed. Action of the relay will close the left

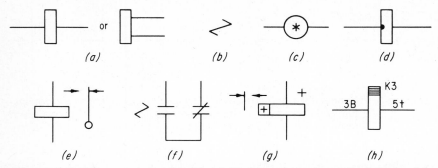

Figure 3·7 Relay symbols. *(a) (b) (c) (d)* Approved relay coil symbols. *(e)* and *(f)* Relays with transfer contacts. *(g)* Polarized relay with transfer contact. *(h)* Slow-release relay with letter-number designations.

contact and open the right one. The term *slow-release* (Fig. 3·7*h*) is only relative. This relay closes one contact before it activates another, but the total action happens very quickly.

3·7 The resistor

Two approved symbols for the resistor are shown in Fig. 3·8*a*. The older, zigzag symbol is probably used the most. The rectangle symbol, originally used in the electrical-controls field, has been adopted by a number of companies in other areas. Made with a 60° angle between adjacent lines, the zigzag symbol needs only three points on each side, unless extra taps or other special features require more. The asterisk within the rectangle means that identification should be placed within or near the rectangle. Typical values would be 210 (ohms) or 20 kΩ (20,000 ohms, often 20k, 20kΩ, or 20,000Ω in drawings).

Resistors may be fixed, variable (rheostat), or tapped, with either fixed or variable taps (see Fig. 3·8*d* and *e*). They are usually linear. If not, a special nonlinear symbol is available. Resistors are used for such purposes as dividing voltage, dropping voltage, developing heat, and minimizing current and voltage surges. The shaft of the arrow in Fig. 3·8*b* is drawn at about 45° in this and other symbols that require variability.

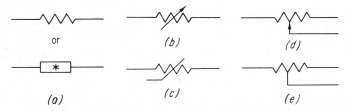

Figure 3·8 Symbols for resistors. *(a)* General. *(b)* Variable or adjustable. *(c)* Nonlinear. *(d)* With adjustable contact. *(e)* Tapped.

3·8 The switch

The purpose of the switch is to open or close circuits. The words *break* or *make* are often used instead of *open* and *close*. Mechanical-switch symbols are usually combinations of contact symbols, and they may be fixed, moving, sliding, nonlocking, etc. They are shown in the position in which no operating force is required. This is sometimes called the *normal,* or *initial,* position, in which the circuit is not energized. Figure 3·9*f* is not really a switch symbol; it simply defines the switching function. If, instead of a bar, an *X* were shown, we would have an open (break) contact. Many other devices—diodes, transistors, tubes, and cryotrons—also perform switching functions.

Rotary-switch symbols are viewed from the end opposite the control knob and with the operational sequence in the clockwise direction. The projection on each segment of the "wafer" represents the moving contact. When several functions are performed, the tabular form of presenting information, as in Fig. 3·10, is preferred. Here, dashes link the terminals that are connected. In position 2, for example, terminals 1 and 3 are connected (not terminal 2) and terminals 5 and 7 and 9 and 11 are connected.

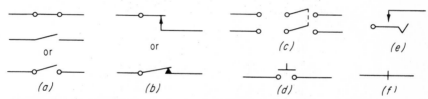

Figure 3·9 Symbols for switches. *(a)* Single-throw, general: closed and open. *(b)* Nonlocking, momentary: opening (break). *(c)* Double-throw, two-pole. *(d)* Pushbutton: closing (make). *(e)* Locking: closing (make). *(f)* Switching-function symbol: closed contact (break).

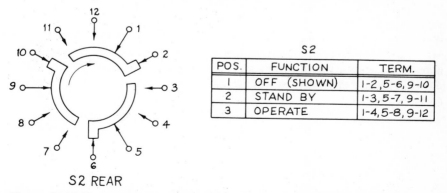

POS.	FUNCTION	TERM.
1	OFF (SHOWN)	1-2, 5-6, 9-10
2	STAND BY	1-3, 5-7, 9-11
3	OPERATE	1-4, 5-8, 9-12

S2 REAR

Figure 3·10 Position-function relationships for rotary switches. *(From American Standard "Drafting Manual," ANS Y14.15, "Electrical and Electronics Diagrams." Used by permission of the publisher, The American Society of Mechanical Engineers, 345 E. 47th St., New York, NY 10017.)*

3·9 The transformer

The symbols used for inductors are also utilized in the drawing of transformers. The American National Standard Y32·2 approves the use of either the helix symbol or the more rudimentary symbol. Except for one instance, ANSI/IEEE Y32E shows only the simpler symbol. We shall use the simpler symbol in most of our drawings. Transformers are made with air cores (usually found in high-frequency circuits) or with iron or laminated cores (found primarily in low-frequency ac circuits). The standard, however, does not require that the two parallel lines be drawn for magnetic or metallic cores. The polarity symbols in Fig. 3·11e are placed so that the instantaneous direction of current into one polarity mark corresponds to current out of the other polarity mark. An IEC symbol having round dots near each line is shown in the Appendix.

The power transformer supplies power (usually in different amounts) to two or more circuits. The secondary windings are often drawn with different numbers of loops, or cusps, to suggest the relative voltages going to these circuits. Sometimes the secondary circuits have additional taps.

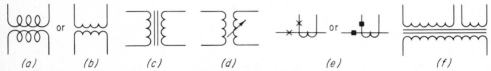

(a)   *(b)*   *(c)*   *(d)*   *(e)*   *(f)*

Figure 3·11 Symbols for transformers. *(a)* and *(b)* General. *(c)* If it is desired to show a magnetic core. *(d)* With one winding having adjustable inductance. *(e)* Current transformer with polarity markings. *(f)* Power transformer.

3·10 New device symbols

In an expanding area such as electronics, new devices for which no symbols exist will be forthcoming. Two methods of solving the problem of portraying these devices are in common use: (1) a new symbol for each device may be invented or designed, or (2) a rectangular block with appropriate identification may be inserted in the diagram at the place where the new device would be placed. Figure 3·12 shows a symbol which was designed by someone to show a cryogenic switch.

Figure 3·12 A symbol invented to show a cryotron.

3·11 Size of symbols

Theoretically, size of a symbol is not important. But the relative sizes of symbols are an important matter. These relative sizes are shown in the standards, in

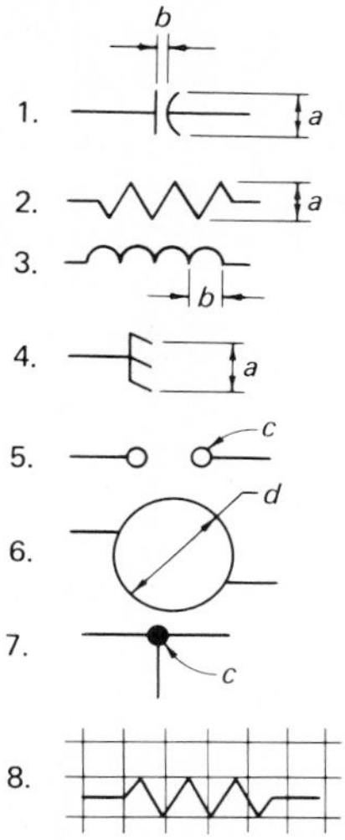

| | Measurements in fractions of an inch | | | | | | | |
| | minimum | | | | maximum | | | |
	a	b	c	d	a	b	c	d
Capacitor	.25	.06			.40	.10		
Resistor	.15				.30			
Inductor		.15				.25		
Chassis	.25				.35			
Terminals			.06					.10
Transistor envelope				.60				.80
Connection			.06				.12	

Figure 3·13 Suggested sizes for electrical-device symbols. (The lowest figure shows a resistor symbol that has been drawn on cross-ruled paper.)

Fig. 3·13, and in Appendix C of this book. Figure 3·13 suggests minimal and maximal sizes for several commonly used symbols. The minimum sizes are for small diagrams or for larger diagrams that are filled with many devices. The maximum sizes are for large drawings, say those on paper larger than 12 × 18. Other factors may dictate what size symbols should be drawn. A student or drafter using a template to make symbols is pretty much limited to the sizes that can be made with the template. Crowded conditions may require that the drafter draw symbols smaller than the suggested sizes. Lines are usually of medium weight—the same weight (width) used to draw the connecting paths and other lines in a drawing. However, symbols are sometimes made with heavy, thick lines for purposes of emphasis.

3·12 Drafting aids

There are at least four ways in which symbols can be made more quickly than by drawing them with conventional drawing instruments. These methods use:

1. Templates
2. Preprinted symbol cutouts
3. Typesetting equipment with symbols
4. Computer graphics

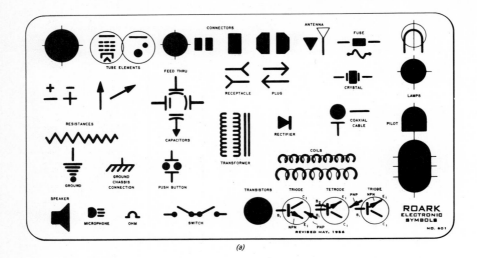

(a)

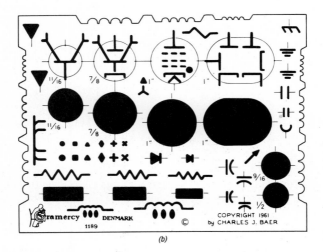

(b)

Figure 3·14 Templates and appliqués (pressure-sensitive adhesives). *(a)* and *(b)* Templates (the black areas represent open spaces). *(c)* Appliqués. *(Tech Tac and Bishop Graphics, Inc.)*

There are many designs of electrical templates on the market, and no one template will make all the symbols that are in use today. Great care should be used in the selection and purchase of such a device. A template should be used with a T square or drafting machine, as described in Chap. 1. Some templates are so thin that they tend to slip under the edge of the T square and hence are not entirely satisfactory. Some users prefer to raise the template off the

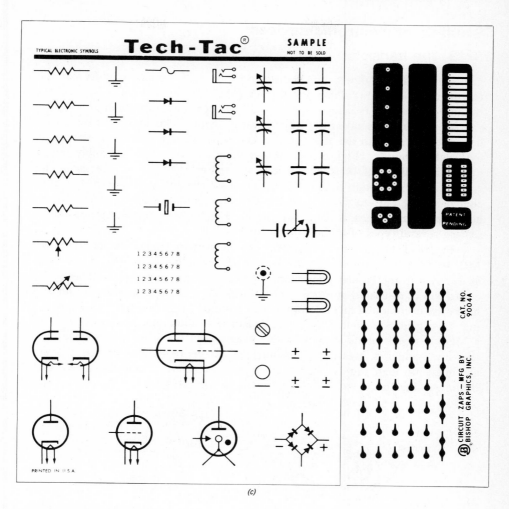

Figure 3·14 *(Continued)*

paper. One method is to place the template over a triangle and use the grooves that are over the open part of the triangle.

Preprinted symbols are manufactured as contact (pressure-sensitive) adhesives, or appliqués. These are cut out or lifted off a sheet of preprinted symbols and then positioned on the drawing in the correct place.

Certain equipment including the typesetting machine, interactive computer graphics devices, and automatic drafting machines can be programmed to produce many symbols. A detailed discussion of their use is beyond the scope of this book.

Symbols of electronic devices

3·13 The transistor

Called a *solid-state device*, the transistor is made of semiconducting material (germanium or silicon) which has characteristics that lie midway between those of good conductors (such as copper) and insulators (such as glass).

Most transistors have three leads; some have four. The three outside connections are connected to the base, emitter, and collector in a junction type transistor, such as bipolar. Figure 3·15 shows details of transistor assembly and a hydraulic analogy of the way a triode transistor performs. A change in the current flowing through the low-impedance base circuit causes a corresponding change in the current flowing to the collector circuit. The collector circuit, having high impedance, is the output circuit. Figure 3 · 16 shows symbols representing several types of bipolar transistors.

Figure 3·17 shows the symbols for field-effect transistors (FETs) and MOSFETs, which are transistors made with the metal oxide (MOS) process, as opposed to the bipolar transistors shown in the previous figure. (These processes will be discussed at more length in Chap. 7.) These transistors are finding heavy use in computer and microprocessor circuits. The letters are for the following: G = gate, S = source, D = drain, and U = substrate. The active area of a MOS device is at its surface, where a gate electrode applies a voltage to a thin layer under it, creating a "channel." Contacts to the end of this channel are made through the N or P regions, and these are called the *source* and *drain*. An idealized cross section of a N-channel MOSFET is shown in Fig. 3·18.

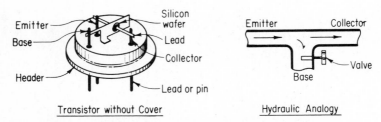

Figure 3·15 Assembly and action of bipolar transistor.

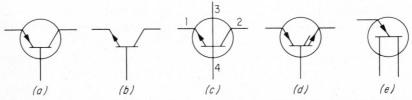

Figure 3·16 Symbols for bipolar transistors. *(a)* PNP type. *(b)* NPN type. *(c)* Tetrode. *(d)* PNPN type. *(e)* Unijunction with P-type base.

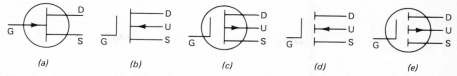

Figure 3·17 Field-effect transistors (FETs). *(a)* N-channel junction-type FET, or JFET. *(b)* N-channel depletion-type metal-oxide FET, or MOSFET. *(c)* P-channel depletion-type MOSFET. *(d)* N-channel enhancement-type MOSFET. *(e)* P-channel enhancement-type MOSFET.

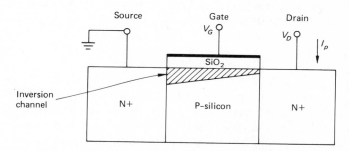

Figure 3·18 Idealized cross section of N-channel MOSFET.

Several points should be made about transistor-symbol construction:

1. The circle (envelope) does not have to be drawn if no confusion arises or if no leads are attached to the envelope.

2. Orientation, including a mirror-image presentation, does not change the symbol meaning.

3. For the NPN and PNP transistors, the base symbol is drawn about one-third of the way "up" if the envelope circle is drawn.

4. Collector and emitter lines are drawn at about 60° to the base symbol, and the arrowheads do not touch the baseline.

5. The vertical line in the junction-type FET, called the *channel*, is drawn through the center of the circle.

Although the envelope circle is not required, many engineers and drafters prefer or recommend that it be drawn. For this reason, we have shown most of the transistor symbols in Figs. 3·16 and 3·17 with envelopes. In Fig. 3·16c, the leads have been numbered, starting with the emitter, which is the standard order when numbering is desired. The student should become familiar with the symbols and current-flow characteristics of bipolar PNP and NPN transistors. Figure 3·19 supplies this information. The circular symbol with ~ inside indicates a signal source. We have only briefly touched upon the electron action of the transistor, a fascinating subject. There are several good books that go into this subject fully. Several are listed in the Bibliography.

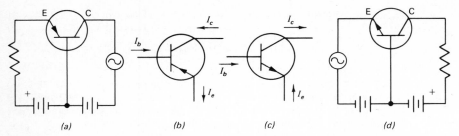

Figure 3·19 Transistor biasing. *(a)* Biasing of PNP transistor in an amplifier circuit. *(b)* and *(c)* Electron flow in PNP and NPN transistors. *(d)* Proper biasing of NPN transistor.

3·14 The diode

Another often-used semiconductor device is the diode. Figure 3·20 shows symbols for several different kinds or arrangements of diodes. The basic rectifier symbol is usually not encircled, but it—as well as the other symbols shown—may be enclosed in a circle if it is desired. This symbol points in the direction of conventional current, sometimes called the "easy" direction. The bar part of the symbol corresponds to the cathode of an electron tube. Forward-biased diodes produce a "square" wave. Diodes are not always forward-biased, however. Diodes are used for different purposes such as rectification, detection, amplification, and switching. The tunnel diode, for example, has a negative-resistance characteristic that permits it to be used as an amplifier or as a switch. The bridge rectifier provides an easily smoothed dc output with about twice as much average voltage as the full-wave bridge.

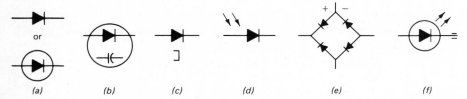

Figure 3·20 Diode symbols. *(a)* General. *(b)* Capacitive (varactor). *(c)* Tunneling (tunnel). *(d)* Photosensitive type. *(e)* Bridge. *(f)* Light-emitting (photoemissive) type (LED).

The location and installation of semiconductors requires the following procedures:

1. Don't exceed their maximum voltage, current, or power ratings.
2. Use adequate heat sinks for power devices.
3. Avoid prolonged exposure to heat during soldering.

4. Don't locate sensitive circuit devices adjacent to heat-producing power devices.

5. Don't solder semiconductors into an electrically live circuit.

The *silicon controlled rectifier* (SCR) will be discussed in Chap. 8.

3·15 The electron tube

A vacuum tube consists of an evacuated envelope, a cathode which supplies electrons, and an anode which collects the electrons; it may have one or more grids. (See Fig. 3·21.)

When the cathode is heated, it releases electrons, which flow to the anode (or the *plate*) when the latter has a positive potential with respect to the cathode. A grid, which is a fine mesh or helix that offers little obstruction to the electron flow if it is at the same or slightly lower potential than the cathode, may be placed between cathode and anode. However, if this grid is attached to a circuit that receives certain signals, the grid potential may rise and fall, causing alternating changes in the flow of electrons toward the positively charged plate. Because of this throttling, or valvular, action, the electron tube is sometimes referred to as a valve.

Production of electron tubes has dropped since the advent of transistors and integrated circuits. Only General Electric and one or two others still manufacture tubes.

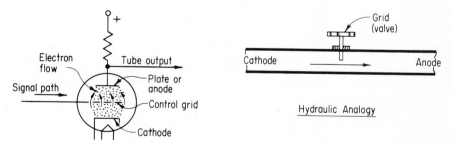

Figure 3·21 Action of electron flow in vacuum tube.

3·16 Tube elements

The triode tube (whose action is described in Fig. 3·21) is so named because it has three electrodes—a plate, a cathode, and a grid. It also has a heater symbol, but the heater in this case is not called an electrode. In this instance, the cathode is the indirectly heated type—it consists of a thin metal sleeve which surrounds a heater of a tungsten-alloy wire filament. Sometimes the heater symbols are shown in a separate heater-circuit diagram and thus do not appear on the tube symbol. Also, a tube may be directly heated, in which case the heater element itself is the cathode. If it is the heater symbol shown in Fig.

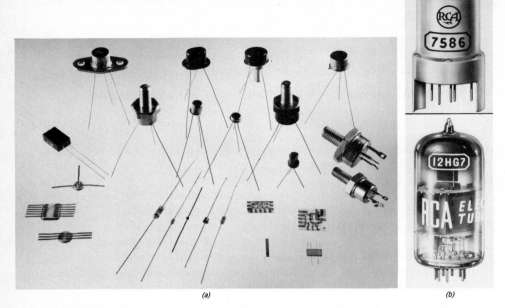

(a)　　　　　　　　　　　　　　　　　　　　　　(b)

Figure 3·22 Electron tubes and solid-state (semiconductor) devices. *(a)* Transistor, diode, and integrated semiconductor circuits. In the upper part of the photograph, the devices having three leads are transistors. The smaller devices in the lower part are diodes, and most of the flat objects are integrated semiconductor circuits. *(General Electric Co.) (b)* Electron tubes. Model 12HG7 is a frame-grid-type sharp-cutoff pentode in a T9 envelope. Model 7586 is a medium-mu Nuvistor triode in ceramic-and-metal construction. *(RCA, Electronic Components and Devices Division.)*

3·23*a*, it is drawn without the cathode symbol. The included angle in this symbol is 90°; the angle of the x-ray, or target, electrode is 45°. The electrodes shown in Fig. 3·23 are from ANSI/IEEE Y32E and are reproduced in the back of the book along with many examples of different kinds of electron tubes. The circular tube symbol is usually drawn a little larger than the transistor

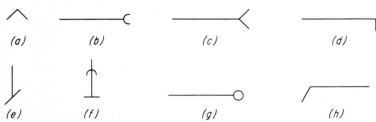

Figure 3·23 Symbols for tube electrodes. *(a)* Directly heated cathode, or heater. *(b)* Photocathode. *(c)* Deflecting electrode. *(d)* Excitor. *(e)* Target, or x-ray anode. *(f)* Dynode. *(g)* Cold cathode. *(h)* Ignitor, or starter.

symbol. In many commercial drawings, it is made about 1 to $1\frac{1}{2}$ in. in diameter. Sometimes the circle is made with a heavy line to make the tube (or transistor) stand out.

3·17 Tube envelopes

In most cases, a tube envelope is drawn as a circle. There are numerous instances in which this is not possible or desirable; in such cases, it is drawn as an elongated or split envelope. Figure 3·24 shows examples of elongated and split envelopes. The pentode can be drawn in either the circular or the oblong envelope, but it is less crowded in the oblong. Figure 3·24d and e shows different methods of splitting tube symbols. The tube represented by Fig. 3·24c and d is a twin triode.

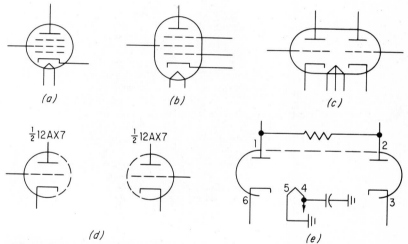

Figure 3·24 Tube-envelope symbols. *(a)* Pentode in circular envelope. *(b)* Pentode in elongated envelope. *(c)* Twin triode in elongated envelope. *(d)* Multiunit triode in split envelope. *(e)* Multiunit diode in split envelope.

3·18 Picture tubes (Cathode-ray, kinescope, or video tubes)

There are two types of video or cathode-ray tubes, electrostatic and electromagnetic. Both types have three main parts to their system—an electron gun, deflecting devices, and a fluorescent screen. The electrostatic system is used primarily in oscilloscopes, while the electromagnetic system is used in most television tubes and for radar. Picture tubes are generally portrayed symbolically, as shown in Fig. 3·27. In Fig. 3·27a, the deflecting electrodes are shown, while they are omitted in Fig. 3·27b. In this figure, however, dashed lines signifying conductive coatings on the glass are shown. Accelerating voltage is applied to the

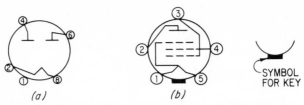

Figure 3·25 Base terminal markings. *(a)* Small pins. *(b)* Large pins.

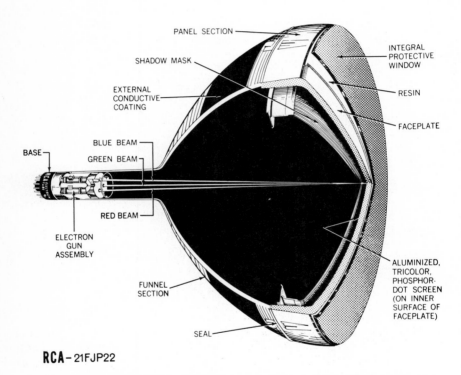

RCA– 21FJP22

Figure 3·26 Photograph of a cutaway section of a 70° color picture tube. *(RCA, Electronic Components and Devices Division.)*

inside coating through a high-voltage connector on the outside of the tube. The outside coating, grounded to prevent radiation, forms a capacitance with the inside coating to filter high-voltage pulses applied to the inside of the tube. In this symbol, the deflecting electrodes are not shown; presumably they are shown elsewhere, separately—a not unusual practice. Figure 3·27*c* and *d* shows another way of drawing picture-tube symbols. Most diagrams indicate the tube shape more or less as it is shown in Fig. 3·27*a* and *b.*

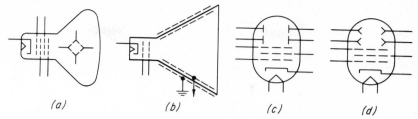

Figure 3·27 Picture-tube symbols. *(a)* and *(b)* Electromagnetic. *(c)* Electrostatic. *(d)* Electromagnetic.

3·19 Gas-filled tubes

The action of gas-filled electron tubes is considerably different from that of vacuum tubes. In the gas-filled type, many more electrons reach the anode than leave the cathode. This so-called *gas amplification* occurs because atoms of the gas become ionized. In one much-used tube, called the thyratron, a triggering or firing action takes place, following which the tube acts as a rectifier. A large dot in a tube symbol indicates that it is gas-filled.

3·20 Nomenclature

Tubes, and to a lesser extent semiconductors, can be classified according to the number of electrodes as follows:

Diode—two electrodes	Pentode—five electrodes
Triode—three electrodes	Hexode—six electrodes
Tetrode—four electrodes	Heptode (pentagrid)—seven electrodes

3·21 Qualifying symbols

There are many qualifying symbols that, when added to a device symbol or conductor, indicate that that special characteristic is important to the function of the device. We have shown a few of these in Figs. 3·2, 3·8, 3·9, and 3·20. Because their number has grown so much in the last two decades, we have shown many of the qualifying symbols (listed in ANSI/IEEE Y32E) in Fig. 3·28.

3·22 Integrated-circuit modules

The symbols shown thus far have been for individual devices. The wide use of integrated-circuit packages has made it necessary to somehow portray the entire circuit, which may have the equivalent of hundreds of transistors or other devices, as a part of a larger circuit. Present practice includes two symbols

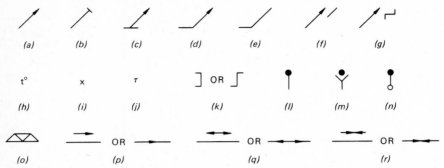

Figure 3·28 Qualifying symbols. *(a)* Adjustability, general. *(b)* Preset adjustability. *(c)* Linear, extrinsic. *(d)* Nonlinear, extrinsic. *(e)* Nonlinear, intrinsic. *(f)* Adjustable, continuous. *(g)* Adjustable, in steps. *(h)* Temperature dependence. *(i)* Magnetic-field dependence. *(j)* Storage. *(k)* Break-down, overvoltage absorber. *(l)* Test-point recognition symbol, general. *(m)* Test point for a test jack. *(n)* Test point for a circuit terminal. *(o)* Electret (without electrodes). *(p)* Direction of power or signal flow, one way. *(q)* Either way. *(r)* Both ways, simultaneously.

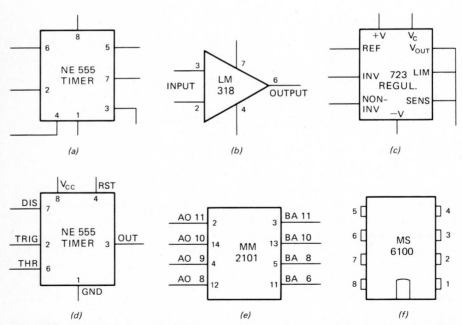

Figure 3·29 Symbols for integrated-circuit modules. *(a)* Rectangle with pin numbers. *(b)* Triangle, usually used for amplifier circuits only. *(c)* Rectangle with functions assigned to each pin. *(d)* Rectangle with pin numbers and assignments. *(e)* Rectangle with pin numbers and circuit identification. *(f)* Standard numbering of pins on 8-pin dual-in-line package.

that are employed to show the integrated circuit. They are the rectangle and the equilateral triangle, as shown in Fig. 3·29*a* and *b*. The triangle is usually reserved for ICs that are amplifiers because it is so listed in the standards. However, since many ICs are not amplifiers, the rectangle is widely used.

Unfortunately, the use of the rectangle is not spelled out in the standard. Usually, the manufacturer's number and the function are listed, as has been done in Fig. 3·29*a, c,* and *d.* Pin numbers are also generally shown. More information, such as the ground or input voltage, is helpful. Circuit destination or identification, as shown in Fig. 3·29*e,* is also helpful. Going on the assumption that a schematic or wiring drawing should be easy and quick to read, the authors favor the methods shown in Fig. 3·29*d* and *e.* Figure 3·29*f* shows the typical pin arrangement for an 8-pin dual-in-line package. The reader will notice that the other examples shown do not follow this pin sequence. This is because the pins have been rearranged (on the drawing) to facilitate making the physical circuit. The manufacturer of the integrated circuit provides the information about the pin assignments. Packages for LSI (large-scale integration) chips have various numbers of leads. The largest commercially available package has 64 leads, but the package sizes are growing.

SUMMARY

Nearly every electrical and electronics device has a standard symbol which can be used to represent the device in a diagrammatic drawing of a circuit. These symbols are drawn with medium-weight lines but can be made with heavier lines, if it is necessary to highlight the symbol. Transistor symbols may or may not include the circle envelope, while tube symbols must have the envelope symbol. Theoretically, there are no particular sizes to which symbols should be drawn. Practically, though, there are minimal and maximal sizes for any symbol or set of symbols, and, within a drawing, symbols should be drawn in correct relative sizes. Usually a device is represented by a symbol of one size throughout a drawing. Not more than two sizes of a symbol are recommended for a drawing. There are certain basic electrical and electronics devices with which a reader or drawer of electrical drawings should become familiar. Orientation of a symbol depends upon the direction of the conductor path along which it is placed. Symbols for the most commonly used devices or functions have been standardized on a U.S. and international basis.

QUESTIONS

3·1   What authority (publication) or authorities would be good source material for the selection of symbols to be used in an electrical drawing?

3·2   How would you go about determining the size to make a symbol for a circuit breaker in an electrical drawing?

3·3   What determines the position (orientation) of a symbol in a drawing?

3·4   When would you put the polarity markings of a battery on its symbol in a drawing?

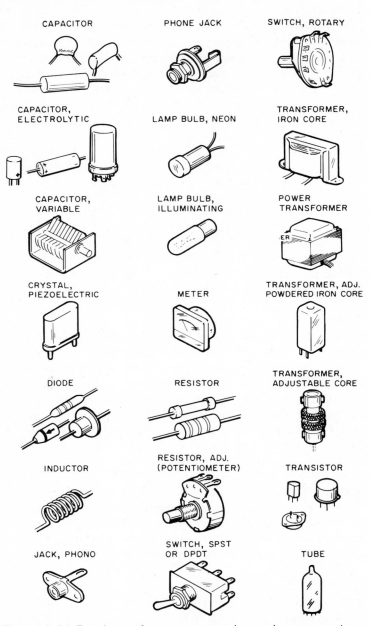

CAPACITOR

PHONE JACK

SWITCH, ROTARY

CAPACITOR, ELECTROLYTIC

LAMP BULB, NEON

TRANSFORMER, IRON CORE

CAPACITOR, VARIABLE

LAMP BULB, ILLUMINATING

POWER TRANSFORMER

CRYSTAL, PIEZOELECTRIC

METER

TRANSFORMER, ADJ. POWDERED IRON CORE

DIODE

RESISTOR

TRANSFORMER, ADJUSTABLE CORE

INDUCTOR

RESISTOR, ADJ. (POTENTIOMETER)

TRANSISTOR

JACK, PHONO

SWITCH, SPST OR DPDT

TUBE

Figure 3·30 Drawings of some commonly used components. *(Artwork courtesy Heath Company.)*

3·5   What would be a typical designation that would be placed in the rectangular resistor symbol? In a circular relay-coil symbol?

3·6   How many "points" would you make in drawing the zigzag resistor symbol? How many turns would you make on the helical inductor symbol? On the other inductor symbol?

3·7   Where more than one symbol is approved for an element (or component), how would you go about determining which symbol to use on an electrical drawing?

3·8   Assume that you are required to draw the line representing a signal path of a circuit approximately $\frac{1}{2}$ mm wide. How wide a line would you use to draw the symbols in that circuit drawing? Why?

3·9   What determines whether a ground symbol or chassis symbol should be used in an electrical drawing?

3·10 A statement in the standard indicates that not more than two sizes can be used for any one symbol in the same drawing. Describe a situation that, in your opinion, would require two different sizes of the same symbol in a drawing.

3·11 The IEC-approved capacitor symbol of two parallel lines has a disadvantage. What is it?

3·12 Name five different kinds of switches or devices that can be used for switches in electric circuits.

3·13 How would you show a device in a circuit drawing if no known symbol for that device exists?

3·14 What does the dot signify when placed within the tube circle symbol?

3·15 What is the difference between a pentagrid and a pentode tube?

3·16 What is the difference between a diode and a triode tube? Between a tetrode and a diode?

3·17 Name some typical types of tubes that would lend themselves to representation by the elongated envelope symbol when graphically portrayed.

3·18 What is the drawing size of the circular part of a transistor symbol as compared to the drawing size of an electron-tube circle symbol?

3·19 Symbolically, how is a polarized capacitor shown in a circuit path?

3·20 What does the bar in the diode symbol represent?

3·21 What do the letters V_{GG} and V_{DD} stand for?

PROBLEMS

The problems below will require that the student make use of the American National Standard or some standard which indicates the correct symbols. The symbols shown in Appendix C are taken from several American standards. It would also be advantageous for the student to look at some of the schematic drawings shown at various places in the text, especially in Chap. 6, before beginning to solve the problems below.

Cross-section paper with four or five divisions to the inch may be helpful, especially if the symbols are to be sketched freehand. It is not an absolute requirement, however. Any type of detail or tracing paper or film will be adequate. Exercises can be drawn on $8\frac{1}{2} \times 11$ paper, except where otherwise indicated.

3·1 Draw lightly three horizontal lines 9 in. long and 3 in. apart. At 2-in. intervals (starting $\frac{1}{2}$ in. from the left end), draw the symbols for the following elements:
   a. Capacitor (fixed)
   b. Resistor (general)
   c. Battery (multicell)
   d. Speaker
   e. Ammeter
   f. Transformer (general)
   g. Relay coil
   h. Fuse
   i. Contact (normally open)
   j. Motor
   k. Transformer (current)
   l. Switch (pushbutton, make)
   m. Potential transformer
   n. Inductor (continuously adjustable)
   o. Antenna
   Add the name below each symbol.

3·2 Follow the instructions given in the preceding problem, but show the following symbols:
   a. Shielded capacitor
   b. Chassis connection
   c. Transformer with core
   d. NPN transistor
   e. Contact (normally closed)
   f. Inductor with two taps
   g. Polarized relay
   h. LED
   i. Adjustable capacitor
   j. Pushbutton switch
   k. Variable resistor
   l. Switch, double-throw
   m. Tetrode transistor

  n. Split-stator capacitor

  o. Inductor with ceramic core

3·3 Follow the instructions given in Prob. 3·1, but show the following symbols:

  a. Capacitive diode

  b. PNP transistor

  c. N-channel depletion-type MOSFET

  d. Delay function

  e. Pool-type vapor rectifier

  f. Incandescent lamp

  g. Field-effect transistor, N-type base

  h. Thermistor

  i. Pentode tube

  j. Semiconductor thermocouple, current-measuring

  k. Light-emitting diode

  l. Thyratron tube (gas-filled triode)

  m. PNPN transistor

  n. P-channel enhancement-type MOSFET

  o. PNIN transistor, with ohmic connection

3·4 Follow the instructions given in Prob. 3·1, but show the following symbols:

  a. Dipole antenna

  b. Schmitt trigger

  c. Separable connectors

  d. Slow-operate relay

  e. Breakdown diode, unidirectional

  f. Capacitor with split stator

  g. Pickup head, recording

  h. Switching function, transfer

  i. Thermistor, general

  j. Adjustable resistor

  k. Field-effect transistor, N-type base

  l. Twin-triode tube

  m. Four-conductor shielded cable

  n. Tunnel diode

3·5 Sketch or draw the following tube symbols, placing the name under each tube. Use indirectly heated cathodes.

  a. Diode

  b. Tetrode

  c. Thyratron (gas-filled triode)

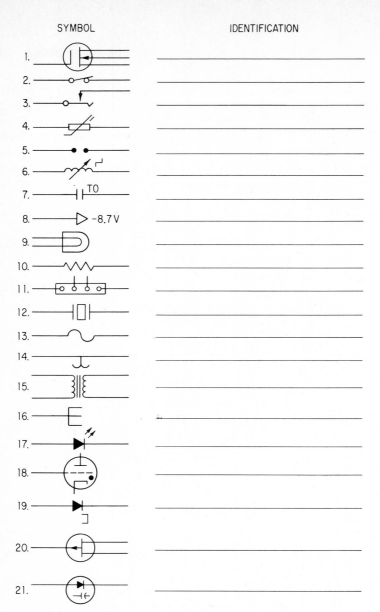

SYMBOL   IDENTIFICATION

1. _____

2. _____

3. _____

4. _____

5. _____

6. _____

7. _____

8. _____

9. _____

10. _____

11. _____

12. _____

13. _____

14. _____

15. _____

16. _____

17. _____

18. _____

19. _____

20. _____

21. _____

Figure 3·31 (Prob. 3·6) Symbol identification exercise.

d. Twin triode with common cathode
e. Gas-filled ignitron (with exciter and control grid)
f. Triode
g. Pentode
h. Twin diode
i. Pentagrid or heptode

3·6   Redraw the symbols shown in Fig 3·31 at about twice the size they appear in the book. (Use one 11 × 17 or 12 × 18 sheet, or two 8½ × 11 sheets.) Then identify each device symbolized at the right. (*Suggestion:* Draw a horizontal line to the right of each symbol, as shown in Fig. 3·31, and letter the name along each line.) Identification, in many cases, should include more than just the name. If it is a transformer, for example, what *kind* of transformer it is; if a switch, what type it is and what its operating condition is.

3·7   Draw the circuit shown incompletely in Fig. 3·32, and put the symbols in the places shown, correctly positioned. Where figures 1, 2, and 3 appear, draw resistors. At 4, draw a capacitor (fixed). At 7, draw a variable capacitor. At 5, draw a PNP transistor, with emitter at *e,* and so forth. At 8, draw an inductor. Place identifying lettering at terminals as follows: 11, *J*10; 15, *GRD;* 16, +5; 12, *J*15; 13, *J*20; 17, *J*21; and 14, −5. All lines in the completed diagram should be the same weight. If necessary, go over lines that are too light. Use 8½ × 11 paper.

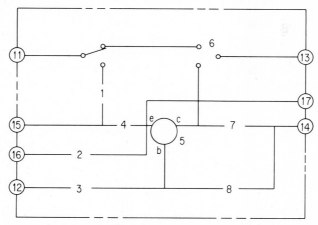

Figure 3·32 (Prob. 3·7) Symbol exercise. Transmit-standby Apollo color camera. (No-dot system is used.)

3·8   Complete the circuit shown in Fig. 3·33 by adding the symbols listed below at the places where the numbers appear. Rectifiers CR4 through CR7 may be redrawn as the bridge in Fig. 3·20*e*.

Locking switch (make): 1

Multicell battery: 2

Polarized capacitor: 3

Solid-state rectifier: 4

Resistor: 5

Diode: 6

Resistor: 7

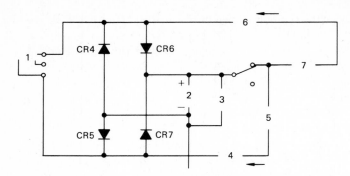

Figure 3·33 (Prob. 3·8) Calculator charging circuit.

The completed drawing should have all lines the same weight. Use $8\frac{1}{2}$ × 11 paper.

3·9   Draw the circuit shown in Fig. 3·34 about twice as large as it appears there. Then add the symbols as indicated by their standard letter-number designations. (These can be found in Mil Std 16B, shown in Appendix A and elsewhere in the book.) The transistor is of the NPN type, with base, emitter, and collector leads identified. Use the terminal symbol at the input and output. C_1, C_3, C_4, and C_6 are variable. L_1 and L_2 have fixed taps. The lines on the completed drawing should have the same weight. Use $8\frac{1}{2}$ × 11 paper.

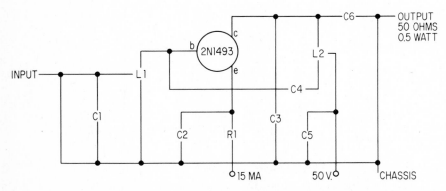

Figure 3·34 (Prob. 3·9) Symbol exercise. Power-amplifier circuit.

3·10 Each symbol in Fig. 3·35 is incorrect or not American National Standard. First, identify the symbol that is being depicted. Then draw the correct symbol and label it, using the same general layout as in Fig. 3·35. Use $8\frac{1}{2}$ × 11 or 11 × 17 paper.

3·11 Identify some symbols assigned by your instructor. Use the format assigned.

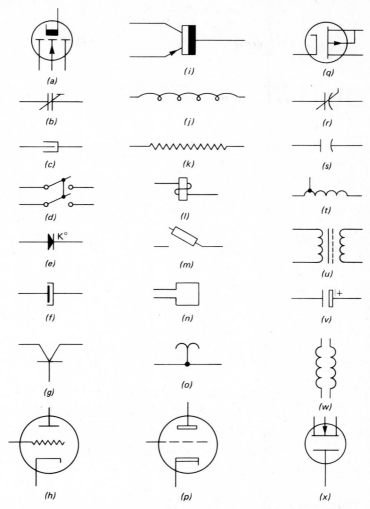

Figure 3·35 (Prob. 3·10) Symbol exercise. Incorrect symbols.

Chapter 4
Production drawings

Before an electronics/electrical package is built, instructions must be given to the production workers who are going to have to build or assemble the unit. Drawings convey these instructions very well, although they sometimes must be accompanied by other information. There are several different kinds of drawings used for this purpose. Correct or recommended practice for most of these drawings is set down in certain ANSI, IEEE, and Military Standards. Some of the more relevant standards are listed below:

ANSI Y14 "Drafting Manual," which includes

14.1 "Drawing Size and Format"

14.2 "Line Conventions, Sections and Lettering"

14.3 "Multi and Sectional Drawing Views"

14.4 "Dimensioning and Tolerancing for Engineering Drawings"

14.6 "Screw Threads"

14.15 "Electrical and Electronics Diagrams"

Mil Std 242 "Electronic Equipment Parts (Selected Standards)"

Mil Std 196 "Joint Electronics Type Designation System"

Mil Std 429 "Printed Wiring and Printed Circuits Terms and Designation"

Mil Std 681 "Identification, Coding and Application of Hook Up and Lead Wire"

Company standards, if any, that may be more specific or restrictive than ANSI, IEEE, or Military Standards

In this chapter we will discuss the following types of drawings that are commonly used for production:

Wiring or connection diagrams

Cabling diagrams

Harness diagrams

Sheet-metal layouts

Assembly drawings

Printed circuit layouts

Connection Drawings

4·1 Connection, or wiring, diagrams

In order to connect the various component devices (resistors, transistors, etc.) in an instrument such as an amplifier, it is the usual custom to follow a *connection diagram*. The older name for this type of drawing is *wiring diagram*, and, because it is so widely used, we shall use these two terms interchangeably.

In a manufacturing company these drawings are used by the time-and-motion, quality-control, and estimating sections, as well as by individual workers. Wiring diagrams are also used in the maintenance area by persons doing checking, troubleshooting, and modification. Sometimes they are used with schematic diagrams (explained in Chap. 6) to provide additional information.

In mass production, a connection diagram may be used at first, then put away after the worker has memorized the steps and connections. In some instances, companies with active methods sections are able to produce without such a drawing.

4·2 Types of connection diagrams

These diagrams may be classified according to their general layout as follows:

1. Point-to-point
2. Highway, or trunkline
3. Baseline, or airline
4. Straight-line

There are also other ways of labeling wiring diagrams, as the reader will discover. Such titles might include *cabling, harness,* and *interconnection* diagrams.

4·3 Point-to-point diagrams

This is the oldest type of wiring drawing. Figure 4·1 shows a pictorial form of the diagram in which each part and terminal are drawn about as they appear to the viewer. This type of drawing has met with great success in the do-it-yourself kit industry and other places.

Figure 4·2 is another type of pictorial point-to-point diagram, but it is not as close to what the actual object looks like as is the first drawing. Many of the symbols are pictorial, but their physical relationship to each other is not accurate. Furthermore, the lines representing the wires (conductors) are drawn either horizontally or vertically, not as they actually are placed in the chassis or frame of the automobile. A distinctive feature of this drawing is the color code and identification system required when there are as many wires as are presently utilized in motor vehicles. The wire in the upper right corner, for example, is *L*7-18 *BK*. This means that this is conductor *L*7, which is of size 18 wire (see p. 432) and is black in color.

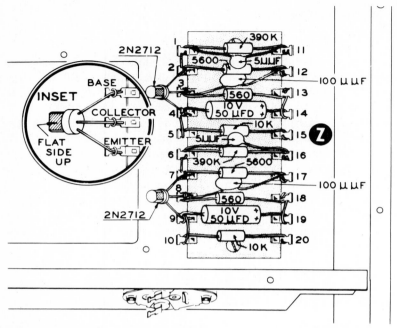

Figure 4·1 A pictorial production drawing for wiring an electronic assembly. *(Heath Company.)*

In the two drawings discussed so far, the component parts have been drawn pictorially. In most wiring diagrams, these parts are shown by means of symbols or by simple geometrical figures such as squares, rectangles, and circles. This will be the case with the connection diagrams that follow, including the aircraft wiring diagram of Fig. 4·3.

Figure 4·3 shows only a small part of the original drawing, but it contains enough details to give a fair picture of what such a drawing looks like. Note the switch symbols $S8$, $S9$, etc., the battery symbol, $BT1$, and the circuit-breaker symbols, $CB1$, $CB2$, etc. Note, also, the uniform spacing of wire conductors, drawn about $\frac{1}{4}$ in. apart, with their rounded "corners." Actually, most of the wires are routed in bundles (harnesses), and the physical routing of wires may or may not bear close resemblance to the way they are drawn here. Item $J9$, below center and right, is a female receptacle into which male connector $D7$ is plugged. The letters represent the pins in the connector. Readers will have a difficult time tracing the paths of many of the wires in this drawing because only part of the drawing is here. They can trace line $H27$ from switch $S8$ to $CB5$ and line $H37$ from $S8$ to $J9$, however. In Figs. 4·2 and 4·3, medium-weight lines have been used to represent wires. This is common practice, although some manufacturers draw all or certain conductors more heavily.

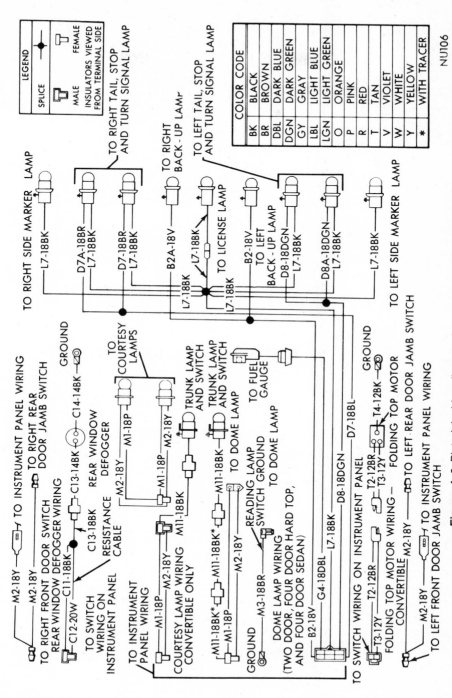

Figure 4-2 Pictorial wiring diagram for an automobile. *(Chrysler.)*

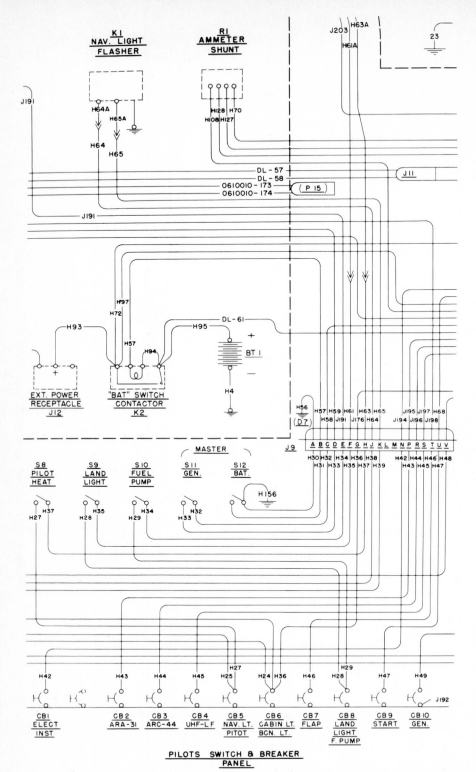

Figure 4·3 Connection (wiring) diagram for a small aircraft. *(Cessna.)*

4·4 Interconnection (external-wiring) diagrams

Figure 4·4 is another example of a point-to-point wiring diagram. It consists of conductors (wire or cables) connecting items of equipment or unit assemblies. Internal connections are customarily omitted in such a drawing. Note that the conductor lines are either horizontal or vertical and have sharp corners. Note, also, the shielded and twisted wires running horizontally through the center of the diagram. A newer twist symbol is shown in the Appendix.

4·5 Highway, or trunkline, type of wiring diagram

Practically any electrical assembly or system can be shown wired or connected graphically by the point-to-point method. But the same assembly or system can also be connected graphically by other methods. Sometimes the point-to-point diagram gives the clearest picture and fastest reading, yet, at other times, one of the other methods might give the best picture and provide easier and quicker reading. One of these other methods is the highway type of connection diagram. This type differs from the point-to-point style in that *conductors* (*conductor paths* might be better) are merged into long lines called *highways* (or *trunklines*) instead of being drawn as separate, complete lines from terminal to terminal. The short lines leading from the terminals to the highways are called *feed,* or *feeder, lines,* and must have some sort of identification near the highway so that the conductor path can be followed by the reader.

One or more highways may be shown, depending upon the wire routing necessitated by the physical (or graphical) arrangement of components. These trunklines do not have to conform to actual bundles or harnesses, if such are used. Figure 4·5 shows the rear view of a control panel of a copy-reproducing machine. In this figure, the highways are located so that the lengths of the feed lines are short and so that there is a minimum of crossing of highways and feeders. (Highways may cross, and feeders may cross each other and highways, but the number of crossings should be kept to a minimum.) Each conductor is given a number, and that number appears periodically on the drawing so that the reader can follow the conductor path. Line 1, for instance, can be traced from *P*2 to *TB*3 and on to *TB*4, which does not appear on the drawing. Line 2 can be followed from *P*2 to the control terminal. It must be remembered that there is a separate wire for line 1 and another wire for line 2 from *P*2 to *TB*3. The highway is just an imaginary thing. The direction a conductor path takes when it is in the highway is indicated by the direction of the quarter-circle arc where the feed line enters the highway. For example, when leaving the control terminal, line 11 goes to the right and line 14 goes to the left. Short 45° lines, as shown in the inset of Fig. 4·5, are also approved by the ANSI/IEEE Standard.

A different method of identifying circuit paths is in wide use. This ANSI-recommended method suggests that each feed line have this designation near the point where it joins the trunkline: (1) component destination, (2) terminal or component part, (3) wire size or type (if necessary), and (4) wire color. In

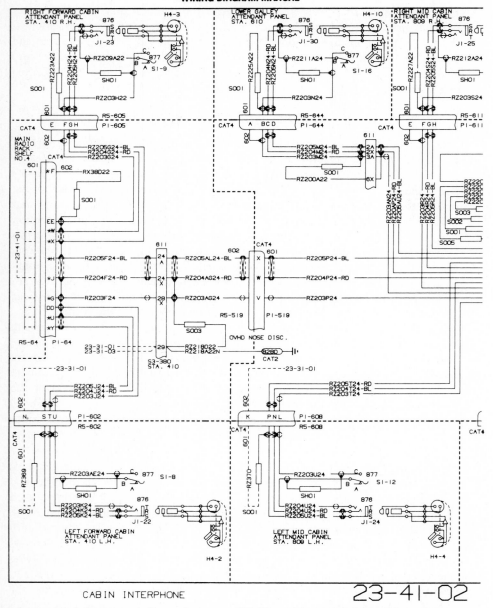

CABIN INTERPHONE

23-41-02

Figure 4·4 Interconnection diagram. DC-10 cabin interconnection system. *(McDonnell-Douglas.)*

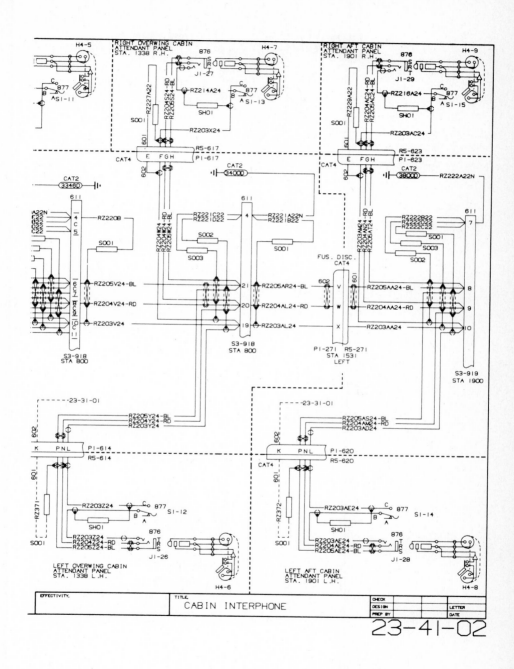

CABIN INTERPHONE

23-41-02

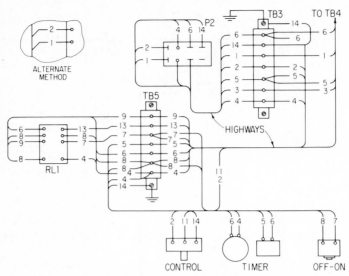

Figure 4·5 A typical highway type of connection diagram.

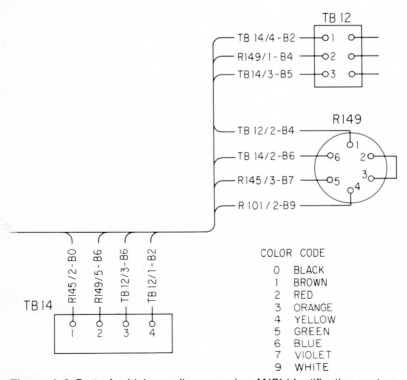

COLOR CODE

0	BLACK
1	BROWN
2	RED
3	ORANGE
4	YELLOW
5	GREEN
6	BLUE
7	VIOLET
9	WHITE

Figure 4·6 Part of a highway diagram using ANSI identification system.

Fig. 4·6, the top feeder line at terminal board 12 carries the following identification: *TB*14/4-*B*2. Broken down, this would be: *TB*14—destination; 4—terminal at destination; *B*—size of wire (say, No. 22 hookup wire by a prearranged code); and 2—color (red, by Mil Std 122 color code shown on the drawing and in Appendix B). Going to *TB*14 and terminal 4, we find the other end of this connector, and here it has the reverse identification, *TB*12/1-*B*2.

The choice of location and number of highways may depend on several factors. Separate highways for different cable groupings may be desirable. If a group of wires needs to be segregated or shielded, a separate highway for that group may be desirable. Such a highway might be drawn close to, and parallel to, another highway. Occasionally situations may arise where certain wires, called *critical wires,* should be drawn from point to point and not merged into a highway. Special identification for these wires may be necessary.

The highway type of connection diagram is particularly suited for showing the wiring of large panels where there are many terminals and conductors. Figures 4·5 and 4·6 are both panel-wiring diagrams. A panel-wiring diagram generally shows the back, or wiring, side of the panel. Terminal strips, switches, breakers, etc., are usually shown in exact or approximate position with regard to each other. Sometimes the panel is drawn as a scale drawing showing all terminals, lights, lettering, and switches, and the wiring is shown diagrammatically elsewhere on the drawing sheet.

4·6 Airline, or baseline, diagrams

These wiring drawings, with a rather misleading name, are similar in some ways to highway diagrams. The *airline,* or *baseline,* as it is sometimes called, is an imaginary, usually horizontal or vertical, line conveniently located so that short feed lines may be drawn from component terminals to it.

The heavy dark line in Fig. 4·7 is the airline in this drawing. The feed lines are drawn at right angles to it, and there are no curves where they join the airline, as there are in highway diagrams. As in the case of highway diagrams, good identification of feed lines is necessary. In this figure, a simplified version of the American National system is used. Each feeder has the number of the destination component part and the color of the wire. The drawing is read as follows: Take the conductor, which is identified as BK-R 13, leading from part 15. Now go to component 13 and look for the black-red feeder. It is quickly found with the identification BK-R 15. So we have an identification and reading procedure similar to that used in highway diagrams. The major difference between airlines and highways is that a highway must go from one component to the destination component, sometimes making several bends and turns to get there, but an airline may stop at any convenient place and another airline may be drawn near the destination component. In our example this does not happen, but in some airline drawings having many component parts there are many airlines spotted at different places close to groups of parts. In such cases, a feed line may enter one airline and come off another airline. This makes for

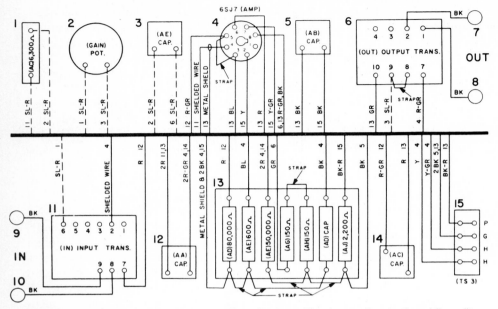

Figure 4·7 An airline type of connection diagram. The heavy line is the airline. (It is often not made heavy.)

slower reading and tracing of conductor paths, but for systems or packages containing hundreds of wires and terminals, the airline type of diagram may be less cluttered and confusing than point-to-point and highway drawings are.

Identification is important in two respects. First, the component parts themselves must be labeled with large block numbers in or near their upper left corners. And second, proper letters should be used for colors where a numbered color code is not used. One-letter designations are adequate for some colors such as yellow and red, but two letters are necessary for others such as black (BK), blue (BL), brown (BR), green (GN), gray (GY), and slate (SL). Figure 4·7 follows this concept fairly well except that GR is used for green, which is a bit confusing, although SL (slate) is used for gray. (ANS Z32.13, "Abbreviations for Drawings," specifies G for green and GY for gray.)

Frequently, so many components are present in a large system that if they were shown as one column or strip, the airline diagram would be too long and narrow for practical purposes. To avoid this, the diagram may have to be laid out in several columns side by side. In some cases, a feed line may represent more than one wire, as witness the feeder coming from terminal 8, component 4. Some of the feeders in the left part of the diagram are shown with dashes. They happen to be part of a feedback circuit which is being emphasized. This is not very common practice. In both highway and airline diagrams, straps or "pigtail" leads are often shown. These go from one terminal of a component to another and are often identified as *straps* or *PTs*.

4·7 Straight-line diagrams

Another type of wiring or interconnection diagram is the straight-line type that is shown in Fig. 4·9. Instead of drawing the components or terminals as rectangles, circles, etc., each component number is listed at the top with a vertical line directly underneath. Conductors that go to each component are shown with a connection dot and the terminal number close to the dot. Wire colors are placed above, often near the center of the "span."

Component numbers usually start with the lowest at the left and increase, with the highest at the right. Since there is no nine in Fig. 4·8, the point-to-point version, the number 9 has been omitted in Fig. 4·9. Numbers 2, 8, and 11 have been assigned to the connectors and ground as indicated. This type of diagram has similarities with the wiring list shown in the next section of this chapter.

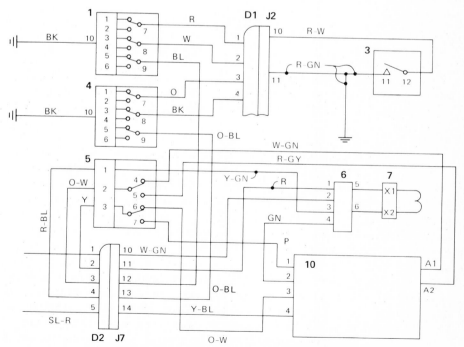

Figure 4·8 A point-to-point connection diagram.

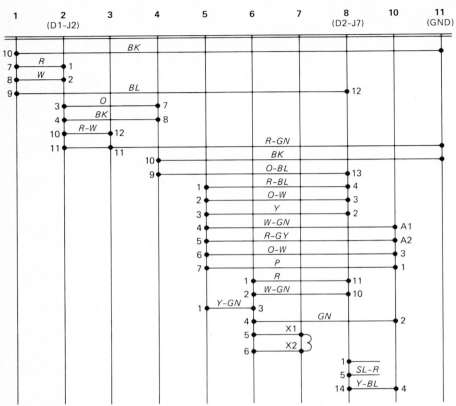

Figure 4·9 A straight-line diagram of the point-to-point connection diagram shown in Fig. 4·8.

4·8 Other examples of connection diagrams

Frequently, it is desirable to show the wiring diagram and the schematic (elementary) diagrams on the same drawing. Figure 4·10 shows these two diagrams for a motor controller. In this panel-wiring drawing, two different line widths are used for conductors. Narrow lines, say $\frac{1}{100}$ in., are used for the control, or pilot, section of the circuit; and wide lines, say $\frac{1}{50}$ in., represent the line-voltage part.

A *cabling diagram* of an industrial TV installation is shown in Fig. 4·11. Actually, it is just a variation of the interconnection diagram. The corners of the connecting conductors have been rounded to simulate cable. If drawn to scale, this could also be called a *location diagram*. Any wiring diagram that is drawn to scale, or approximately so, may be classified as a *location diagram*. This term is not used as extensively in the United States as it is in the British

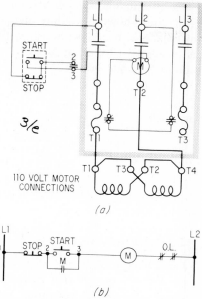

(a)

(b)

Figure 4·10 Connection *(a)* and elementary *(b)* diagrams of a single-phase starter. *(From Thomas E. French and Carl L. Svensen, "Mechanical Drawing," 6th ed., McGraw-Hill Book Company, New York, 1957.)*

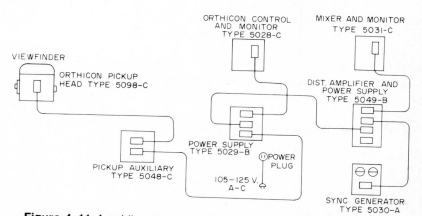

Figure 4·11 A cabling diagram of an industrial television installation.

Commonwealth. Additional information, such as sizes and types of cables and types of plug connectors, is often shown in cabling diagrams.

For very complex installations, it may be necessary to compile *wiring lists* or *to-and-from* diagrams in addition to, or in place of, connection diagrams. A typical heading for such a list might be:

Item No.	Symbol	Color	Size	From		To	
				Component	Terminal	Component	Terminal

Additional information, such as routing (conduits, ducts, etc.), function, and Military type number, may be shown.

4·9 Line spacing and arrangement

The reader may or may not have noticed the neat, uniform line spacing in some of the examples—particularly the point-to-point diagrams. This spacing and layout were not achieved accidentally, but rather by careful planning and good use of drafting aids. Two especially helpful aids are (1) the preliminary freehand sketch (or sketches) and (2) a commercially produced or office-made undergrid.

When making the original sketch, it is advisable to consider laying out the components so that the simplest, neatest wiring diagram will result. This may include changing the location of the components with regard to each other (if such juggling is permitted) and changing the order of the terminals. Rather elementary graphic examples are shown in Fig. 4·12. The shorter and more direct the lines are, the quicker and easier it is for the reader to trace the paths. Keeping the crossings to a minimum also facilitates reading.

Spacing of parallel lines is most often at $\frac{1}{4}$ in. This permits use of a prepared undersheet having $\frac{1}{4}$-in. grids, which, if a transparent drawing medium is used, greatly facilitates the making of the final drawing. For various reasons (lettering requirements, reduction of drawing) other spacings may be required.

4·10 Wiring harness or local cabling

A large portion of the expense in producing complex electronics assemblies can be laid to the wiring. The wiring operation can be optimized (made simpler and cheaper) by study of the chassis and connection drawings, followed by the drawing of a harness diagram. This is generally done in a series of steps:

1. Study of the mechanical assembly, noting the general path which the harness should take and any obstacles which might cause difficulty.
2. Making a drawing of the outline of the harness. This can best be done if a suitable full-scale chassis or assembly drawing exists. A sheet of tracing

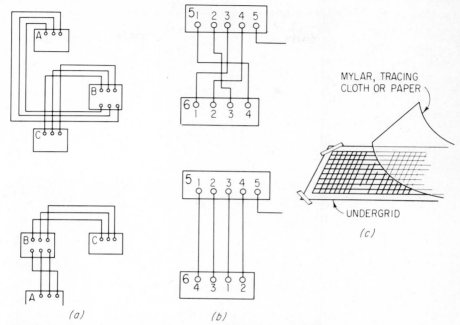

Figure 4·12 Layout of wiring diagrams. *(a)* Experimenting with location of components. *(b)* Changing terminal locations (lower part). *(c)* Using preprinted undergrid.

paper or other translucent drawing medium is then placed over the drawing, and the harness outline done on the top piece.

3. Completion of the harness (sometimes called *local-cabling*) diagram itself, using the assembly drawing and the wiring drawing for the assembly. This may include "breakout" points where individual wires (or several wires) break out of (or enter) the harness and where nails or pegs will be driven into a board or jig on which the cabling will be assembled.

4. Proper identification of each wire on:

   a. The drawing itself.

   b. A table or listing of wires by color, etc.

Figure 4·13 shows a typical local cabling, both by itself and as it later appears connected to the assembly. There is additional cabling and wiring on the same assembly.

Figure 4·14 shows how such a diagram may be begun, and Fig. 4·15 shows the final full-scale drawing of the harness with breakout points and lines showing where the wires are stripped. For example, the wires in the upper left corner extend beyond the line marked 104, 5 but are stripped back to this line, and the bare wire is cut so that $\frac{3}{4}$ in. remains beyond the line. The approxi-

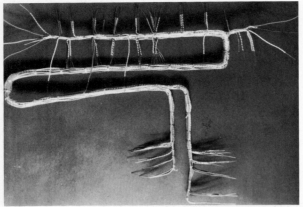

(a)

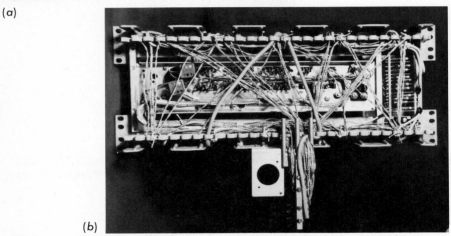

(b)

Figure 4·13 Photograph of a local cabling harness. *(a)* Harness by itself. *(b)* Harness attached to chassis. *(Western Electric Co.)*

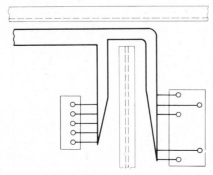

Figure 4·14 Laying out a harness drawing. Terminals and other parts of chassis are shown with light lines.

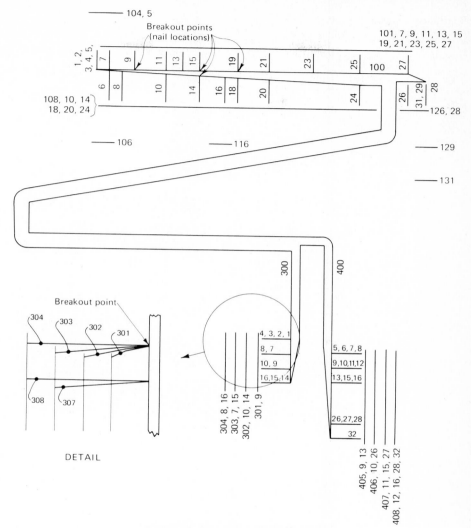

Figure 4·15 A harness drawing.

mate thickness of the cabling is shown and is based on the number (and thickness) of wires that are in the harness at these places. If 26 wires are in the center part, the approximate thickness there would be about $\sqrt{26} \times \frac{1}{20}$ (thickness of insulated No. 22 wire), or about $\frac{1}{4}$ in. plus, which is only slightly less than what the finished harness measures at this point.

Table 4·1 shows the tabular form of wire identification and routing. It is often placed on the same sheet as the harness drawing. This system uses a sort of "station" method whereby leads numbered in the 100s are at one area

Table 4·1 Harness-wire routing

Color	Start	Break Out At	Finish	Terminal Data	
BL	101		119		CAP(*C*34)
BL	120		114		TERMS
BL	111		121	103	1
BL	118		427	104	2
R	408		120	105	INDR
R	109		412		(*L*31)
R	428		114	106	CAP(*C*31)
R	115		111		REL(AL)
R	110		316		TERMS
R	304		128	107	77*R*
R	129		121	108	BOT 3, 4
BK	107		103	109	TOP 1, 2
BK	106	108,118		110	REL (DB)
		120,123	415		

NOTE: This tabulation is part of drawing shown in Fig. 4·15.

of the harness and those numbered in the 300s and 400s are at other areas, or "stations," along the harness. Other numbering systems are sometimes used. Figure 4·15 is one of a set of drawings made for the assembly of the electrical equipment. Other drawings include a schematic diagram and connection diagrams.

Other types of drawings may be necessary or desirable. Figure 4·16 shows three drawings that one may encounter in the cabling, or interconnection, area. Figure 4·16*a* is an orthographic assembly of a multiple-conductor cable. This drawing may be accompanied by a parts list which identifies the connectors (2), clamps (2), sheath (1), and individual wires with their length, color, and pin number or letter. Figure 4·16*b* shows a ribbon connector for a microprocessor board which is mounted on a chassis that includes, among other items, a keyboard. Details of a female ribbon connector for edge connections to a PC card or board are shown in Fig. 4·16*c*. The flat-edge connectors are common on many cards and boards, as shown on the single-board computer in Fig. 7·31.

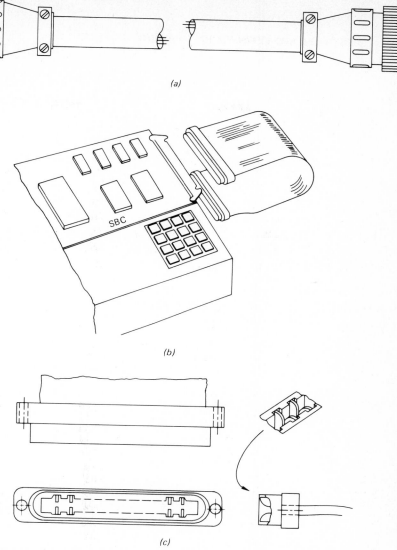

Figure 4·16 Connectors and cables. *(a)* Cable detail of circular female connector. *(b)* Pictorial drawing of a ribbon connector for microprocessor assembly. *(c)* Details of female edge connector for printed circuit board. Ribbon and strip connectors having 40 and 50 connections are common.

Construction and Assembly Drawings

4·11 Sheet-metal layouts

In order that the frames, chassis, shields, and other such parts of electrical or electronics assemblies can be correctly and economically manufactured, drawings must be made that will tell the builder exactly how the item is to be made. Figure 4·17 shows the pattern of a shield for some electronics equipment of a satellite. It is laid out flat; the lines on which it is to be folded are shown

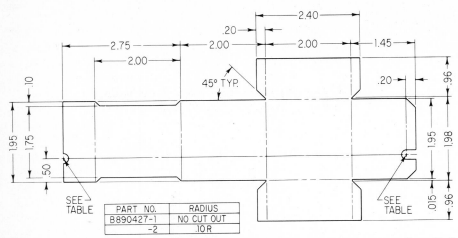

PART NO.	RADIUS
B890427-1	NO CUT OUT
-2	.10 R

Figure 4·17 The pattern drawing of a shield for a communications satellite. *(Bell Telephone Laboratories.)*

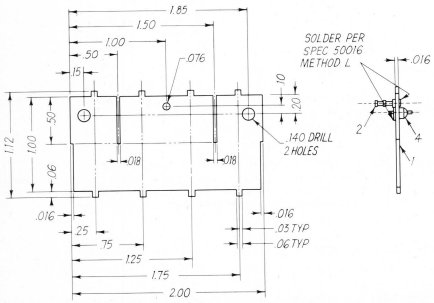

Figure 4·18 Construction drawing of center fin, including terminals. *(Bell Telephone Laboratories.)*

with double dashes. In addition to having complete dimensions, the drawing tells exactly what kind of material is to be used and how the material (an aluminum alloy) is to be finished. Although it does not tell how to fold the metal, a little study will reveal that this will make a boxlike enclosure if the sides are folded at 90° to each other. Chassis diagrams are treated in the same way, unless they have too many dimensions to make the folding-out graphical concept practical. Figure 4·17 uses the standard two-decimal system of dimensioning, which has become quite popular in U.S. industry. All except the most critical dimensions are rounded to the nearest hundredth of an inch (then usually to the nearest even hundredth). Note the 0.015 dimension which the designer considered to be more critical. The aligned system (guidelines parallel to dimension lines) is employed.

Figure 4·18 shows the same type of drawing for another piece of satellite equipment, except that a different dimensioning system is used. Here, horizontal location dimensions are given from the left edge because of the critical distances to the square projections, which must fit into mating recesses on another piece. This method of dimensioning to a well-defined datum plane (it should be on a finished surface) is often the best way for numerical-control (N/C) manufacturing. It does require two straight edges, however.

4·12 Chassis manufacture

Figure 4·19 shows only part of a construction drawing for a chassis like that shown in Fig. 4·13, with the wiring harness. The complete drawing has four views, plus supplemental drawings showing the exact shapes of some of the holes and other cutouts. Because of size limitations, we have shown only the left quarter of two adjacent views and some of the cutout details, which are placed around the edge of the sheet. The cutout dimensions are given in separate details to avoid cluttering the main views with too much information and to provide a practical way to give the correct tolerances for mounting holes and their components in many situations (Figs. A, C, and D).

For complete manufacture of this chassis, a separate set of instructions, entitled Manufacturing Layout and Time Rate, is issued to the manufacturing section. Some of the 22 steps included in this Layout and Time Rate are shown in Table 4·2. Note that even the tools are specified.

This chassis drawing is a typical engineering drawing made to full scale (1 in. = 1 in.). It can be used later on as a basis for an assembly drawing (for putting the rest of the chassis together) and for the harness, or local-cabling, drawing. The advantages of making it to full size are now fairly evident. However, chassis for some miniaturized packages cannot be drawn to full scale because they are too small. They must be drawn larger than actual size in order that details and dimensions can be appropriately shown.

More and more layout drawings will use metric dimensioning as the United States slowly converts to SI (Système International d'Unités) in which the base

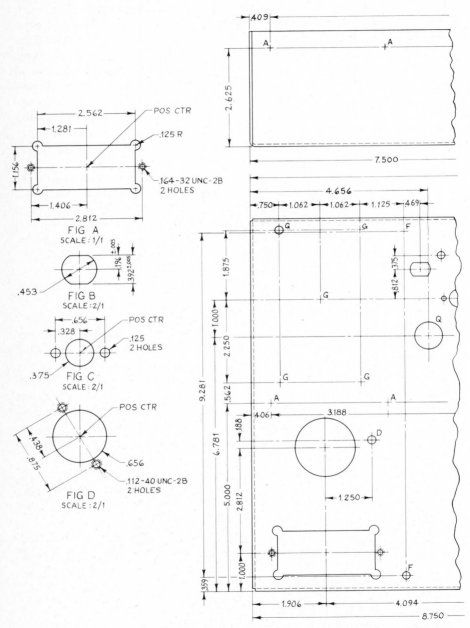

Figure 4·19 Part of a drawing for the construction of a main chassis. *(Western Electric Co.)*

Table 4·2 Manufacturing Layout and Time Rate*

RM 512261	Brass sheet, .063″ × 24″ ×84″ Grade A.
(Stock No.)	(5) parts per sheet.
	(20) sheets per 100 parts.

Operation Description	Machines, Tools, Gages, Test Sets, etc.
metal parts	
1. Shear strips to 15.923 ± .010″ × 24″	Niagara shear (11005)
a. Gage part from front measuring scales on	Use front gaging.
shear (backup).	18″ vernier caliper
b. Place parts on pallet.	
2. Trim parts to 22.674 ± .010″ × 15.923	Niagara shear (11005)
a. Gage part from front measuring shear (backup).	Use front gaging.
	24″ vernier caliper
3. Perforate blank complete except that (4) corner notches (cutouts) are not to be blanked at this time.	#5 Minster punch press (62221)
(1) setup	P&D C-729515
(2) handlings	Bolster C-673918-9
4. Omitted	
5. Deburr part.	Bench
a. Place part on pallet.	Pneumatic sander
6. Omitted	
7. Tap (2) holes .164–32 for "B" cluster	Snow tapper (79003)
(1) Setup	(1) H.S.S. tap, .164–32
(2) Strokes	(2) Flute plain pt.

* Five of twenty-two steps are described.

unit of length is the meter. Figure 4·20a shows a bracket that is dimensioned in SI and the unidirectional placement of figures, in which all guidelines are drawn horizontally and all numbers and letters are read from the same direction. The unidirectional system has long been approved by ANS Y14·4, as has the aligned system exhibited in the previous three examples. Many companies are employing dual dimensioning in the manner shown in Fig. 4·20b, with millimeters added above the inches. Dimensioning to the nearest millimeter is accurate enough for most manufacturing facilities and equipment. Sometimes more accuracy is needed, in which case dimensions are given to the nearest 0.1 mm.

The detailer or designer must be certain that the manufacturer has metric drills. Otherwise he or she will have to specify standard U.S. twist drills. The nearest U.S. drill to a 5.00-mm (0.1968-in.) hole is No. 9 (0.1960 in.), and the closest drill to a 9.00-mm (0.3543-in.) hole is size T (0.358 in.). Both the metric and the older U.S. twist-drill tables are in the Appendix.

4·13 Hole and terminal data

Chassis and wiring boards may be dimensioned according to standard drafting practice. There are several methods in use. Two systems that would be quite

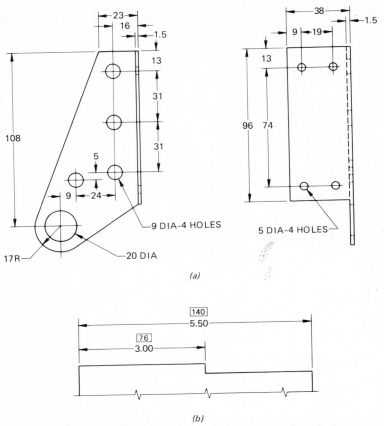

(a)

(b)

Figure 4·20 Metric and dual dimensioning. *(a)* Using millimeters as the standard units. *(b)* Dual dimensioning with inches below and millimeters above.

suitable for parts having many holes are shown in Fig. 4·21. These methods can be used for manual operation of standard drill presses or for N/C machine tools. Figure 4·21*a* illustrates the type of drawing that is ideal for a programmer to use. Various sizes of holes are indicated by different letters, and the holes within a letter group are numbered sequentially. The *xy* reference point is located close to a corner, in this case at the center of hole A1. The programmer will be able to write the program (which will later—through tape—give instructions to the machine tool about how to do the job) in a very efficient manner. Holes that are the same size will be drilled one after another. After these holes have been drilled, the drill bit will be changed (manually on some tools, automatically on others) and then the next series of holes will be made.

The table of Fig. 4·21*a* is so constructed that the *absolute* method of N/C can be easily performed. However, the other method, called *incremental,*

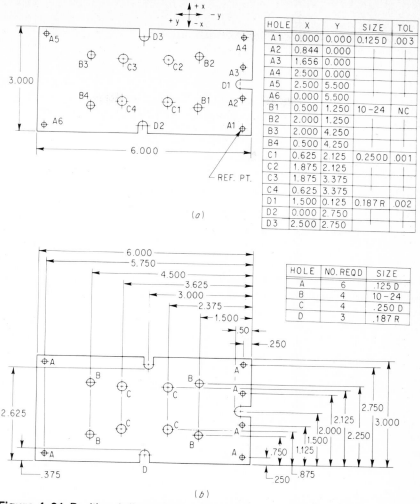

HOLE	X	Y	SIZE	TOL
A1	0.000	0.000	0.125 D	.003
A2	0.844	0.000		
A3	1.656	0.000		
A4	2.500	0.000		
A5	2.500	5.500		
A6	0.000	5.500		
B1	0.500	1.250	10-24	NC
B2	2.000	1.250		
B3	2.000	4.250		
B4	0.500	4.250		
C1	0.625	2.125	0.250D	.001
C2	1.875	2.125		
C3	1.875	3.375		
C4	0.625	3.375		
D1	1.500	0.125	0.187 R	.002
D2	0.000	2.750		
D3	2.500	2.750		

(a)

HOLE	NO. REQD	SIZE
A	6	.125 D
B	4	10-24
C	4	.250 D
D	3	.187 R

(b)

Figure 4·21 Positional dimensioning (drilling layouts). *(a)* Coordinate system. *(b)* Baseline system.

may also be used by subtracting x and y values of one hole from the next. Figure 4·21*b* also lends itself to both absolute and incremental programming approaches. Actually, it is not necessary to draw the holes on this type of drawing. Some companies follow the practice of showing centerlines only. Holes B1 to B4 are, of course, threaded. They are No. 10 taps of the coarse-thread (NC or UNC) series having 24 threads per inch. (See table, Fig. 4·21*a*.)

Standard threads can be identified in tables such as are found in Appendix B. The unified (UNC, UNF, etc.) series includes standard U.S. threads in inches or fractions. In this table, we find the No. 10 coarse (UNC) thread, with 24 threads per inch, and the fine (UNF) thread, with 32 threads per inch. Either type of thread could be used; the coarse is more common. Another table, which identifies standard coarse and fine metric threads, is also included in the Appendix.

4·14 Assembly drawings

Figure 4·22c shows the required components drawn in place on a board. Some reasons for locating the components on such an assembly drawing are:

1. To fit the parts into the space available
2. To satisfy wiring requirements, such as accessibility for connections and short wiring paths
3. To achieve a pattern without crossings if printed circuitry is to be used
4. To satisfy electrical requirements such as shielding, built-in capacitances, etc.
5. To satisfy manufacturing requirements such as assembly by N/C equipment.

Such a layout is made by (1) studying the original schematic, or elementary, diagram to determine what components have common connections, where common lines (e.g., ground, B +) are, etc.; (2) drawing one or more point-to-point wiring diagrams freehand in order to get an optimal arrangement; (3) further refinement, if justified or necessary, which might include cutting out outlines of parts on heavy paper and shifting them into different arrangements; and (4) drawing the final assembly drawing, such as is shown, to scale. Each component is given a designation, such as $R1$ (resistor No. 1) and $C5$ (capacitor No. 5), or the value, such as 100 for a 100-Ω resistor. The assembly of Fig. 4·22c was derived from the schematic diagram of Fig. 4·22a and then by experimentation in the location of components. Figure 4·22b represents one start in the layout of components, with the switch and volume control at the front, wires to the speaker and automobile radio at the back. Because the volume control is so bulky and the circuit board and chassis so small (less than 4 in. long), the final layout (Fig. 4·22c) includes a recessed portion for $R7$. For reasons that will be explained shortly, the layout was made with a 0.1-in. (2.54-mm) undergrid. Note that all holes (for connection to printed circuitry on the other side) are on the intersections of the grid lines. A photograph of the PC-board assembly is shown in Fig. 4·23.

This drawing can be used for two or more purposes: (1) to give a parts-location layout, along with a list of parts and catalog numbers, to production and service personnel, and (2) to provide the location of holes for a drilling drawing such as Fig. 4·21.

Another type of assembly drawing is that used for the assembly of a chassis itself. Figure 4·24 shows one of three views of such an assembly. The drawing also lists the various parts as follows: (1) chassis (also shown partially in Fig. 4·19), (2) panel, (3) and (4) angles, (5) brackets, (6) and (7) rivets, (8) grommet, (9) Penn fastener, and (10) self-clinching fastener. (Numbers 7 and 8 are shown on another view.)

As in the case of Fig. 4·19, this drawing is accompanied by a Manufacturing Layout and Time Rate sheet which lists each assembly step and the tools with which the operations are to be performed. Chassis come in various sizes and

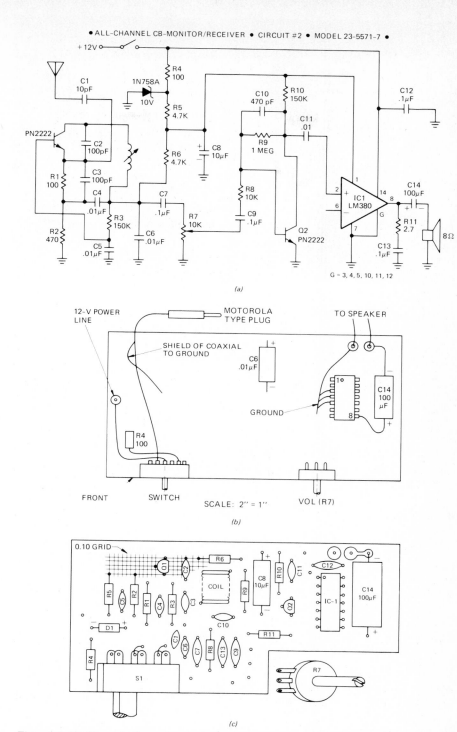

Figure 4·22 Steps in the making of a components layout drawing of a CB receiver for automobiles. *(a)* Schematic diagram. *(b)* Trial layout of components (incomplete). *(c)* Final component assembly, or components location drawing.

99

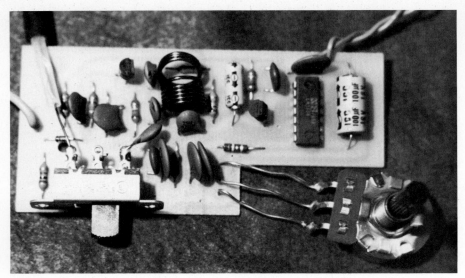

Figure 4·23 Photograph of the CB-receiver circuit which has been drawn in Fig. 4·22. *(Kantronics, Inc.)*

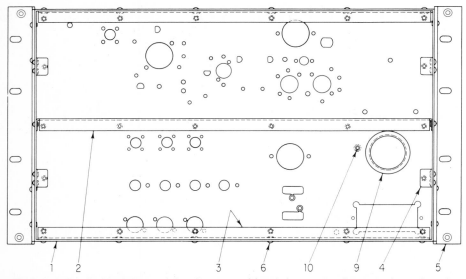

Figure 4·24 A drawing used for the assembly of the parts of a complete chassis. *(Western Electric Co.)*

shapes. Their main purpose is to hold, or support, component parts, but they also often furnish shielding and rigidity and even act as parts of electric circuits.

4·15 Photodrawing

An interesting development in the graphics area is photodrawing. Figure 4·25 shows a photograph of an assembled radio with the accompanying letter-number designation of parts. These photographs perform a function similar to that of

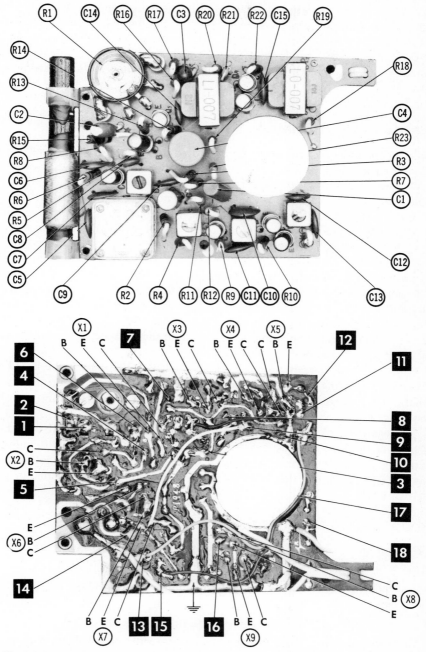

Figure 4·25 Photodrawings of a radio assembly, both sides. *(A Howard W. Sams Circuitrace Photo.)*

Fig. 4·22. In some ways a photograph is clearer than a drawing. Certain things, like disk capacitors and coils, show up better in a photograph than in a drawing, especially if the assembly is crowded. Electrical installations that have been extensively modified can be shown better by photographs than by well-worn drawings that have had many changes made on them. Numbers for parts identification, alignment symbols, and connection points are drawn directly on the photograph. A complete list of parts usually accompanies a photodrawing, such as is shown here.

Printed Circuit Boards

In general, there are two systems employed to get a plated (printed) circuit on a circuit board. One method is to use a copper-clad laminate (board) and etch away the unwanted copper foil, leaving the desired circuit pattern. In this system a plating resist is applied in the image of the desired pattern. The board is subjected to the etching process, wherein that portion of the copper that is not covered by the resist is removed. One of several methods is employed to get the resist pattern onto the board.

The other process is the *additive* one, in which the conductor pattern is deposited on the board. Usually this is copper. Tin lead is then deposited on the copper by the electroplating process. After large mounting holes, if any, are made and gold connection tips, if any, are added, the circuit board undergoes a *reflow* process. During this process the copper and tin paths become solder, and the component connections are soldered to their respective conductor paths at their pads or plated-through holes.

Often a solder-resistant plastic is printed over the side of the board that has the solder and covers those areas that are to have no solder. This *solder mask* prevents solder from bridging closely spaced conductors and splashing on other unwanted places.

Regardless of how the printed circuit board is to be manufactured, the layout of the final artwork is pretty much standard. Considerable thought and drafting go into the layout. A study of the calculator board (Fig. 4·26) should indicate to the reader that careful layout of components and conductor paths is required.

4·16 Making drawings for a printed circuit

The first step in making a set of drawings for a printed circuit is to study the elementary diagram. Such a study will provide such information as:

1. What groups of components have common connections
2. What the peak potential differences will be
3. Grounds, voltage supply lines, etc., that are required
4. Ground paths, large and heavy, and heat sinks that will be necessary
5. Other pertinent information

(a)

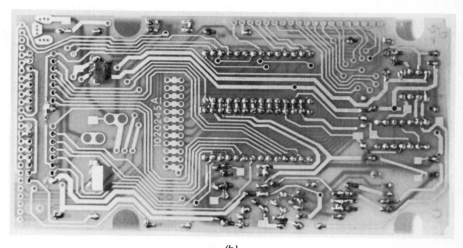

(b)

Figure 4·26 Both sides of the main printed circuit board for a scientific calculator. *(a)* Component side. *(b)* Foil side. *(Texas Instruments, Inc. Photograph: Precision Arts, Ltd.)*

The next step is to make a *preliminary sketch* from the schematic diagram, such as that shown in Fig. 4·22a. The components involved in this circuit are mostly capacitors ($C1$, $C2$, etc.), resistors ($R1$, $R2$, etc.), two transistors (PN2222) which we shall call $Q1$ and $Q2$, and an integrated amplifier circuit $IC2$. In the sketch (or sketches, because there may be several trials before the most likely pattern results), all components are located and circuit paths are sketched in. This is in reality a rough wiring diagram. Such a diagram was made before the component layout of Fig. 4·22c could be made. Now this sketch can be checked to see if:

1. Crossovers are zero or a minimum.

2. Bypass and grid lines are as short as possible.

3. Longer lines, such as ground and high-current, are placed near or around the edge of the board.

4. The design is neat and compact.

5. Other specifications (see Art. 4–19) are met.

Figures 4·22 and 4·27 show the final component layout as seen looking at the component side of the CB-receiver board. The components, as mentioned before, have been drawn so that their leads will pierce the board at the intersection of the 0.10-in. (2.54-mm) grid lines. Figure 4·28 shows two grids that are frequently used. Not shown is a 0.05-in. (1.27-mm) grid system.

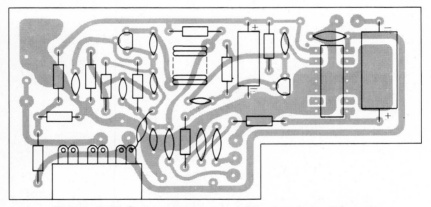

Figure 4·27 Component layout of CB printed circuit board.

In order to position components properly it is necessary to know their size, how far apart their leads must be placed, and how close together the components may be positioned on a board. Table 4·3 and Fig. 4·29 supply this information for many of the more frequently used components.

The difference between Figs. 4·22c and 4·27 is that the conductor paths have been drawn in the latter. Fig. 4·27 is really 4·22 with this work added. The ground path is shown extending from the upper left corner, going down in the left center part, and then going up to the integrated circuit and capacitor, upper right. It is extra wide and heavy at each end, and thus acts as a heat sink in those areas. (A heat sink removes heat by radiating it away from heat-producing elements.) This board is to be a single-sided circuit board with components on one side and printed wiring on another. The circuit paths are therefore drawn with light shading because they are to be on the other (far) side. As in the case of many similar products, the circuit paths will be made by the etching

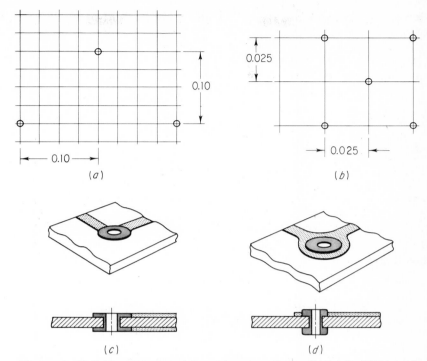

Figure 4·28 Grid systems and through connections. *(a)* 0.1-in. (2.54-mm) grid. *(b)* 0.025-in. (0.635-mm) grid. *(c)* A plated-hole connection. *(d)* A plated eyelet.

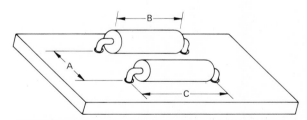

Figure 4·29 Axial lead spacing for resistors and tubular-shaped components. See Table 4·3 for center-to-center distances. *(King Radio Corp.)*

process, in which a thin layer of conducting material (called a foil) is bonded to one side of the board and then etched away so that only the conductor paths remain.

The great bulk of PC boards are made from a paper-base phenolic or a glassy epoxy laminate, called G-10, which is superior to the phenolic in that

Table 4·3 Component Spacing on Printed Circuit Boards

	Dim B	Dim C	Dim A
Resistors			
130 Family			
$\frac{1}{8}$ W	0.140	0.300 or 0.400	0.100 (2.54)
$\frac{1}{4}$ W	0.250	0.400 (10.2)	0.100
$\frac{1}{2}$ W	0.375	0.550 (14.0)	0.150 (3.81)
1 W	0.562	0.750 (19.1)	0.225 (5.72)
2 W	0.687	0.850 (21.6)	0.325 (8.25)
136 Family			
13-xxx-2	0.281	0.425 (10.8)	0.100
13-xxx-22	0.343	0.500 (12.7)	0.100
13-xxx-12	0.560	0.725 (18.4)	0.150
Diodes			
D-41	0.200	0.300 or 0.400	0.100
D-26	0.450	0.500	0.300 (7.62)
D-14	0.300	0.400	0.100
D-13	0.350	0.500	0.300
Tubular Capacitors			
0.01 μf			
Small	0.625	0.800 (20.3)	0.300
Large	1.000	1.300 (33.0)	0.450 (11.4)
0.10 μf			
Small	0.875	1.100 (28.0)	0.500
Large	1.625	1.950 (49.5)	0.650 (16.5)
Ceramic Capacitors			
Small		0.250 (6.35)	0.200 (5.08)
Medium		0.350 (8.89) or 0.400 (10.2)	0.200
Large		0.400 or 0.500 (12.7)	0.200 or 0.300 (7.62)

Dual-In-Line Integrated-Circuit Packages
Ceramic and Plastic

Number of Pins	Distance between Adjacent Pins	Distance between Rows of Pins
8	0.100 (2.54)	0.300 (7.62)
14	0.100	0.300
16	0.100	0.300
18	0.100	0.300
20	0.100	0.300
22	0.100	0.400 (10.2)
24	0.100	0.600 (15.2)
28	0.100	0.600
40	0.100	0.600

NOTE: Dimensions at left are in inches. Those in parens are in millimeters.
Resistor data from King Radio Corp.
DIP data from Texas Instruments, Inc.

it has a higher resistance to warping. They are of several standard thicknesses ranging from $\frac{1}{32}$ (0.79) to $\frac{1}{4}$ in. (6.35 mm). Copper foil comes in several thickness or weights as shown in Table 4·4. The cross section of the conductor should be large enough to carry the required current with a certain temperature rise. Table 4·5 is a partial table that provides some of this information for a 40°F rise above the ambient temperature (that which exists in the immediate environment).

Table 4·4 Weight and Thickness of Copper Foils for Printed Circuit Boards

Foil Weight		Foil Thickness	
oz/ft²	mg/cm²	in.	mm
0.5	16.74	0.00068	0.0173
1.0*	33.5	0.00135	0.0343
2.0†	66.95	0.0027	0.0686
3.0	100.4	0.0042	0.107
5.0	167.4	0.0068	0.173

* Often referred to as 1-oz foil
† Often referred to as 2-oz foil

Table 4·5 Width of Conductor Paths in Relation to Current for a 40°C Temperature Rise above Ambient

Current (A)	1-oz 0.00135 (0.0343) Copper	2-oz 0.0027 (0.0686) Copper
1.5	0.015 (0.381) Wide	0.008 (0.203) Wide
2.5	0.031 (0.787)	0.015 (0.381)
3.5	0.062 (1.58)	0.031 (0.787)
4.5	0.125 (3.18)	0.062 (1.58)

NOTE: Minimum spacing between conductors is 0.031 in. (0.787 mm) for voltages up to 150 V.

Figure 4·30 shows the *final artwork* or *master layout* of the CB-receiver circuit board. It was made by placing Fig. 4·27 upside down, then putting a standard acetate medium over it and applying commercially available tape and printed-circuit patterns. This single-sided printed circuit board has served as a typical example. Later, a double-sided board layout will be presented.

4·17 Use of tape and other contact-adhesive aids

Making a PC-board layout with tape is not quite as easy as it looks to the beginner. The preprinted appliqués (contact-adhesive patterns) are usually easy enough to separate from their protective backing. However, getting them on the exact desired location and oriented correctly is difficult for the person who has no experience or correct tools. (The latter should include a very sharp X-

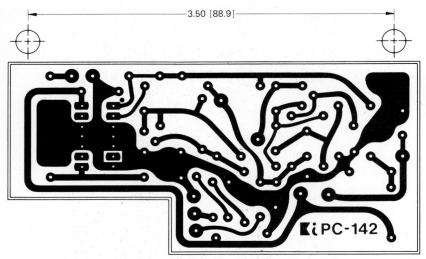

Figure 4·30 Final artwork (master layout) of CB receiver, foil side. This was made with commercially available printed-circuit tapes and aids. *(Kantronics, Inc. and Bishop Graphics, Inc.)*

acto knife or equivalent and a thin, flat spatula-type tool for manipulating the pattern onto the drawing medium.) Getting a straight line with tape also requires experience. One way is to hold the tape down at its starting point (usually on a pad which has already been emplaced), pull it slightly at the other end, and then lay it down and cut it to the desired length. One must try to allow for the stretching of the tape. Unfortunately, the tape is not supposed to be stretched, and the manufacturer's instructions so indicate. The key to this approach is the word *slightly.* For, as the authors and many others have discovered, it is extremely difficult to get a straight line if the tape is not stretched.

Pads and registration marks are usually put down first, then conductor paths, grounds, and heat sinks (if any) are added. Mylar for this artwork is available with grid lines spaced 0.050 (1.27), 0.100 (2.54), or 0.125 in. (3.18 mm) apart. Such a grid may be placed under a clear sheet of Mylar, which is to be the final artwork (master layout), or a grid with blue or brown "drop-out" lines[1] may itself be used for the master layout. When the tape is laid down, it should overlap onto the adjacent part of the lands, elbows, etc. (see Fig. 4·31).

It is very convenient to have 90° elbows available commercially, but not all corners are 90°. Other bends can be made with tape if one crimps it on the inside radius; or they can be done with ink, in which case one makes a line the same width as the tape. It is also possible to obtain 45° elbows commercially.

---

[1] These lines will not appear on the photographic reproduction.

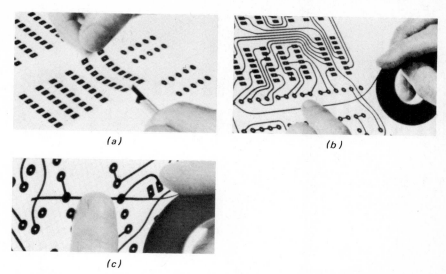

Figure 4·31 *(a)* Emplacing a large stick-on pattern on a PCB layout. *(b)* and *(c)* Applying tape for short circuit-path and longer circuit-path distances. *(Courtesy of Bishop Graphics, Inc.)*

The contact-adhesive aids shown in Fig. 4·32 deserve some comment. First, the dimensions listed are full, or actual, size. (The original drawing was reduced to about half size to fit in this book.) Second, the aids come in a number of different sizes (widths, in the case of lines). Conductor tapes are available 0.015 (0.38), 0.020 (0.51), 0.026 (0.66), 0.031 (0.787), 0.040 (1.02), 0.046 (1.17), 0.050 (1.27), 0.062 (1.57), 0.090 (2.29), 0.100 (2.54), and up to 2 in. (50.8 mm) wide. Some patterns are such that they can be cut to desired lengths (Fig. 4·32*d*), and some come in fixed sizes (Fig. 4·32*e* and *f*). Figure 4·32*i* represents a set of lands which have the same center-to-center spacing as Fig. 4·32*e*. But the pads are small enough that a small conductor can be placed between them as shown—a situation that is sometimes unavoidable. Otherwise the larger pads are preferred because they provide better soldering and are more stable if a lead or component has to be removed and replaced. The integrated-circuit connector pattern, Fig. 4·32*e*, comes in different sizes and shapes. An optional arrangement is to buy flatpack patterns which have long fingers that look much like Fig. 4·32*d*. The outline of the IC package can be drawn over the pattern, which can be trimmed to the exact size and shape with a small knife.

Target patterns, Fig. 4·32*g*, are designed so that when superimposed any slight mismatching can be seen and corrected. Land or donut patterns, Fig. 4·32*h*, come in many sizes and combinations of I.D.s and O.D.s. Careful attention must be given to inner diameters because different wire sizes are often present in any single printed circuit. Thus, wires to speakers, potentiometers, and outside power, for example, are usually larger than the leads to typical components on the board.

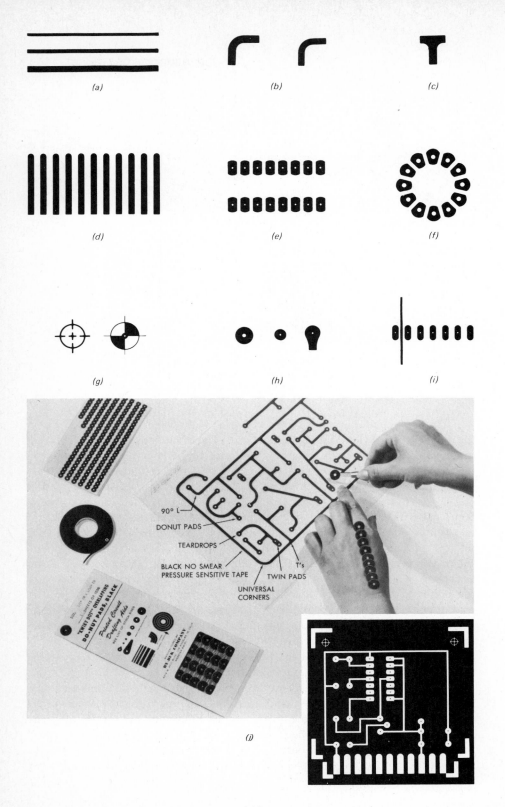

(a)

(b)

(c)

(d)

(e)

(f)

(g)

(h)

(i)

90° L

DONUT PADS

TEARDROPS

BLACK NO SMEAR
PRESSURE SENSITIVE TAPE

TWIN PADS

T's

UNIVERSAL
CORNERS

(j)

Our master layout was drawn to a 2:1 scale (twice size). Scales of 2:1 and 4:1 are common, but 10:1 and even 100:1 have been used where extreme accuracy has been required. Tapes and aids can be obtained for most of these scales. For instance, a 0.062-in. (1.57-mm) tape on a 2:1-scale layout will produce a line that is actually 0.031 in. (0.787 mm), which is large enough to accommodate most currents (Table 4·5).

The check list that follows later provides other specifications for good PC-board layout. Figure 4·30 includes two reduction targets with the required final dimension provided so that the finished product will be the desired size.

4·18 Other aspects of PC-board layout and construction

More accurate layouts can be made with the coordinatograph, Fig. 4·35, which has a plotting accuracy of 0.001 in. (0.0254 mm). Another system for drawing accurate conductor paths uses a scribe that has two accurately spaced points that cut sharp, precise lines on a peel-coat polyester film. The area between the lines represents the conductor path and remains after the rest of the film is peeled away.

Orientation of components is important from the production standpoint. Figure 4·34 shows in diagrammatic form the PC board moving along the waves of the wave-soldering equipment. The solder just touches the locations on the foil (solder) side of the board where leads from the components go through the board by way of the holes which have been drilled through their pads or lands. This makes small, neat, strong solder connections. In order to minimize unavoidable "bridging" of solder, dual-in-line packages, flatpacks, etc., are positioned with their long dimensions parallel to the direction in which the board moves across the solder waves. This is often the long dimension of the PC board, but that depends upon the production equipment and methods in use.

Figure 4·32 Some commercially available contact-adhesive aids for printed-circuit drawings. *(a)* Conductor tapes, 0.046 (1.17), 0.050 (1.27), and 0.100 in. (2.54 mm). *(b)* 90° elbows, 0.125 (3.18) and 0.100 in. (2.54 mm) wide. *(c)* Tee, 0.125 in. (3.18 mm). *(d)* Connector pattern, 0.20 in. (5.08 mm) ctr. to ctr. *(e)* Sixteen-lead DIP patterns 0.20 in. (5.08 mm) ctr. to ctr. *(f)* Twelve-lead TO-5 pattern 0.74 (18.9) O.D., 0.075 in. (1.91 mm) I.D. *(g)* Target or registration marks. *(h)* Pads or lands, 0.281 (7.14) O.D., 0.062 (1.58) I.D.; 0.187 (4.75) O.D., 0.040 in. (1.02 mm) I.D.; teardrop 0.281 (7.14) O.D., 0.031 in. (0.787 mm) I.D. *(i)* Seven-lead pad pattern, 0.20 in. (5.08 mm) ctr. to ctr., with 0.031-in. (0.787-mm) conductor-path tape between two lands. *(j)* Two drawings (final artwork) that have been made with tapes and other adhesive aids. *(These and other aids can be obtained from several companies, including Bishop Graphics, Inc., P.O. Box 5007, Westlake Village, CA 91359; and BYBUK Company, 4314 W. Pico Blvd., Los Angeles, CA.)*

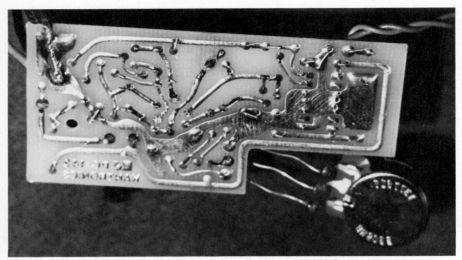

Figure 4·33 Photograph of foil side of CB-receiver printed circuit board. The shield of coaxial cable can be seen soldered to the ground at upper right. Note the large wires to the potentiometer at bottom. These require larger pads and holes than most of the other component leads. *(Photograph: Precision Arts, Ltd.)*

(In the additive system of manufacture, the solder bath is not used. Interconnections are formed during the reflow process. If this system is used, the orientation of components to reduce bridging is not important.)

Part of a N/C machine which automatically puts components on a board and clinches them into position is shown in Fig. 4·36c. This method requires that the pattern be limited to the *X* and *Y* axes as shown in Fig. 4·36a. The components are first placed on tapes in the order in which they are to be inserted

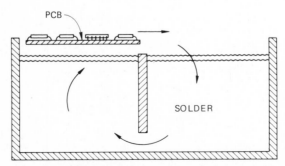

Figure 4·34 Diagram of wave-soldering process. The bottom of the printed circuit board touches the solder waves. *(King Radio Corp.)*

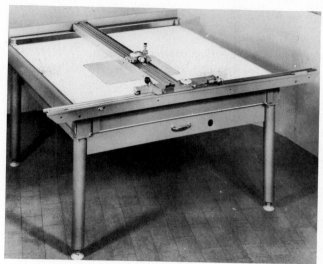

Figure 4·35 The coordinatograph is often used for precision drafting of printed circuits. *(Keuffel & Esser Co.)*

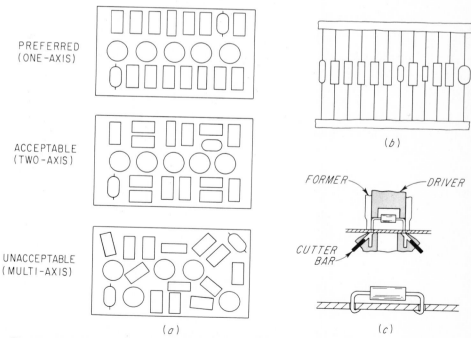

PREFERRED
(ONE-AXIS)

ACCEPTABLE
(TWO-AXIS)

UNACCEPTABLE
(MULTI-AXIS)

(a)

(b)

FORMER

DRIVER

CUTTER
BAR

(c)

Figure 4·36 Orientation of components for automated assembly of printed circuit boards. *(a)* Preferred and unacceptable designs for N/C assembly. *(b)* Components assembled on tape. *(c)* Schematic drawing of N/C insertion machine action.

on the board. The tape (Fig. 4·36*b*) moves through the machine much as an ammunition belt is fed through a machine gun. Components are then placed on the board one after the other in a line. The idea is to minimize the number of passes that the board and components have to make through the machine.

Transistors and capacitors are available with preformed "standoff" leads that facilitate their assembly on the board. DIP elements can be mounted directly on the board with their leads projecting through, or they can be separated with a thin spacer to eliminate possible shorting. They can also be plugged into sockets which are fastened to the board as other components are. This facilitates removal and replacement of ICs, microprocessors, etc. Completed boards are often sprayed with a material that inhibits the deleterious effects of humidity, fungus, etc. This is called *conformal coating.*

4·19 Some specifications for good printed-circuit-board design

The following are some minimum design standards prepared by a company that produces thousands of printed-circuit boards a year:

1. All components will be oriented on the *X* or *Y* axis wherever possible. Dual-in-line IC packages will be oriented on one axis and keyed in the same direction wherever possible.

2. Components will be mounted on only one side of the PC board.

3. Maximum board size shall be 5.80 (147) × 11.70 in. (297 mm), if possible, to allow wave soldering and programmed assembly. (The sizes may vary from company to company, depending on the equipment used. This specification will not apply if the electroplating process is used instead of the solder bath.)

4. Components should be located so any component can be removed from the board without removing any other part.

5. When large areas of ground plane are required, crosshatch or solder mask to avoid excessive buildup and warping during soldering.

6. All boards must have at least two holes referenced to the artwork. These holes, which may be mounting or extractor holes, must be 0.062-in. (1.57-mm) diameter or larger.

7. Component pad centers will conform to standard spacing and should be maintained on even 0.025-in. (0.635-mm) graduations.

8. Minimum spacing shall be maintained, as shown in Table 4·3 and Fig. 4·29.

9. Components that dissipate more than 2 W should not be mounted on a PC board, but should be heat sunk directly to the chassis.

10. Determine pad size as follows:

    a. Maximum misregistration allowed: 0.002 in. (0.051 mm).

    b. Minimum working tolerance: 0.008 in. (0.203 mm).

c. Plated-through holes require a minimum of 0.006 in. (2 × 0.153 mm) for through-plating.

d. Allow 0.002 in. (0.051 mm) for etching tolerance per ounce of copper.

e. The above plus the minimum required annular ring multiplied by two for each side of hole.

As an example, we might have a finished hole size of 0.031 in. (0.787 mm), misregistration 2 × 0.002 for 0.004 in. (0.101 mm), working tolerance 0.008 in. (0.203 mm), through-plating 0.006 in. (0.304 mm), etching tolerance 0.002 in. (0.051 mm), minimum land 2 × 0.005 for 0.010 in. (0.254 mm), for a total of 0.061 in. (1.57 mm), which would be the minimum pad size.

11. Component bodies will be at least 0.05 in. (1.27 mm) from the edge of the board.[1]

12. No conductor path shall be closer than 0.025 in. (0.635 mm) to the edge of the board.[1]

13. No component pad perimeter will be closer than 0.05 in. (1.27 mm) to the edge of the board.[1]

14. Conductor paths should be oriented to run parallel to the longest axis on the solder side of the board. (This reduces bridging during wave soldering.)

15. Conductor paths should be oriented to the *XY* coordinate system. Necessary deviations should be at 45°.

16. All final artwork will be done at a 4:1 scale minimum. Any board requiring very critical tolerance may be done at 10:1.

17. Wherever possible the following markings will be added to the component side of the board:

a. Termination numbers or wire colors

b. Polarity for capacitors, diodes, etc.

c. Test-point identification

d. Adjustment identification

4·20 Double-sided and multilayer board layout

Figure 4·37 is the schematic diagram of the display-board circuit for a small radar unit. The components consist of three integrated circuits *I*1, *I*2, and *I*3; three cold cathode display lamps *DS*1, *DS*2, and *DS*3; two transistors *Q*1 and *Q*2; two resistors *R*1 and *R*2; and a connector *J*1. After some analysis of the schematic, an assembly drawing (or component layout) was made. Figure 4·38 shows a satisfactory arrangement of the devices. Note that the cathode lamps and integrated circuits are positioned with *I*1 close to *DS*1, etc. The

---

[1] This is a product design specification. It may or may not apply to PC boards designed by other firms.

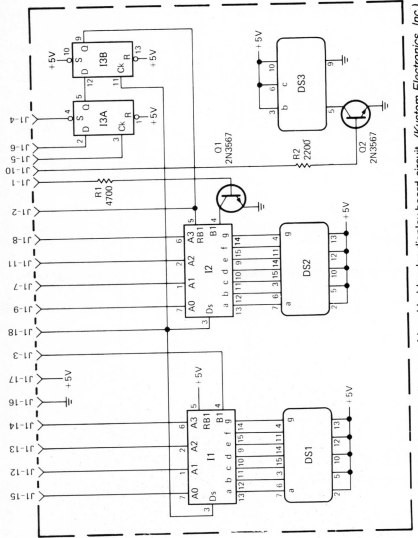

Figure 4·37 Schematic diagram of hand-held-radar display-board circuit. *(Kustom Electronics, Inc.)*

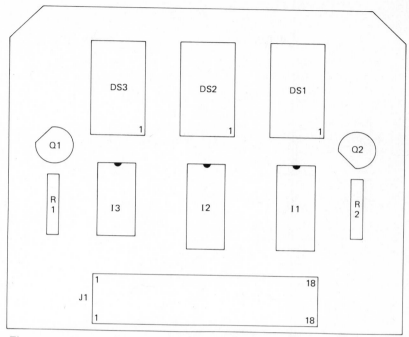

Figure 4·38 Assembly drawing (component layout) for hand-held-radar circuit board.

schematic diagram shows seven connections between $I1$ and $DS1$; therefore, if they are located close to each other, crossovers will probably be minimized. If one allows space for some circuit paths between the devices and lays out to scale (double or 2:1 in this case), the size of the board that results is about 3 (76) $\times$ 2$\frac{3}{4}$ in. (70 mm).

Using this layout, one or more rough sketches of possible circuit patterns have been made. One such sketch is shown in Fig. 4·39. This was made on a sheet of tracing paper placed directly over the layout of Fig. 4·38. A logical step is to run ground and power paths where they can conveniently be connected to components. Such a line starts at terminal 9 of $DS3$, connects at e of $Q1$, and goes to $J1$-16, ground. Other logical procedures are to sketch paths between components in one direction, say vertically such as between 7 of $DS2$ and 13 of $I2$, and between 4 of $DS2$ and 14 of $I2$. Note that there are few crossovers in the area between $DS2$ and $I2$ on the left and $DS1$ and $I1$ on the right. Also note that it is possible to have conductor lines going under the ICs and cathodes. This is possible even on the component side because the bottoms of the devices, when mounted, are raised almost a millimeter off the board.

Careful layout and preliminary sketching have not been able to eliminate quite a few crossovers of the radar circuit. This indicates that a double-sided board is the most logical and economic construction. (It is possible to treat a

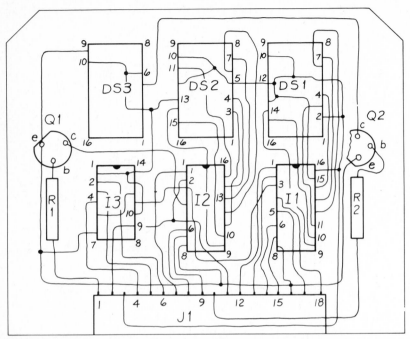

Figure 4·39 Preliminary sketch of connections for radar circuit board. Such preliminary drawings are the link between the schematic diagram and the master artwork.

crossover as a component on a one-sided board by making a "jumper" out of wire. But the economics of modern PC-board manufacture suggests that such crossovers be held to a bare minimum.) Therefore it was necessary to make more sketches for the printed paths on the component side and for the printed paths on the other side. One approach is to put all the vertical paths (as in Fig. 4·39) on one side and the horizontal ones on the opposite side. It usually is not this simple, but this is one way to begin.

The final results of such preliminary sketching are shown in the two master layouts of Fig. 4·40. Each layout was drawn double size. The distance between pads was 0.2 in. (5 mm) the paths were laid down with 0.026-in. (0.66-mm) tape, and the same figure was used for minimum space between paths. Most paths are horizontal or vertical, but there are some paths in congested areas (between I2 and I3 for instance) where it was necessary to use angled lines in order to meet minimum spacing requirements.

The master layout must be made on a dimensionally stable medium to a scale larger than the final product. Registration is important, and two (in some cases, three or more) registration points quite far apart are required. The two points used in Fig. 4·40 are such that the small squares in the centers remain white when both sides are exactly aligned with each other. The heavy corner

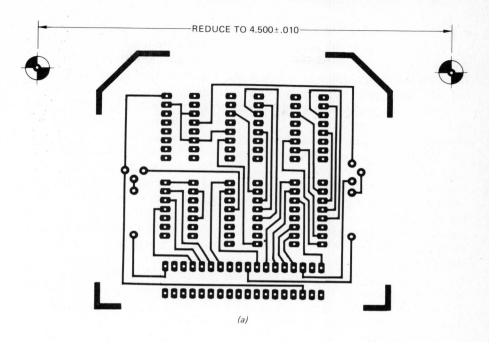

(a)

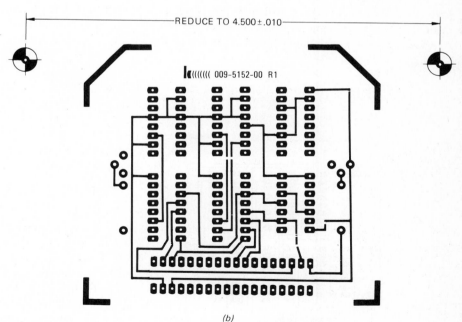

(b)

Figure 4·40 *(a)* Master layout for near side (component side) of printed circuit board. *(b)* Master layout (final artwork) for far side of printed circuit board. *(Kustom Electronics, Inc.)*

markers establish the board outline on the inside of these markers, as shown in Fig. 4·41. Lines in the master artwork must be solid black, unless a colored taping system is used. If the color system is used, blue tape is used for conductor paths on the component side and red tape is used on the opposite side. These specially made tapes can both be put on one sheet of clear polyester film. With the use of filters, the photographic process can "hold" one color while "dropping" the other color, and vice versa. Thus two circuit patterns, one for each side of the PC board, can be made from the two-color layout.

Once the master artwork is completed, it can be used to make other drawings, such as the drilling drawing (Fig. 4·41), the marking drawing (Fig. 4·42), and solder masks. Marking drawings include such items as polarity markings, serial number, vendor's number, and the name of the company using the board. (Many equipment manufacturers have their PC boards made by firms that specialize in PCB production.) Marking drawings are usually made with commercially available adhesive lettering and markings. Care should be taken in locating the markings and lettering so that they are not hidden by the components. Lettering and registration marks should be at least 0.015 in. (0.381 mm) thick, and letters should be high enough to read after reduction to final board size. The number 1 is used to indicate where pin 1 of the IC is to be located. The dashed lines for component outlines are not actually shown.

If the density of components (and wiring) that are to be in a circuit is high, two sides of a single board may not be enough to accommodate all the printed circuitry. In such an instance, more than one board may be necessary. A photograph of a multilayer board is shown in Fig. 4·43. This high-density

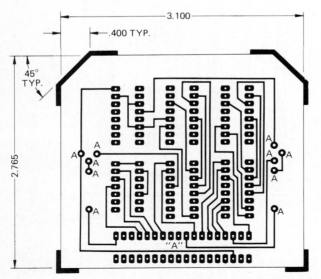

HOLE INDEX		
SYM.	QTY.	DESCRIPTION
A	46	.040 DIA.
B	94	.031 DIA.

Figure 4·41 Drilling drawing for PC board.

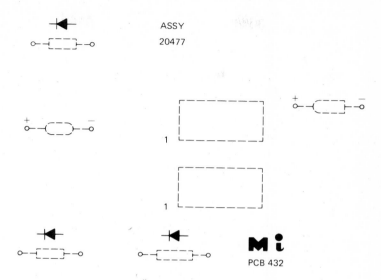

Figure 4·42 Marking drawing for a small printed circuit board.

Figure 4·43 Photograph of a four-layer board accommodating a time-base logic circuit. At least two layers of printed circuit paths can be seen in the lower right sector of the board. *(Kustom Electronics, Inc.)*

board with a lot of integrated circuits has four layers of printed circuitry printed on two pieces with a thin layer of what is called *prepreg* material in between. In the lower right-hand sector, just below the last integrated circuit and to the right or left, can be seen three colors (or shades) of lines, which represent circuit paths in various layers. Boards with as many as 14 layers have been manufactured. Sometimes a number of circuit boards interconnect with (plug into) such a multilayer board. In these cases such a board is called the *mother* board. Figure 4·44 shows a cross section of a six-layer board. After the board patterns have been made from the artwork, they are aligned with precision and the layers of prepreg material are placed between the boards. Then the assembly is placed in a heated press under high pressure. A six-layer board made of three 0.012-in.- (0.30-mm-) thick laminates might have a total thickness of 0.062 in. (1.57 mm), including substrate, prepreg, and thicknesses of printed wiring.

The design and layout for multilayer boards is beyond the scope of this book. Considerable experience is desirable. For one thing, a decision must be made about which circuits are to be placed in which layers. Sometimes ground planes and voltage buslines are the only items in a single layer. Accuracy of registration and manufacture are very important. Computer-aided layout is often used.

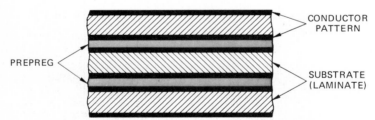

Figure 4·44 Cross section of a six-layer circuit board. Total thickness is about 0.062 in. (1.57 mm).

4·21 Use of the computer in printed-circuit-board layout

As the authors have suggested several times previously, the precision requirements for high-density printed-circuit work are such that automated drafting should be considered. Figure 4·45 shows two sides of a board layout made by an automatic drafting system (ADS), which was part of a computer-assisted design (CAD) system. More details of this system are provided in Art. 12-6. Because of the extreme precision available, this layout was drawn at a 1:1 scale. The thin lines are 0.01 in. (0.25 mm) wide and 0.02 in. (0.51 mm) apart. The three lines going horizontally through the center of the board (Fig. 4.45b), representing ground and voltage (V_{cc}) busses, are much wider, as is usually the case.

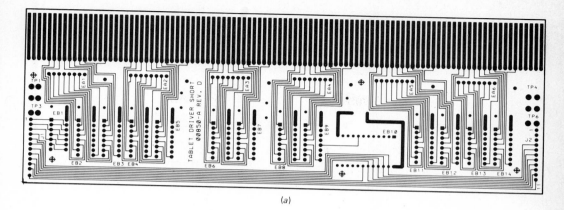

(a)

(b)

Figure 4·45 A circuit-board layout that was made by a computer-assisted design (CAD) system. *(a)* The component side with circuit paths. *(b)* The far (foil) side.

Two other aspects of CAD and layout of PC boards might be mentioned. First, it is possible to write computer programs that will determine the optimum (best) paths for circuit paths. In such a program the computer might be fed information for the radar circuit as follows: $J1$–1 to $R1$ to $Q1b$; $Q1c$ to $I2$–4; and $DS3$–10 to $DS3$–6 to $DS3$–3 to $DS2$–13; etc. Other information will be necessary for the computer to determine the best circuit-path layout. Second, a rough PC-board layout may be made by a person. Then, using a digitizer in which the cursor (Fig. 4·46) is positioned on the corners and ends of each path (drawn mechanically or freehand over a grid not shown in the illustration), the operator processes this information into a computer. The computer has been programmed to expedite the final artwork on a flat-bed or drum plotter so that all lines are neatly and accurately drawn horizontally or vertically, as was the case of Fig. 4·45, and/or sometimes in other directions if the situation requires. An overview of computer-aided design and drafting is presented in Chap. 12.

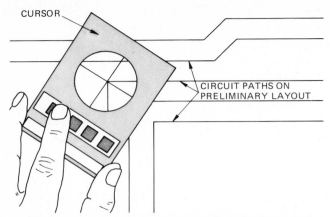

Figure 4·46 Using a digitizer to make a computer layout of a printed circuit. The cross hairs of the cursor are at a turn in a circuit path. Pushing the button will provide the coordinates of this location.

SUMMARY

There are many types of production drawings for the construction and assembly of electrical/electronics equipment. Some examples are connection diagrams, panel diagrams, chassis drawings, cabling diagrams, and assembly drawings. Connection diagrams may be classified as point-to-point, straight-line, airline (baseline), or highway. Careful identification of conductors must usually be made; this may include such items as color, component and terminal destination, size, shielding, and function.

The objects being connected are sometimes shown in symbol form and sometimes in elemental (rectangular or circular) form. Interconnection and cabling diagrams are variations of the connection diagram.

Chassis drawings can be used as the basis for wiring diagrams and parts-assembly drawings. Special problems in dimensioning may arise, requiring rather unique treatment. These drawings are often supplemented by complete parts lists and sets of manufacturing instructions giving each step and tool required. The method of manufacture often dictates how such a drawing is to be dimensioned.

Artwork for printed-circuit boards must be accurate, and therefore is often drawn two or four times the size of the finished conductor pattern. It may be done in ink on a stable medium such as Mylar, or with tape and preprinted pads, bends, etc. Companies that make or use PCBs usually have extensive lists of specifications that must be followed by the drafter or designer. Some of the drawings that are often included in the production of a PC board are:

1. Preliminary sketches for component location and conductor paths
2. Component layout
3. Master layout, final artwork
4. Drilling drawing
5. Marking drawing
6. Solder mask

Many scales are in use for PC-board layout. Both 2:1 and 4:1 are common. Some computer-assisted drawings are drawn full size. Scales of 10:1, 20:1, and 100:1 have been used. Printed-circuit paths are often drawn horizontally or vertically. Sometimes electrical requirements, such as resistive pinching and inductive coupling, dictate what directions adjacent lines should go. Production methods and economics (minimum use of materials) also must be considered in circuit-board layout.

QUESTIONS

4·1 What is the purpose of the solder mask in printed-circuit-board production?

4·2 What data should be included in a wire-harness table?

4·3 What is the difference between an airline, or baseline, diagram and a highway diagram?

4·4 When is it desirable to lay out a wiring diagram in a pictorial (such as isometric or perspective) view?

4·5 If a connection diagram is to show elements (such as tubes, resistors, etc.) of a circuit, to what source would you refer in order to portray those elements?

4·6 Assuming a circuit could be laid out with either a point-to-point or highway type of connection diagram, which type of diagram would probably require more lettering? Why?

4·7 Are connection diagrams sometimes accompanied by other types of drawings or electrical diagrams? If so, what might the other drawing be?

4·8 What letters would you use to indicate the following colors—black, red, white, yellow, gray, brown, green, red-blue?

4·9 In what situations would you use different line widths when drawing a connection diagram?

4·10 What is the sequence or order recommended in ANS Y14.15 for identification of feed lines in a connection diagram?

4·11 How is the viewing direction of a chassis layout selected?

4·12 Is it true that airlines (baselines) are usually drawn horizontally? Is it true that highways are usually drawn horizontally?

4·13 It may be said that a cabling diagram is a variation of a certain type of connection diagram. What particular type would this be?

4·14 What are the steps that are used in making a local-cabling diagram?

4·15 What elements make up a Manufacturing Layout and Time Rate sheet for construction of a chassis?

4·16 Briefly describe, by sketching, two or more different systems used for dimensioning holes on a chassis or board.

4·17 What drawings may be required for the production of a rather complex printed circuit?

4·18 What are some typical widths for printed conductor paths?

4·19 What is a good minimum distance between the outer edge of a PC board and an outside conductor path?

4·20 What is the first step in making drawings for a printed circuit?

4·21 How is the master layout used in the production of a printed-circuit board?

4·22 Show, by means of sketches, the following: eyelet, board outline, pad, elbow, and registration mark for use in PC-board construction.

4·23 What is included in a marking drawing?

PROBLEMS

4·1 Redraw the installation shown in Fig. 4·47 as a highway type of connection drawing, with at least three highways. For each lead use standard identification as follows: (1) component destination, (2) terminal, and (3) color of wire. For example, the lead from terminal 17 of component 3 would read D1/1/Y and would be placed near component 3. Drawing sheet size: 11 × 17 or 12 × 18.

4·2 Redraw the installation shown in Fig. 4·47 as a straight-line connection diagram. Use standard abbreviations for wire colors. Use 11 × 17 or 12 × 18 paper.

4·3 Redraw the connection diagram of Fig. 4·48 as a highway type of connection diagram. Use at least three highways. For each lead, use the standard identification as follows: (1) component, (2) terminal, and (3) wire color. For example, near component 1 the identification of the lead attached to terminal 1 would have the identification 15/6—O—W, meaning that it is attached at the other end to terminal 6 of component 15. Use 11 × 17 or 12 × 18 paper.

4·4 Redraw the connection diagram of Fig. 4·48 as a straight-line connection diagram. Use appropriate abbreviations for color designations. Use 11 × 17 or 12 × 18 paper.

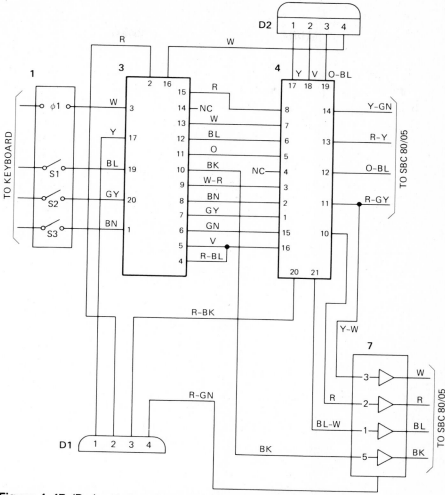

Figure 4·47 (Probs. 4·1 and 4·2) Point-to-point connection diagram of keyboard-to-microprocessor interface.

4·5 Draw the wiring diagram of the Honda CB 750 as it appears in Fig. 4·49. This may be improved by using standard symbols for contacts, diodes, etc., and abbreviations for wire colors, such as GN for green and BL for blue. Switching schedules may be included or omitted, as your instructor indicates. Use 11 × 17 or 12 × 18 paper.

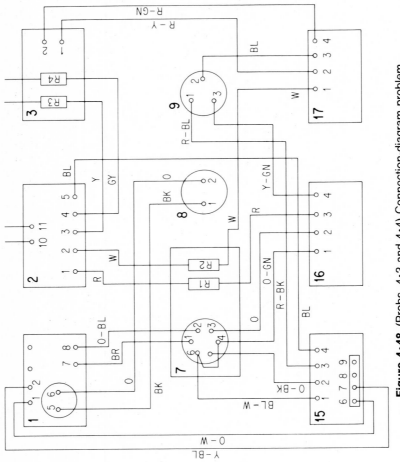

Figure 4·48 (Probs. 4·3 and 4·4) Connection-diagram problem.

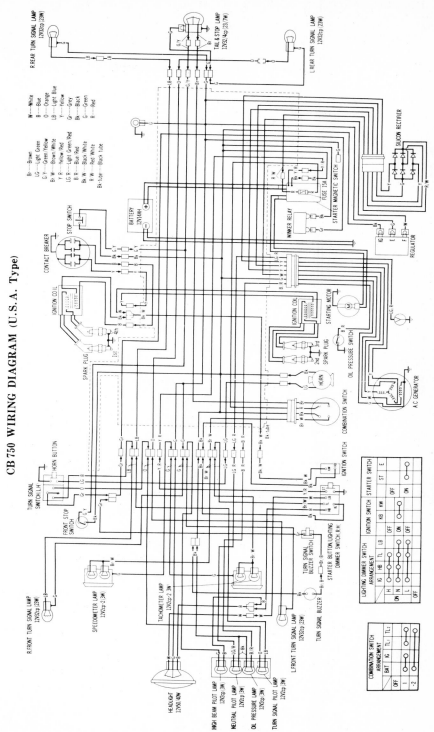

CB 750 WIRING DIAGRAM (U.S.A. Type)

Figure 4·49 (Prob. 4·5) Motorcycle wiring diagram.

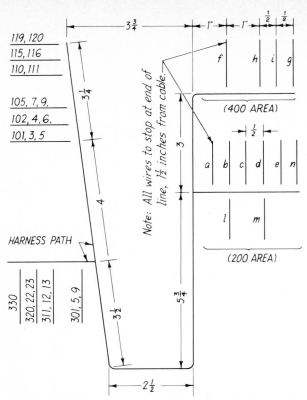

Figure 4·50 (Prob. 4·6) Cabling-harness problem.

4·6 The rudimentary diagram of a local-cabling harness appears in Fig. 4·50. Complete the diagram using the following information:

Start	Finish	Area	Start	Finish	Area
101, 3, 5	f	400 (call 401, 3, 5)	301, 5, 9	n	200
102, 4, 6	e	200	311, 12	i	400
105, 7, 9	d	200	313	m	200
110, 111	c	200	321, 22	h	400
115, 116	b	200	323	l	200
119, 120	a	200	330	g	400

Since starting-line 101 terminates at f in the 400 area, call the termination 401; call the termination of 103, 403; etc. Then add a wire-routing diagram to your drawing. At different places the harness should have different thicknesses, which you should estimate.

4·7 Figure 4·51 shows four views of a subchassis for an amateur-radio receiver. Not shown are a cover and anodized outside front panel. The front view has dual English-metric dimensioning, whereas the top and side assembly

130

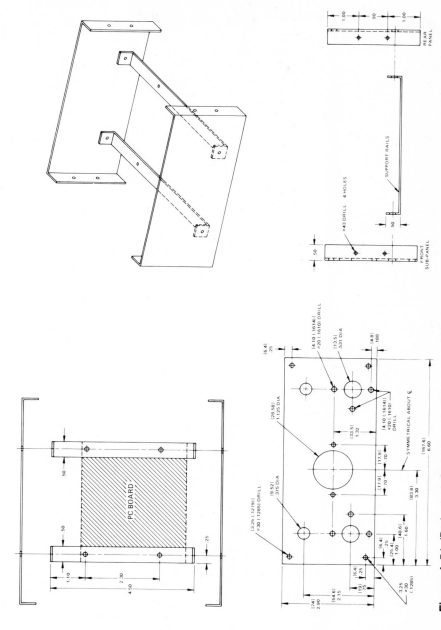

Figure 4·51 (Prob. 4·7) Orthographic and pictorial views of a subchassis for an amateur-radio receiver. Front view is dual dimensioned. *(Kantronics, Inc.)*

views are dimensioned in inches. Draw to full scale three or four views of this chassis. Use metric, dual, or English system of dimensioning, as your instructor directs. Without dimensioning this will fit on 11 × 17 paper; with dimensioning, on a C-size (17 × 22) sheet.

4·8 The interconnection diagram of Fig. 4·4 is point-to-point. Redraw it as a highway diagram. For the purposes of this problem, cables and other conductors may be merged into the same highway(s) if desired. Use 11 × 17 or 12 × 18 paper.

4·9 Redraw the diagram shown in Fig. 4·4 as a straight-line connection diagram. Numbers and some colors may have to be added to components and lines. Use 11 × 17 paper.

4·10 Draw the 2- × 2-in. circuit board shown in Fig. 4·52b to four times actual size. Then complete the location of the components for the schematic diagram shown in Fig. 4·52a. Show the necessary wiring (as hidden lines on the other side of the board) to agree with the schematic diagram. Add notes as required by your instructor. An alternative solution would be to draw a mirror image of the board and put the components in as hidden parts and the wiring in as solid. If it is desired to make a parts list, the values of the components can be found in Fig. 6·6. Use 11 × 17 or 12 × 18 paper. See below for typical component sizes.

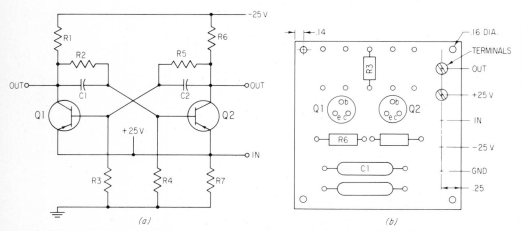

Figure 4·52 (Probs. 4·10 and 4·11) A wiring diagram or printed-circuit drawing problem. *(a)* Schematic diagram. *(b)* Proposed board, partially laid out.

4·11 Do the work indicated in Prob. 4·10, but make a printed-circuit-board drawing. Use 11 × 17 or 12 × 18 paper.

4·12 The circuit shown in schematic form in Fig. 4·53 is to be wired for a prototype, or experimental, assembly. Using a pattern similar to that shown in Fig. 4·53, design a component assembly that will fit on as small a

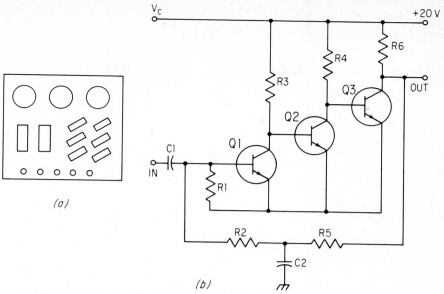

Figure 4·53 (Prob. 4·12) DC-amplifier connection-diagram problem.

Typical Component Sizes

	Transistor Diameters	
a. $\dfrac{0.370}{0.360}$	b. $\dfrac{0.650}{0.550}$	c. $\dfrac{0.345}{0.322}$

Resistor Sizes

	Length	Diameter	Max. Resistance
$\frac{1}{4}$ W	0.375	0.093	22 MΩ
$\frac{1}{2}$ W	0.375	0.138	22 MΩ
1 W	0.562	0.225	22 MΩ
1 W	0.715	0.237	22 MΩ
2 W	0.688	0.318	22 MΩ

Capacitor (Tubular Paper) Sizes

	200 V dc W	400 V dc W
0.01 μF		
Small	$\frac{1}{4} \times \frac{5}{8}$	$\frac{1}{4} \times \frac{5}{8}$
Large	$\frac{3}{8} \times 1$	$\frac{3}{8} \times 1\frac{1}{8}$
0.1 μF		
Small	$\frac{7}{16} \times \frac{7}{8}$	$\frac{15}{32} \times 1\frac{1}{8}$
Large	$\frac{9}{16} \times 1\frac{5}{8}$	$\frac{5}{8} \times 1\frac{5}{8}$

card (or board) as possible, leaving about 0.30 clearance from the parts to the edge of the board. Use the $\frac{1}{2}$-W resistor, small 0.01-μF capacitor, and 0.370/0.360-diameter transistor shown in the data given on page 133. The wiring should be on the opposite side of the components. If drawn four times actual size, the drawing may fit on $8\frac{1}{2} \times 11$ paper, but it would be safer to use 11×17 or 12×18 paper.

4·13 Figure 4·54 shows a pictorial view and several other partial views of a chassis. The hole sizes are as given in the table below. Make a complete set of drawings for the construction of this chassis, which has one panel labeled A, two panels on each side marked C (each having two holes), and one panel in front labeled B. There is no panel in the rear. Panel B is swung out in the pictorial view to show how C is folded. B is actually vertical. Dimensions are in inches.

Hole	Letter or Number Drill	Diameter (in in.)
D	N	0.302
E	See detail	See detail
F	#38	0.1015
G	K	0.281
H	#27	0.144
J	#20	0.161

The chassis may be drawn in two ways. Several orthographic views, with dimensions, of the chassis in its completed folded form would be satisfactory. Showing the single piece of metal laid out flat, with fold lines and a small pictorial view of the folded chassis, would also be satisfactory.

All burrs should be removed. The 20-gage steel should be degreased per Spec. 5160 and coated with clear varnish per Spec. 5160. Tolerance is ±0.02 in. Dimensions may be converted to the metric system. Sheet size: 11×17, 12×18, or larger.

4·14 Figure 4·55 shows a pictorial view and partial views of a bracket, plus a schedule for hole sizes. This bracket is part of the assembly shown in Fig. 4·24. Make a complete drawing for the construction of this part. The bracket is to be made of steel sheet, cold-rolled and commercial quality (CRCQ). The holes and ovals are to be punched out, and "C" holes countersunk for a 0.190–32 flat-head machine screw. The part is to be degreased per Spec. 51606 and zinc-plated for 289A finish. Tolerance is ±0.016 in. Dimensions may be converted to the metric system. Use 11×17 or 12×18 paper.

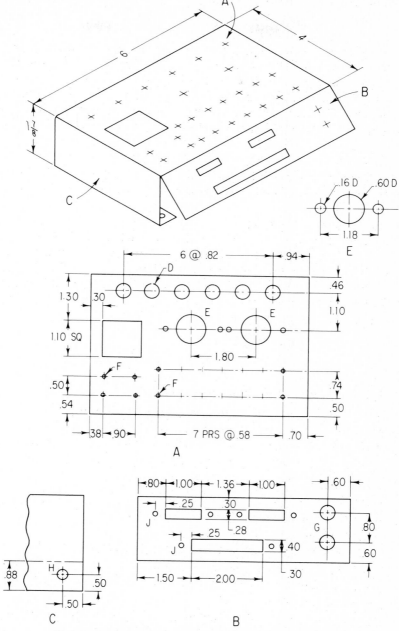

Figure 4·54 (Prob. 4·13) A chassis drawing problem.

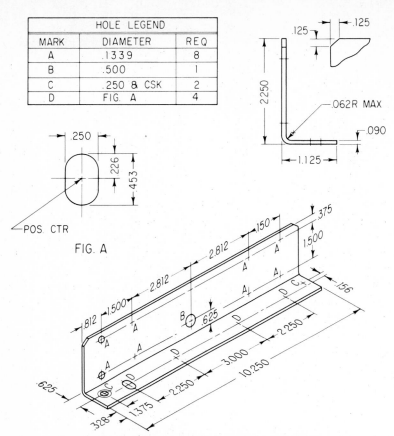

HOLE LEGEND		
MARK	DIAMETER	REQ
A	.1339	8
B	.500	1
C	.250 & CSK	2
D	FIG. A	4

POS. CTR

FIG. A

Figure 4·55 (Prob. 4·14) Bracket for TD-2 chassis.

Figure 4·56 (Prob. 4·15) Pictorial view of TD-100 chassis.

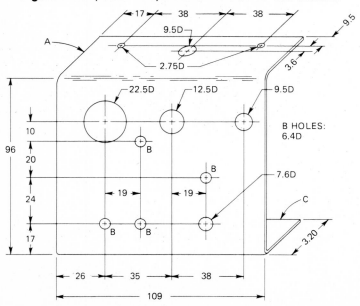

4·15 Make a three-view drawing of the chassis shown in Fig. 4·56. Flanges A and C are identical. The chassis is to be made of 20-gage (0.08 mm) CR steel and degreased. Finish is to be gray enamel, baked 525A, 0.025 mm thick. Dimension it according to one of the methods prescribed in this chapter. One of the views may be a flat foldout, or development. This would permit dimensioning for N/C drilling of the chassis. Dimensioning may be done in the dual or metric system. Dimensions shown are in millimeters. Use 11 × 17 paper.

4·16 Figure 4·57 shows a preliminary layout of a circuit board that is to contain the sweep-meter circuit shown. Check the layout against the schematic diagram for possible errors or improvement in layout. Make a PCB master

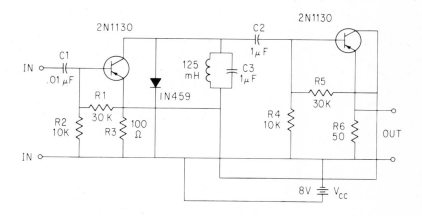

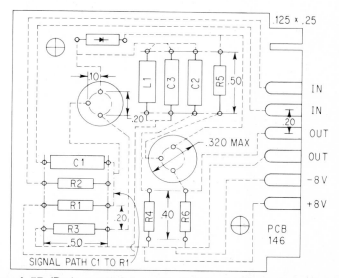

Figure 4·57 (Prob. 4·16) Schematic diagram and preliminary drawing of circuit board for a current-sweep generator.

layout (outside dimensions are 2.20×2.40 in.) to a $4:1$ or $5:1$ scale. Use the 0.10-grid system for locating components, whose sizes are: resistors, 0.25×0.09 in., except $R3$ which is 0.375×0.09 in.; capacitors, 0.422×0.135 in.; $L1$, 0.400×0.15 in.; and diode, 0.275×0.105 in. maximum. Use 0.062 in. conductor paths, with 0.031 in. minimum spacing. Use $8\frac{1}{2} \times 11$ paper.

4·17 Figure 4·58 includes an elementary diagram and a suggested arrangement of parts for a differential amplifier. Make a scale drawing ($4:1$ or $5:1$) of the final master drawing for the pattern. A maximum of four jumpers will be permitted. The lower figure shows the wiring side of one arrangement of components. Transistor diameters are 0.360 ± 0.010 in., and resistor dimensions are 0.093 diameter $\times 0.375$ in. Try to get an acceptable pattern on a board that is no larger than 3.00×2.00 in. and is 0.062

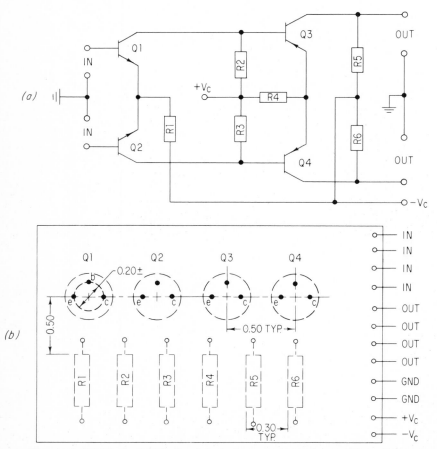

Figure 4·58 (Prob. 4·17) A printed-wiring problem for a differential-amplifier circuit.

in. thick. Freehand preliminary sketches are suggested. If you cannot get an acceptable pattern with the suggested arrangement of components, make your own arrangement. Hole diameters are 0.032 in. Terminal pads are 0.125 in. across. Use 0.062-in. conductor paths with 0.031-in. minimum spacing. Show registration marks and pin #1 (top pin), and label the board PCB 46. Use $8\frac{1}{2} \times 11$ paper. Show one critical dimension.

4·18 A digital timer circuit is shown in schematic form in Fig. 4·59. Make a preliminary wiring sketch, a component layout sketch, and a final artwork for a printed-circuit board for this circuit. Have all the inputs located at one side with fingers 5.08 mm (0.20 in.) apart. Use a 2.54-mm (0.10-in.) grid. Let conductor paths be 1.57 mm (0.062 in.) wide with a minimum spacing of 0.80 mm (0.031 in.) between paths and outside paths and edge of board. Shape and size of board are to be determined by student or instructor, as is scale. Terminal pads are to be 3.18 mm (0.125 in.) in diameter with holes 1 mm (0.040 in.) in diameter. The DIPs, however, will use standard pin spacing of 2.54 mm (0.10 in.). Dimensions of the components shown are in millimeters.

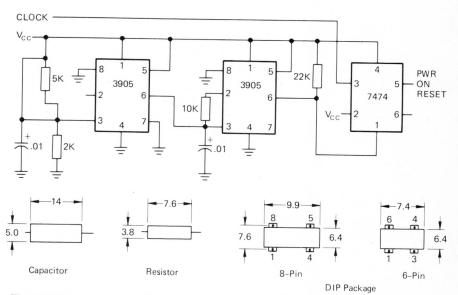

Figure 4·59 (Prob. 4·18) Schematic diagram of a timer circuit with power on reset. Major dimensions of components are given below.

4·19 Figure 4·60 shows the schematic diagram of a video peak-detector circuit. Dimensions of the components are shown in millimeters. Make a sketch of a physical arrangement of components that will accommodate a printed circuit on a minimum-size board. The switch will be attached to the board by wires. It will be mounted elsewhere. Then draw a master layout using

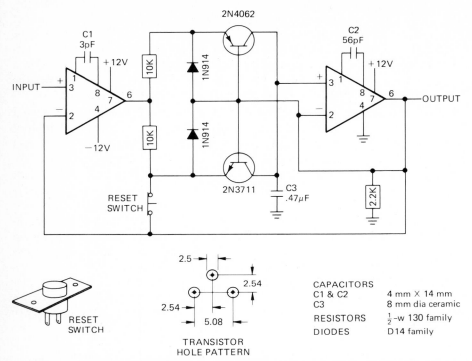

Figure 4·60 (Prob. 4·19) Schematic diagram of a video peak-detector circuit. Major dimensions of most of the components are also shown below.

conductor path widths of 1.57 mm except for ground, which will be 2.54 mm wide. Use a scale of 2:1 for $8\frac{1}{2} \times 11$ paper or 4:1 for 11×17 paper. Make pads 3.18 mm diameter with 1-mm-diameter holes, except for leads to switch, which will require 1.2-mm-diameter holes and a larger-size outer diameter for each wire connection. Refer to Table 4·3 for sizes of components. See Fig. 4·59 for sizes of amplifier DIPs.

4·20 Problems 4·11, 4·12, 4·16, 4·17, 4·18, and 4·19 have to do with printed circuits. Your instructor may require you to make (in addition to the master layout) a drilling drawing or marking drawing, or both. Provision for mounting the board may require the use of mounting holes or brackets. Your instructor may also require a list of material which would include the following for each item on the board: (1) Item number, (2) reference designation, (3) number required, (4) description, (5) manufacturer or part number, (6) remarks.

Chapter 5
Flow diagrams and logic diagrams

A primary drawing is one that shows the function of a circuit or a system in a logical manner. It is usually the first drawing sketched by an engineer or technician. The primary drawing in electronics and industrial control used to be the schematic diagram. But the ascendancy of digital electronics has necessitated that the logic diagram become the primary one, and this is also true for some areas of industrial control. As systems become more complex or miniaturized (using LSI—large-scale integration—components, for example), the block and flow diagrams are becoming more popular and necessary for the explanation of the functions of these systems.

5·1 Examples of block diagrams for electric circuits and systems

Such a diagram may be used to show the operation of a large electronics system. In such a case, a block would represent a complete and removable chassis, such as a preamplifier, a multivibrator, or even a television camera.

However, in a different situation, a block diagram may be used to facilitate the understanding of a radio receiver or a multistage amplifier, for example. In this case, each block would represent a *stage*. This is the case of the diagram shown in Fig. 5·1. At this time it might be a good idea to define the word *stage*. A stage is considered to be that part of a circuit which includes the main device (e.g., transistor, diode, or tube) and the associated devices that go with it, such as biasing resistors, load resistor, voltage dividers, and capacitors. In other words, a circuit may have several stages, hooked together somehow, so that the signal goes first through one stage, then through the next, and so on. Each block represents a major subsystem. The diagram shows major signal and data paths, inputs, outputs, and control points.

Figure 5·1 shows how easy it is to understand a circuit's operation by means of a block diagram. It is clearly shown that the signal comes through the antenna (usually portrayed by a symbol rather than a block) and then progresses through the mixer circuit, through the intermediate-frequency (IF) stages, and finally to the output stage and speaker. The oscillator, which is an

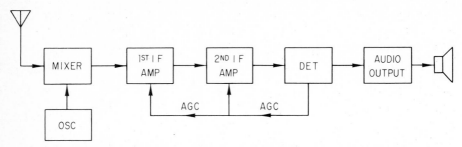

Figure 5·1 Block diagram of a typical transistor radio receiver circuit.

auxiliary circuit, is appended to the main circuit; and, because it is a frequency generator, its output is fed into the signal train as shown by the arrow. A feedback circuit, labeled AGC for automatic gain control, is correctly drawn below the main circuit. If it were desirable to emphasize the AGC circuit, however, it would be appropriate to place it above the main circuit. The purpose of automatic gain control is to prevent fluctuation in speaker volume when the radio signal at the antenna is fading in and out.

5·2 Preparation of block diagrams

A block diagram may use standard symbols for certain elements, but it is predominantly one of blocks (usually, but not always, squares or rectangles). The layout can be facilitated by (1) using freehand sketches in the initial stages, (2) using cutout cardboard blocks of appropriate size and experimental arrangement until the best pattern is achieved, or (3) using cross-ruled paper for an undergrid, thus facilitating the construction of blocks of equal size and uniform spacing between blocks. The size of the rectangles is usually determined by the lettering that goes in them; and, since they are usually drawn about the same size, the block with the most lettering will often set the size of the blocks in the entire diagram.

From Fig. 5·1 and other diagrams in this chapter, certain facts about block diagrams can be deduced, and the following rules for their construction can be listed:

1. The signal path should be made to go from left to right, if possible. In large, complex drawings, the input should preferably be at the upper left and the output at the lower right, if possible.

2. Blocks are usually drawn in one of three shapes: rectangular, square, or triangular. (The triangle represents different items in different types of diagrams. There are also other shapes for certain specialized diagrams, as will be shown later in the chapter.)

3. Once the size and shape of a block are determined, the same size and shape should be used throughout the drawing. The size of a rectangle,

for instance, bears no relation to the importance of the component(s) it represents.

4. A single line, preferably heavy, should be used to show the signal train from block to block. In complex circuits or systems, however, more than one line may have to be drawn leading into or away from a block.

5. Arrows should be used to show the direction of signal flow.

6. Some components, usually terminal ones such as antennas and speakers, are shown by means of standard symbols rather than by blocks.

7. Titles, or brief descriptions, of the components or stages represented should be placed within the blocks.

Aside from the above-listed rules, no standardized procedure exists for the preparation of block diagrams. In Fig. 5·1, for instance, either square or rectangular blocks could have been used. The arrows, which are shown touching the blocks, could have been placed midway between the rectangles if desired.

In Fig. 5·2, both the spacecraft and ground systems are shown. Part of the camera's apparatus is shown pictorially. (It could have been drawn as blocks.) It generates a field-sequential color signal using a rotating color wheel and a single-image tube. The tape recorders compensate for Doppler shift and present real-time information to the scan (color) converter. The second tape recorder is driven by a standard frequency to correct any tape-speed errors. The converter is a storage and readout device.

Figure 5·3 shows the block diagram for a digital watch. Not shown are the batteries (2) and light-emitting diodes. Practically all the circuitry is on a single chip. A three-view photograph of a transparent module without batteries

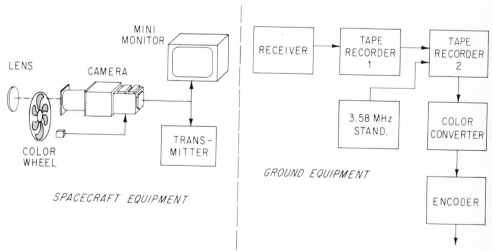

Figure 5·2 Block diagram of the Apollo color-television system. *(Westinghouse Electric Corp.)*

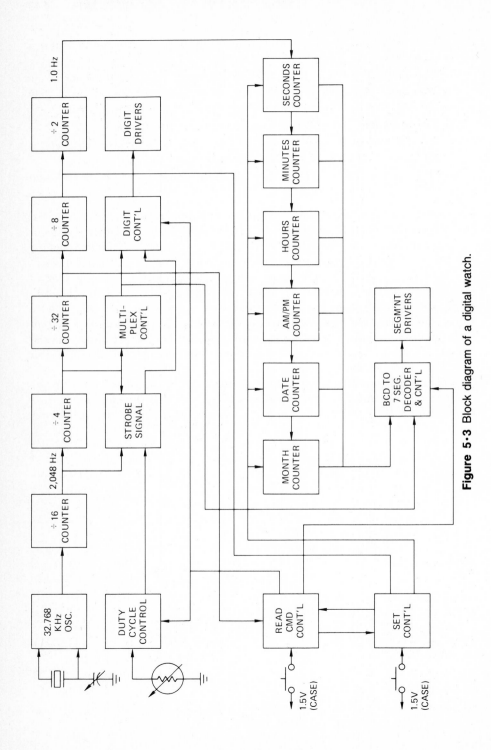

Figure 5·3 Block diagram of a digital watch.

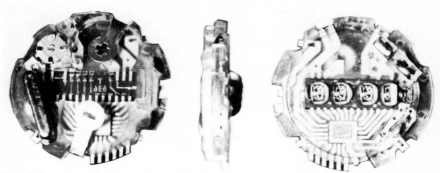

Figure 5·4 Three views of a transparent module of a digital watch. In view at right can be seen the four LEDs (center) and the IC (below center). In view at left can be seen the contacts for the two batteries, the battery separator (T636 in center), the adjustable capacitor (upper left), and the 32,768-Hz quartz crystal (near left edge, inclined at ≈ 75°). *(Texas Instruments, Inc.)*

is shown in Fig. 5·4. Laying out a block diagram for a fairly complex arrangement requires considerable planning, and possibly some experimentation, in order to achieve a neat, well-spaced arrangement of blocks and signal paths.

5·3 Logic diagrams

Beginning with computer design in the 1950s, engineers and manufacturers began to work with logic functions which could be performed by basic circuits. Now people working in many areas—transportation, industry, and communications, for example—find it necessary to make and read drawings that contain symbols representing logic functions. We will describe some of the basic logic functions in this chapter. However, examples will also appear in other parts of the book as various circuits and designs are presented.

The symbols that have been used to represent these functions in the past have not been very well coordinated and standardized. For example, Fig. 5·5 shows six symbols that have been used to represent the AND function. Fortunately the most recent standard, ANSI/IEEE Y32E, has included the revised versions of ANS Y32.14 (1973) and the IEEE and Military Standards that relate to logic symbols. Hopefully they will be accepted by everyone who works with these symbols so that only the distinctive and rectangular shapes will be in use. Inasmuch as the distinctive shapes are quicker to follow, they are widely used, and we shall use them for most of our examples. Figures 5·6 and 5·7 show the distinctive shapes for the more commonly used functions. As is the case with other symbols, no exact size has been specified. However, the authors have noted that if the units that are specified in Fig. 5·7 are drawn as millimeters, the symbols will be just about the right size for most large-size drawings.

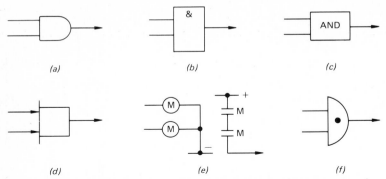

Figure 5·5 Graphical representation of the AND function or gate. *(a)* ANSI-IEEE–approved distinctive-shape symbol. *(b)* ANSI-, IEEE-, and IEC-approved rectangular symbol. *(c)* Old symbol. *(d)* NEMA symbol. *(e)* JIC symbol. *(f)* Old distinctive-shape symbol. Symbols *(a)* and *(b)* are recommended.

Now for a description or definition of the functions. AND: If a signal is impressed at *A,* and at the same instant a signal is impressed at *B,* there will be a definite output signal at *C.* The "high" signals are often referred to as the 1-state or the ON-state, and the "low" signals are often referred to as the 0-state[1] or the OFF-state. Hence, the truth (logic) tables in Fig. 5·6 show the numbers 1 and 0. The table for AND indicates that there is a high (sometimes referred to as HI) signal at the output only when there are simultaneous high signals at A and B. The OR table shows that a HI signal or pulse at A or B or both will produce a HI output signal. We believe the other tables in Fig. 5·6 are self-explanatory, except for the RS flip-flop, which will be described as follows:

> The outputs assume their indicated 1-states when only the S input assumes its indicated 1-state. The outputs assume their 0-states when only the R input assumes its indicated 1-state.

A polarity indicator symbol is shown on the D output of the flip-flop symbol. This denotes that the 1-state of that output is the less positive level. This symbol is placed at the junction of the input or output line and the function symbol and points in the direction of signal flow.

There are other types of flip-flop circuits, but space does not permit their treatment in this text. And there are numerous other functions that have not been described. These are rather well documented in ANS Y32.14, which is now Sec. 14 of ANSI/IEEE Y32.E.

---

[1] The number zero.

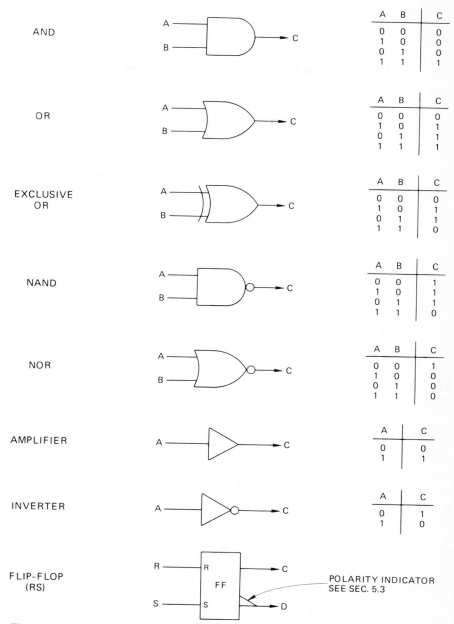

Figure 5·6 Distinctive-shape symbols approved and recommended by ANSI and IEEE. Truth or logic tables for each function are at the right. The number 1, for 1-state, signifies an input or output that is high or "true." Zero indicates that the input or output pulse or signal is zero, low, or "false." *(ANSI/IEEE Y32E.)*

147

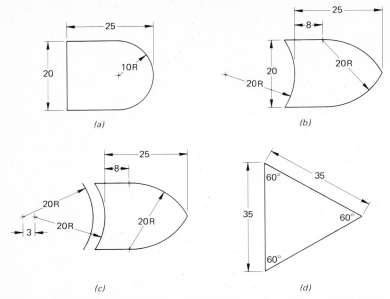

(a)

(b)

(c)

(d)

Figure 5·7 Recommended symbol-outline proportions. *(a)* AND symbol. *(b)* OR symbol. *(c)* Exclusive OR symbol. *(d)* Amplifier symbol. *(ANSI/IEEE Y32E.)*

5·4 Drawing the symbols

Logic symbols, either the distinctive-shape form or the rectangular form, can be drawn with instruments or special templates that are made for this purpose. The standard proportions are shown in Figs. 5·5 and 5·7. (Dimensions in Fig. 5·7 are in millimeters.) In order to provide for situations where many inputs are to be drawn to a symbol, extensions as shown in Fig. 5·8c and *d* may be added.

Sometimes it is desirable to have logic symbols of two or more different sizes on a single diagram. Some of the symbol templates have two sizes of each symbol.

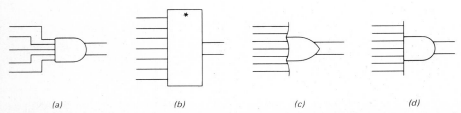

(a)

(b)

(c)

(d)

Figure 5·8 Accommodation for multiple inputs. *(a)* Sometimes used for a limited number of inputs. *(b), (c),* and *(d)* Approved by ANSI/IEEE Y32E. The asterisk is to be replaced by a qualifying letter, symbol, or word, such as OR or A (for AND).

5·5 Negative and mixed logic

The NAND (not AND) function symbol shown in Fig. 5·6 has a small circle at the junction of the symbol and the output line. This circle is a negative indicator symbol. Its presence provides for the representation of the output in terms independent of its physical value. That is, the 0-state of the output is the 1-state of the symbol. If a circle is placed on the input side of a logic gate, the state or polarity of the input signal is reversed before entering the gate or function.

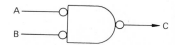

Figure 5·9 An AND function having negative indicators at inputs and outputs. The result of using this negative logic is that the circuit performs the function of an OR circuit.

The result of the negative circle on the output side of an AND gate is that it now performs the NAND function. Fig. 5·9 shows the effect of having negative logic on all leads of an AND function. The truth table is as follows:

Negative Truth Table for AND Function

A	B	C
1	1	1
1	0	1
0	1	1
0	0	0

A glance at the truth table for the OR circuit, Fig. 5·6, will reveal that these two truth tables are the same. That is, the AND circuit performs the same function in negative logic as the OR circuit performs in positive logic.

Instead of the letter C for output, some truth tables show the letters AB (the output for the AND function), $A + B$ (the output for the OR function), $\overline{A}$ (the output for a converter or NOT gate), $\overline{A}B + A\overline{B}$ (the output for an Exclusive OR function), $\overline{AB}$ (the output for a NAND function), and $\overline{A + B}$ (the output for a NOR, not OR, function). There may be, and often are, more than two inputs into a function, in which case the letters will be changed (see Fig. 5·10) or omitted entirely.

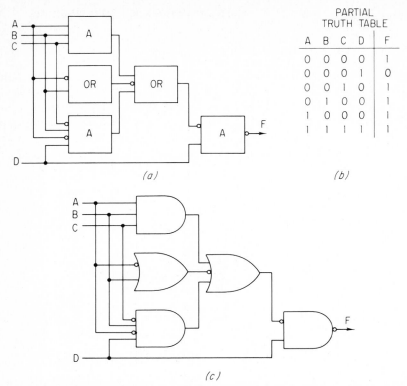

PARTIAL
TRUTH TABLE

A	B	C	D	F
0	0	0	0	1
0	0	0	1	0
0	0	1	0	1
0	1	0	0	1
1	0	0	0	1
1	1	1	1	1

(a)   *(b)*

(c)

Figure 5·10 Example of a logic diagram in which uniform shapes have been used at *(a)* and distinctive shapes at *(c)*.

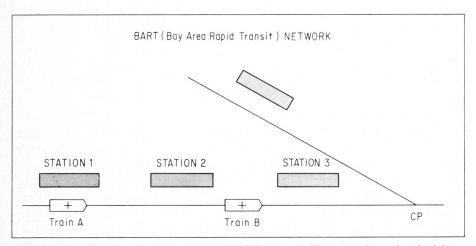

BART (Bay Area Rapid Transit) NETWORK

STATION 1   STATION 2   STATION 3

Train A   Train B   CP

Figure 5·11 Simplified diagram of the BART network illustrates how the decision-table logic matches system conditions with a rule (in this case rule 14). Rule 14 specifies action 1, which is to revise schedules ahead of delayed train to reduce the gap.

Table 5·1 Decision Table for an Automated Rapid-Transit System

Rules	Where Event Occurred*	Is Train Approaching CP at ≤ Minimum Tolerance?	Will First-Come–First-Served Give Correct Sequence?	Will Train Behind Be Delayed?	Train Performance Index†	Can Trains Ahead Be Slowed?	Actions‡
1	0	Yes	Yes	Yes			2
2	0	Yes	Yes	No			0
3	0	Yes	No				3
4	0	No		Yes			2
5	0	No		No			0
6	1			Yes			2
7	1			No			0
8	2				0		0
9	2				1		0
10	2			Yes	2		2
11	2			No	2		0
12	2			Yes	3		2
13	2			No	3		0
14	2				4	Yes	1
15	2				4	No	4

* State 0, between a station and a "merge"; state 1, between a station and a CP that is not a merge; state 2, at least one station before a CP.

† State 0, < 10 s late; state 1, 10 to 30 s late; state 2, 30 to 60 s late; state 3, 60 to 120 s late; state 4, > 120 s late.

‡ Action 0, continue with existing schedule; action 1, revise schedules ahead of delayed train to reduce extended gap; action 2, revise schedules behind delayed train to extend reduced gap; action 3, recommend revised sequence at interlocking; action 4, recommend station run-through.

5·6 The decision table

Somewhat similar to the logic table is the decision table, which is essentially a tabulation of logical relationships consisting of conditions, actions, and rules. Conditions are the variables that influence any decision, while actions are the things to be done once a decision has been made. In the case of BART [San Francisco's rapid-mass-transit system (Fig. 5·11 and Table 5·1)], a computer makes the decision and one of five actions (0 to 4 listed at the bottom of the table) is taken. A CP is a critical point (control location) at which it is especially desirable for trains to be on time.

5·7 Analog-computer programming diagrams

An important step in the writing of a program to be solved on a computer is the making of a flow diagram. Certain standardized blocks are used in making these diagrams. Figure 5·12 shows the basic blocks for analog-computer diagramming. Not shown are such functions as the servo multiplier, division circuit, diode function generator, and servo function generator.

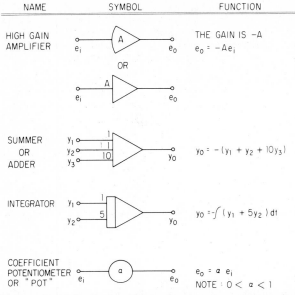

Figure 5·12 Basic functions of general-purpose analog computers and their symbols.

These are the things that can be done with a medium-size, modern, general-purpose analog computer:

1. Add
2. Subtract
3. Multiply:
   a. Easily by a constant
   b. Not so easily by a variable
4. Integrate
5. Generate functions
6. Display or record the results:
   a. By a meter
   b. By an oscilloscope
   c. By a recorder or plotter

The upper limit of the *gain* of good high-gain amplifiers is 10^8. A gain of 5, for example, means that the output signal is five times as strong as the input signal. One problem facing the programmer is that of scaling the output so that it will be properly recorded. In other words, if the voltages used are too large, the curve representing the answer will be too large to fit on the recording graph or oscilloscope. Or, if the voltages used are too small, the answer curve will be too flat to give an accurate answer. Available working voltages are generally between -100 and $+100$ V on large- and medium-size analog computers, and -10 and $+10$ V on small analog computers.

Figure 5·13 shows the flow diagrams which would be used for solving four problems on the analog computer. The first two are quite simple. Problem *a* requires the correct potentiometer setting to achieve an output of -50 V. The answer is 0.04. Problem *b* requires the correct gain to place at the upper input to the second adder to achieve the indicated output. The answer is 2. (The output of the first adder is -3 V.) Problem *c* involves two integrators and pots. If the input of -100 V, later multiplied by 0.322, represents d^2x/dt^2, then the output of the first integrator is dx/dt, which is the velocity v. For scaling purposes, a second potentiometer is set to 0.1, and the integral of $v/10$ turns out to be $-x/10$. Now, if we get an output of 10 V $(x/10)$, $x = 100$ and our answer is 100 ft. Problem *d* involves three integrators in order to solve for x. This diagram shows how the coefficients in the equation can be set up with potentiometers. It also shows initial conditions of zero set into each integrator. Usually, but not always, the initial conditions are zero.

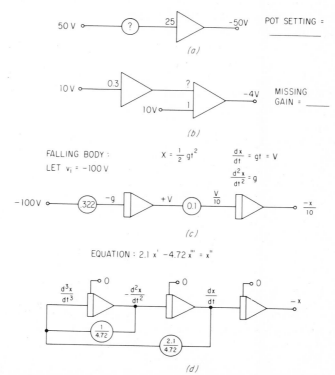

Figure 5·13 Four problems set up by diagrams for solution on an analog computer.

Students who use this text may not have had much calculus instruction. Also, there is more to analog-computer programming than what has been mentioned here. Therefore, the reader may not understand all the problems discussed here. But all readers should appreciate the importance of the flow diagrams in solving such problems. (These diagrams represent the way the computer is wired with patch cords, etc.)

5·8 Digital-computer programming diagrams

Engineers, scientists, and certain technicians in the electrical and electronics fields may find it advisable to use the analog computer for certain problems and the digital computer for other problems. Figure 5·14 shows the basic blocks for flow diagrams in digital programming. We believe that two examples will show fairly clearly how the boxes are used.

The problem shown in Fig. 5·15*a* is that of obtaining the shortest distance between two points in space, given any *X, Y,* and *Z* coordinates for the two points. The problem is programmed so that two sets of coordinates will be read, the distance computed and then printed (on tape, typewriter, or punched card), and the same procedure repeated until there are no more coordinates left. This is shown on the decision block, labeled *n*:0. If *n* is equal to zero *(Yes),* the program is finished; but, if *n* is not equal to zero, there are more coordinates yet to be read and the computer will compute *D* for the next set of coordinates.

Problem *b* consists of solving *p* for *N* values or combinations of *a, b,* and *c.* It has two decision boxes, one for the same purpose as the previous problem. But the first box, labeled *C*:0, is to take care of the situation when a *C* value should happen to be zero. A glance at the equation reveals that *p* will be infinity if *C* equals zero. The computer will try to go to infinity, and precious minutes or hours will be wasted because it cannot reach infinity. To avoid this possibility, the computer can be instructed to do something else whenever a *C* value of

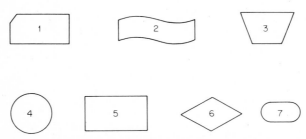

Figure 5·14 Basic symbols for flow diagrams used in digital-computer programming. (1) Punched card. (2) Tape. (3) Input-output. (4) Magnetic tape or termination. (5) Processing, computation or function. (6) Decision (branching). (7) Start or stop.

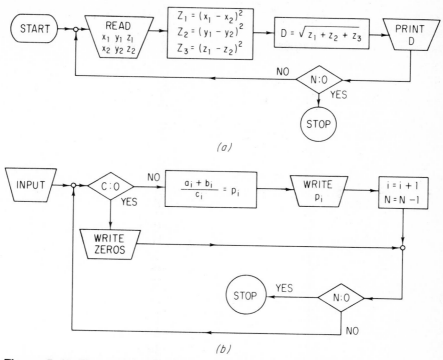

Figure 5·15 Flow diagrams for two problems programmed for solution on a digital computer.

zero shows up. In this case, the instruction is to punch or print a series of zeros, then start over with a fresh set of values.

Flow diagrams may be drawn from left to right and possibly from right to left if there is a second row of blocks below the first. Or they may be drawn from top to bottom. The function boxes are sometimes drawn the same size throughout a diagram, but, because of the varying amounts of material that may be placed in these boxes, they are often drawn in various sizes and proportions.

5·9 Microprocessors and microcomputers

A powerful newcomer to the list of electronic inventions is the microprocessor. When combined with certain other elements, it can be the heart of a microcomputer. A typical single-board microcomputer will have a central processing unit (CPU), memory, and input-output (I/O) elements. These microprocessors and microcomputers can perform as general-purpose computers when combined with peripheral devices such as typewriters, line printers, or magnetic-tape units. Or they may provide process control, such as automating an assembly line. They are now being used for carburetion management of automobiles.

A sound understanding of basic computer operations is necessary for the person who works with microprocessors, particularly in the design, start-up, and maintenance phases. Learning modules are available to assist engineers, technicians, and programmers in learning microprocessor concepts. Figure 5·16 depicts a system of learning modules that include the following:

LCM 1001   microprogrammer
LCM 1002   controller
LCM 1003   read/write memory
LCM 1004   input/output interface

One of the features of microcomputer diagrams is the "bus" system, which may include not only the address bus and data bus shown in Fig. 5·16, but also a control bus and timing bus. More will be written in later chapters about microprocessors and microcomputers.

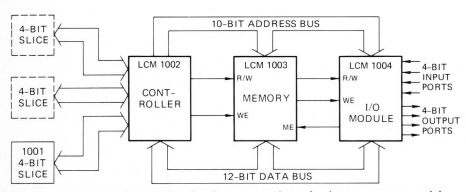

Figure 5·16 Flow diagram for the interconnection of microprocessor modules. *(Texas Instruments, Inc.)*

SUMMARY

Flow diagrams are used for different purposes in different situations. They may be used for depicting an electric circuit or system, for the preliminary design work in computers and other electrical installations, and for programming problems to be solved by computers. A left-to-right direction is sought in planning most block diagrams because this is the normal way in which people read. However, this sequence cannot always be followed. In most block diagrams there is a certain amount of lettering within the blocks. This lettering may determine what the sizes of the blocks will be. Liberal use of arrowheads is made in block-diagram construction. Although the rectangle is widely used in flow and block diagrams, other shapes—and sometimes electrical symbols—

are used as dictated by standard practice. Blocks and flow lines should be evenly spaced if a pleasing drawing is desired.

In recent years, the use of distinctive shapes for logic diagrams has become popular. These shapes facilitate the reading of diagrams for complex systems.

QUESTIONS

5·1 What standard covers the preparation of block or flow diagrams?

5·2 Where are auxiliary circuits, such as feedback, usually placed with regard to the main signal train?

5·3 When would you use a block diagram to show the electrical properties of a packaged circuit?

5·4 In making a block diagram of an electrical system, what shapes would you use for the blocks? Why?

5·5 What direction of flow would you attempt to show in planning a block diagram of an electronic circuit?

5·6 Sketch six different shapes that may be used at one time or another in flow diagrams. Label each shape.

5·7 What are five different functions that might be shown with the rectangle in block or flow diagrams?

5·8 What are three devices that may be shown by means of standard symbols, rather than by boxes, according to customary practice?

5·9 Define the word *stage*.

5·10 With negative logic (circles) used at all outputs, what does a NAND circuit perform? A NOR circuit?

5·11 What is the difference between an Exclusive OR circuit and an OR circuit? (Show by means of a truth table.)

5·12 Add the missing lines in the truth table of Fig. 5·10*b*.

5·13 Add the missing lines of the partial truth table of Fig. 5·10*b* for *A* and *B* only, in the 1-state; *A* and *C* only; and *A* and *D* only, in the 1-state.

5·14 What are the advantages of using distinctive shapes, rather than uniform shapes, in a logic diagram? A disadvantage?

5·15 Is a positive voltage signal which is input into an adder converted to a negative voltage output?

5·16 Is a negative voltage which is input into an integrator converted to a positive voltage output?

5·17 What are the upper and lower limits of a coefficient potentiometer in most analog computers?

5·18 What voltage limits are available in most analog computers?

5·19 Why is a decision box sometimes called a *branching box?*

5·20 What features make the block diagram for a microcomputer distinctive?

PROBLEMS

5·1 Make a simplified block diagram for a digital watch as follows. Use a left-right sequence for the first four blocks, as follows: (1) Quartz crystal, (2) oscillator, (3) frequency divider, (4) wave shaper, (5) battery (with lines flowing to oscillator, frequency divider, and wave shaper), (6) decoder (with lines flowing from battery and wave shaper), (7) digital readout (line flowing from decoder). Make neat uppercase lettering in the blocks, using appropriate abbreviations.

5·2 Construct a flowchart for the basic integrated-circuit fabrication sequence. You may have to turn a corner or two: (1) Crystal growth, (2) wafer slicing, (3) buried-layer diffusion, (4) epitaxial growth, (5) isolation diffusion, (6) open contact windows, (7) interconnection metallization, (8) separate dice, (9) wire bond connections, (10) seal package. Make neat uppercase lettering in each box. Use $8\frac{1}{2} \times 11$ paper.

5·3 Make a block diagram for the following (TR 9–10 BC-SW) radio receiver (follow the instructions given in Prob. 5·1):

    a. External antenna

    b. Mixer

    c. 1st IF amp

    d. 2d IF amp

    e. AF amp

    f. Driver

    g. Output stage

    h. Earphone or speaker jack

    i. Oscillator to feed into the mixer; *and* AGC feedback around the IF stages

Use 11×17 or 12×18 paper unless drawn as two lines, in which case it might fit on $8\frac{1}{2} \times 11$ paper.

5·4 Make a block diagram for the following FM radio receiver (see instructions given in Prob. 5·1):

    a. Antenna

    b. FM RF Amp

    c. FM Converter

    d. 1st FM IF amp

    e. 2d FM IF amp

    f. 3d FM IF amp

    g. AF amp

    h. Output stage

    i. Jack

Use 11×17 or 12×18 paper.

5·5 Arrange the following parts of a digital computer into five blocks with the control unit above the memory unit, both being in the center part of the diagram: (1) Control unit, (2) memory (storage), (3) arithmetic, (4) input-output, (5) control console. Then, in medium lines (or red lines), draw the following lines for control functions: control console to control unit, and vice versa; control unit to input-output; control unit to memory; and control unit to arithmetic. Then, with heavy lines (or black lines), draw the following paths for data flow: memory to control; memory to input-output, and vice versa; memory unit to arithmetic, and vice versa. Use $8\frac{1}{2} \times 11$ paper.

5·6 Complete the drawing of the code converter, Fig. 5·17, by showing distinctive shapes for the NAND and Exclusive OR gates. Draw a border around the entire diagram, including the number 8. Dots are optional. Use $8\frac{1}{2} \times 11$ paper.

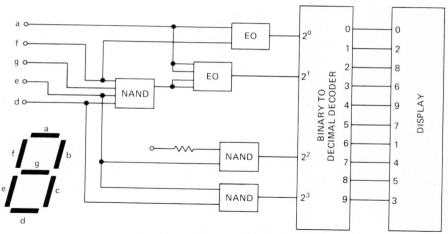

Figure 5·17 (Prob. 5·6) Code-converter circuit.

5·7 Construct a block diagram for a basic regulating system that has the following steps or devices:

a. Power source

b. Regulated quantity

c. Signal-sensing device

d. Error-sensing device

e. Reference

f. Amplifier with feedback

g. Regulator power source

Such a regulating system could be used for the speed control of a motor.

In such a case, the motor would be the regulated quantity and a tachometer would be the signal-sensing device. Use $8\frac{1}{2} \times 11$ paper.

5·8 Make a block diagram of the automobile temperature-measuring and read-out circuit of Fig. 5·18. Block (1) is the forward-feed compensator, (2) capacitor reset, (3) operational amplifier, (4) TTL converter. Above the waveform at (5) letter "1-Hz Square Wave." The diode is a temperature-sensing diode.

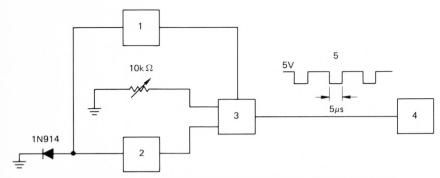

Figure 5·18 (Prob. 5·8) Automobile temperature-measuring circuit.

5·9 Complete the block diagram of the telemetry command and communication interfaces for Apollo–Saturn flight control shown in Fig. 5·19. Add the following lettering at the places indicated by the numbers: (1) Goddard, (2) Manned spaceflight network (MSFN), (3) Houston, (4) Marshall, (5) Kennedy, (6) AF eastern test range (AFTR). Add a suitable title and

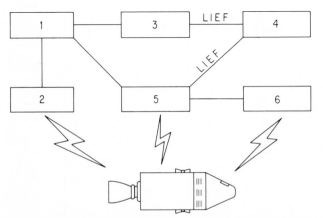

Figure 5·19 (Prob. 5·9) Telemetry system for Apollo-Saturn spaceflight. (NASA.)

the letters LIEF (launch information exchange facility) where indicated. Use $8\frac{1}{2} \times 11$ paper.

5·10 Complete the flow diagram displayed in Fig. 5·20 by adding the following titles to the boxes indicated by the numbers: (1) Video mixer, (2) Video amplifier, (3) Video amplifier, (4) DC restorer, (5) DC restorer, (6) Ground (symbol).

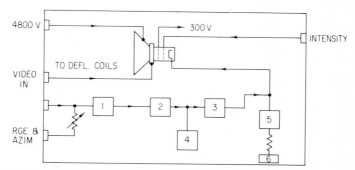

Figure 5·20 (Prob. 5·10) Indicator block diagram for airborne radar.

5·11 Figure 5·21a to f shows simple flow diagrams for analog-computer problems. Supply the missing information as follows:

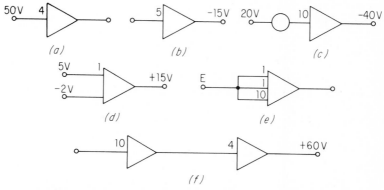

Figure 5·21 (Prob. 5·11) Analog-computer flow diagrams.

a. output _____
b. input _____
c. pot setting _____
d. missing gain _____
e. output _____
f. input _____

5·12 (Parts of this problem require a fair understanding of calculus.) For the diagrams shown in Fig. 5·22, supply the missing information as follows:

a. missing gain _____

b. input _____

c. output _____

d. pot setting _____

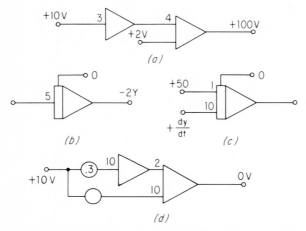

Figure 5·22 (Prob. 5·12) Analog-computer flow diagrams.

5·13 Figure 5·23 shows a flow diagram for a digital-computer program to solve for x and y using Cramer's rule. If $ax + by = p$ and $cx + dy = q$, analyze the program mathematically. Then analyze the program step by step, with a sentence or two describing each block. Can you add something to the flow diagram to make it stop after all values of a, b, c, and d have been read into the computer? (As written, this is not very clear.)

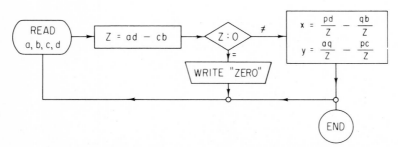

Figure 5·23 (Prob. 5·13) A digital-computer flow diagram.

5·14 Write a flow diagram for a digital computer solution for values of e^x for $x = 1$ through 20. $e^x = 1 + x/1! + x^2/2! + \ldots + x^n/n!$.

5·15 Write a flow diagram for one of the following situations.

    a. The area of a trapezoid

    b. The area under a curve, using trapezoidal or Simpson's rule

    c. The area and volume of a sphere

    d. The sum of all consecutive numbers from 1 to 500

Chapter 6
The schematic
(elementary) diagram

The *schematic diagram*—also sometimes called the *elementary diagram*—is often the primary design drawing of the electronics industry. It is the diagram that shows the functions and relations of the component devices of a circuit by means of graphical symbols. It does not show the physical relationship of those components, however.

In the communications and data processing fields, it is usually referred to as the schematic diagram; in the electrical controls area, it is usually called the elementary diagram.

Such a diagram makes it possible for a person schooled in electronics to trace a circuit with comparative ease. For this reason it is used for design and analysis of circuits, for instructional purposes, and for troubleshooting. The elementary diagram is also used in other electrical areas which are not strictly electronic. One such area is the electric-power field, discussed in Chap. 9.

6·1 Examples of transistors in circuit drawings

In order to give the student a "feel" for the layout of schematic diagrams, we shall present several figures showing popular formats now in use. Then the problem of laying out such a diagram will be discussed. (Without much background in electronics it may not be possible to understand all the authors' comments about the following circuits, but it should be possible to visualize the different drawing patterns that are shown.)

Figure 6·1 shows the three methods of connecting a bipolar transistor in a circuit: *common base* (or grounded base), *common collector,* and *common emitter.* In the common-base circuit, Fig. 6·1*a,* for example, the signal (shown by the ~ in the circle, which indicates an ac input) is introduced into the emitter-base circuit and extracted from the collector-base circuit. The base is thus common to both the input and output circuits. The direction of the arrows shows the electron flow. The voltage or power gain may be in the order of 1,500, and the phase of the signal is not changed.

In the common-collector arrangement, Fig. 6·1*b,* the signal is introduced

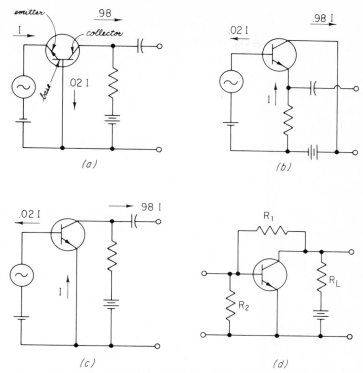

Figure 6·1 Basic bipolar transistor circuits. *(a)* Common base. *(b)* Common collector. *(c)* Common emitter. *(d)* Bias network for common-emitter circuit.

into the base-collector circuit and extracted from the emitter-collector circuit. The power gain is lower than in the other two configurations, and there is no phase reversal. This arrangement is used primarily as an impedance-matching device.

The common-emitter circuit, Fig. 6·1c, can provide power gains of 10,000. The input signal is introduced to the base-emitter circuit, and the output is taken from the collector-emitter circuit. The output signal voltage is 180° out of phase with the input signal. However, this arrangement is the most widely used when more than one stage is required.

Figure 6·1d shows a popular biasing arrangement for a common-emitter circuit, which does away with one of the batteries used previously. (Bias is the difference in potential between, say, the collector and the base.) A voltage-divider network composed of R1 and R2 provides the required forward bias across the base-emitter junction.

In the examples shown, NPN transistors have been used. PNP transistors could be used instead, in which case the battery polarities should be reversed.

6·2 The basic amplifier

Three major functions of transistors and electron tubes are *amplification, oscillation,* and *switching.* Four different *classes* of amplifier service are as follows:

Class A   The collector (or plate in a tube) current flows continuously during the complete electrical cycle.

Class AB   The collector or plate current flows for appreciably more than half the cycle but less than the entire cycle.

Class B   The collector or plate current flows for approximately one-half of each cycle when an alternating signal is applied.

Class C   The collector or plate current flows for considerably less than one-half of each cycle when an alternating signal is applied.

Figure 6·2a shows a basic class A amplifier circuit with a PNP-type transistor. Such amplifiers are used in low-level audio stages, such as preamplifiers and drivers, where "noise" will be at a minimum. Resistors $R1$ and $R2$ determine the base-emitter bias, $R3$ is for emitter stabilization, and the output signal is developed across the collector-load resistor R_L. Capacitor $C1$ bypasses the ac signal around $R3$. (A *loop* has been shown in Fig. 6·2a between $R3$ and the transistor to emphasize the fact that there is no connection at this location. Such loops are not standard and, as a rule, will not appear in drawings in this book. The authors elected to use loops in this figure and in Fig. 6·1b. From now on, except in one or two cases, the standard method of showing crossing, but not connecting, lines will be used.)

A similar amplifier circuit is shown with a triode vacuum tube in Fig. 6·2b. Here the cathode is kept positive by means of the biasing resistor $R1$ and the bypass capacitor $C1$. The capacitor keeps the voltage steady regardless of changes of tube current. R_L is the load resistor.

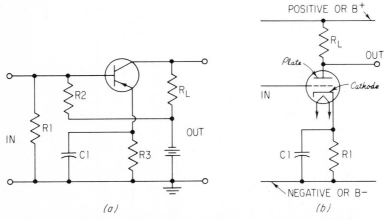

Figure 6·2 Basic amplifier circuits. *(a)* Transistor (triode). *(b)* Vacuum tube (triode).

6·3 Interstage coupling

In the main, there are three methods by which two or more stages (amplifier stages, for example) may be hooked together. These three methods are (1) RC (resistance-capacitance), (2) transformer, and (3) DC (direct coupling). Each has its advantages and disadvantages. Figure 6·3 shows examples of these methods.

Figure 6·3a shows part of a resistance-capacitance-coupled amplifier. C1 is called the coupling capacitor. This method is widely used because of its low cost and the large range of frequencies that can be handled.

Figure 6·3b shows a transformer-coupled network in which one side can be "tuned." (Note that in Fig. 6·3b and c the transistor symbols have not been enclosed by an envelope circle. The reader may recall from Chap. 3 that

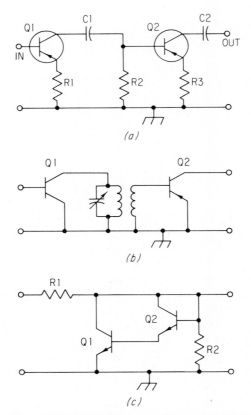

Figure 6·3 Methods of coupling amplifier stages together. *(a)* Part of an RC-coupled amplifier. *(b)* Transformer coupling. *(c)* Direct-coupled (DC) shunt regulator circuit.

the circle is optional. The authors have omitted the circle from these two circuit drawings on purpose to remind the reader that it is not always used. The authors prefer the circle, however, and use it in most examples in the book.) Transformer coupling may also be employed in which both sides or neither side is tuned. Tuning makes possible frequency selection. The use of transformers in coupling has several advantages, but it also is expensive.

Figure 6·3c shows a direct-coupled circuit, which regulates the circuit so that the output voltage is maintained nearly constant. DC amplifiers are normally used to amplify small dc or very-low-frequency ac signals. Typical applications include output stages of series-type and shunt-type regulating circuits, chopper circuits, and differential and pulse amplifiers. Obviously, the cost is low for this type of coupling, and it is, therefore, widely used.

6·4 Patterns for transistor circuits

The nature of the circuitry, including coupling circuits, often indicates a pattern to the engineer or drafter who is making the final sketch or layout drawing of an electronics circuit. The drawing should be planned to allow the pattern to develop in as auspicious a manner as possible.

Figure 6·4a shows a flasher circuit for emergency vehicles, barricades, boats, and aircraft. It will be noted how the transistors are aligned on a horizontal line (invisible), as are the flasher lamps and some other components of the circuit. This type of transistor alignment is possible in a good many circuits, radio and television among them.

Figure 6·4b shows another pattern which develops because of the direct coupling of the transistors. The second and third stages are biased by the preceding stages. A feedback loop containing R2, R5, and C2 adds additional stability to the circuit.

Figure 6·4c shows another pattern which develops when, for instance, PNP and NPN transistors are used in a DC arrangement. A shunt regulator circuit is used to regulate a power-supply *output voltage*. This circuit has a corrective process which ensures the same output regardless of increases or decreases in voltage from the power supply.

Another arrangement of transistors is shown in Fig. 6·5. In a push-pull circuit, such as that shown in Fig. 6·5a, each transistor amplifies half the signal, and these half-signals are then combined in the output (collector) circuit to restore the original waveform in an amplified state. This circuit requires transformer coupling, while that of Fig. 6·5b does not. In this circuit, showing electron current flow by means of arrows, essentially no dc flows through resistor R_L. Therefore the voice coil of a speaker can be connected in place of R_L without excessive speaker-cone distortion.

6·5 Examples of computer circuits

Figure 6·6a shows a saturated flip-flop circuit having one input, called a *trigger*, at *T* and outputs at *A* and *B*. A flip-flop is a memory, or storage, device that

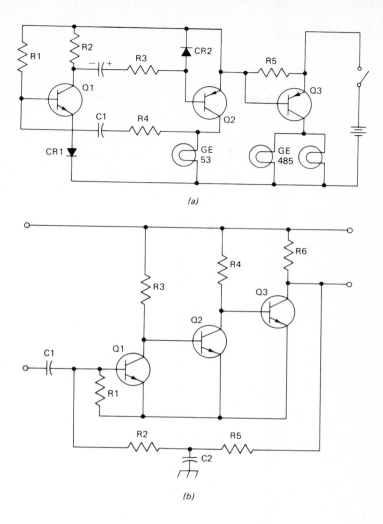

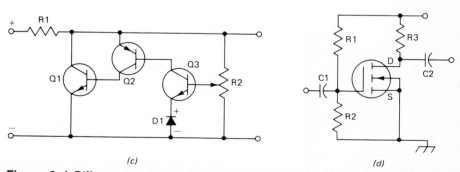

Figure 6·4 Different patterns of transistors in circuits. *(a)* High-power light-flasher circuit. *(b)* Common-emitter DC amplifier. *(c)* Shunt regulator circuit. *(d)* Common-source N-channel enhancement-type MOSFET circuit, similar to common-emitter circuit in *(b)*.

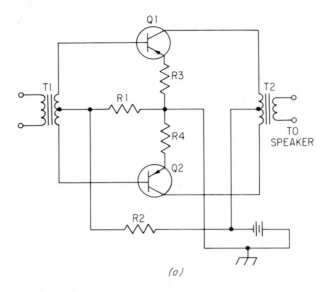

(a)

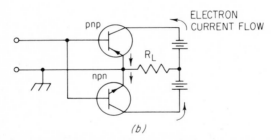

(b)

Figure 6·5 Other patterns. *(a)* A push-pull amplifier circuit. *(b)* Basic complementary-symmetry circuit.

may have one of two states, ON or OFF. It always has two complementary level outputs and may have one, two, or three inputs.

A common feature of flip-flop circuit drawings is the crossing, angled, signal-path lines near the center of the circuit. In most other electrical drawings, the signal paths are drawn as horizontal or vertical lines. It would be possible to draw a flip-flop in this manner, too, but most companies now use the angled lines.

The outputs of this circuit are such that if *A* is +, *B* is −, or vice versa. (We might call *A*+ *and B*− the ON-state and *A*− *and* *B*+ the OFF-state, for instance.) In the *T type,* an input pulse at *T* will change the state of the circuit. If the flip-flop is initially ON, a pulse will change it to OFF, and vice versa.

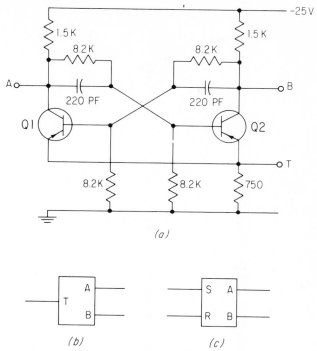

Figure 6·6 A saturated flip-flop circuit. *(a)* Elementary diagram of T type. *(b)* Block diagram of T type and *(c)* RS type. *(From GE Transistor Manual.)*

Another type of flip-flop is the *set-reset,* which is shown by means of a block diagram in Fig. 6·6c. In this circuit, a pulse at the *S* (set) input causes the flip-flop to turn on or stay on, depending on its original state. An input at *R* (reset) causes the circuit to turn off, or stay off.

Figure 6·7 shows several typical logic circuits that are manufactured as integrated circuits (explained in Chap. 7). Figure 6·7a is a direct-coupled circuit, not unsimilar to those of Fig. 6·4. The multiemitter transistor of the TTL circuit (Fig. 6·7d) can be economically fabricated, making the TTL circuit very adaptable to all forms of IC logic. [Note that this drawing uses the no-dot system for showing connections. (See Sec. 3·4.)]

ANSI-IEEE Y32E shows schematic diagrams using both the dot and no-dot systems. The authors have shown several drawings (Figs. 6·7, 6·23, 6·32, and 6·34) without dots in order that the reader may be familiar with this confusing system. Otherwise, most elementary/schematic diagrams in this text will include the connection dots.

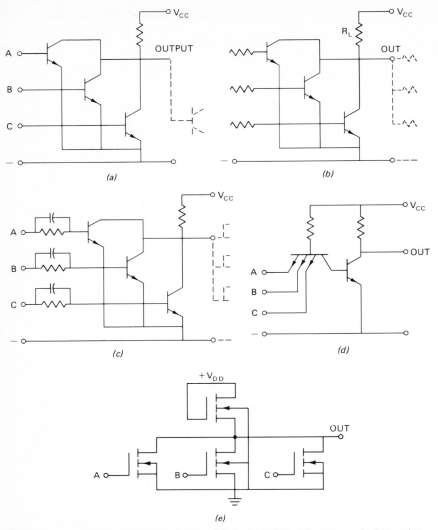

Figure 6·7 Digital integrated circuits (ICs). *(a)* DCTL (direct-coupled transistor logic). *(b)* RTL (resistor-transistor logic). *(c)* RCTL (resistor-capacitor-transistor logic). *(d)* TTL (transistor-transistor logic). *(e)* N-channel enhancement-type MOS field-effect-transistor parallel-drive circuit.

6·6 Reference designations

Symbols of all replaceable parts should be referenced. A reference may be placed above, below, or on either side of its part. ANSI-recommended practice includes giving each part a number, such as resistor $R5$, and its capacity, such as 100 Ω. Thus, we have two lines of designations (combinations of letters and numbers)

172

for most elements in a circuit. Table 6·1 gives letters and examples that are typical of standard practice. Figures 6·10 and 6·15 are good examples.

Table 6·1

Element	Letter	Example
Capacitor	C	C5 10 pF*
Inductor	L	L1 23 mH*
Rectifier (metallic or crystal)	D or CR	D2 or CR2
Resistor	R	R201 270
Transformer	T	T2
Transistor	Q	Q5 2N482 DETECTOR
Tube	V	V3 6AU6 1ST IF AMP

* The abbreviation mH (often MH in drawings) stands for millihenry, a thousandth of a henry. The abbreviation pF (often PF or UUF in drawings) stands for picofarad or micromicrofarad (a millionth of a millionth of a farad).

The letter X is sometimes used for transistors, although Q is recommended by both the Military and American National Standards. $R201$ does not mean the 201st resistor in the circuit. It means that it is the first resistor in the second (200 series) subassembly. Both the transistor and the tube contain a third line—their function. This is optional information insofar as the drawing is concerned.

6·7 Laying out an elementary diagram

Figure 6·8 shows a freehand sketch of a light-flasher circuit. This is the type of sketch that might be made by a design engineer or project chief. The purpose of the sketch is to furnish all the information necessary to make a finished drawing of this circuit and to construct it. If the sketch is very rough and there are many places where improvement is indicated, it may be desirable to make a new sketch.

A drafter should produce a well-balanced work that is pleasing to the eye. In order to do this, it may be necessary to change the configuration of the drawing (reorient some symbols and change some lines), but the circuit must still maintain its original technical significance. One must also strive for simplicity and clarity.

A glance at Fig. 6·8 reveals that, among other things, the transistors could be lined up better, it is a little crowded in the shaded region, and the sketcher

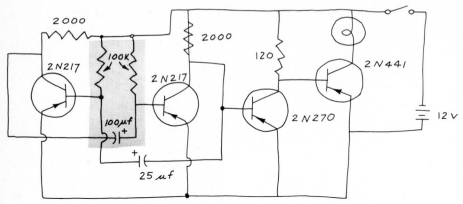

Figure 6·8 Freehand sketch of a light-flasher circuit.

has used loops for crossovers—a nonstandard procedure. Also the components do not have reference numbers.

Figure 6·9 shows how one can go about correcting some of these deficiencies and starting the layout of the diagram. Figure 6·9*a* shows how the spacing between vertical lines in the crowded area can be determined by lettering the

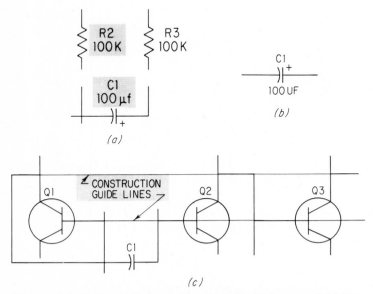

Figure 6·9 Details in laying out an elementary diagram. *(a)* Allowing space for reference designations (identification). *(b)* Alternative arrangement for referencing capacitor. *(c)* Using horizontal guidelines for transistors and other line work.

appropriate reference designations, possibly on a trial basis on a sheet of scratch paper, or on the new sketch if such is being made.

It is obvious that three, and probably all four, transistors can be aligned on one horizontal line to make for a more pleasing and professional-appearing drawing. Therefore a light construction centerline can be drawn horizontally at about midheight through the drawing area. At this time, or possibly later, after more components have been drawn, other horizontal guidelines can be drawn in. One other such line is shown in Fig. 6·9c for the collector outputs.

The transistor envelope circles can be drawn in at this time if the spacing between them can conveniently be determined. Figure 6·9a has provided the spacing around C1, and, by allowing a little more spacing on each side of the vertical lines (on which R2 and R3 are located), we can locate Q1 and Q2 without much fear of having to redraw the diagram. Because there are no components between the other transistors, we can locate them at this time without much difficulty. [Note that this is DC (direct-coupled) circuitry. If it were transformer-coupled, much more work and planning would be necessary.]

After putting in the bases, collectors, and emitters, one can draw more lines, such as appear in Fig. 6·9c, and be well on the way toward completing the circuit diagram. One logical step to perform next would be to draw resistor R4 or R1 (reoriented from the sketch).[1] Then all five resistors could be located at the same level, and the uppermost horizontal line could be located.

Figure 6·10 shows the finished elementary drawing of the flasher circuit.

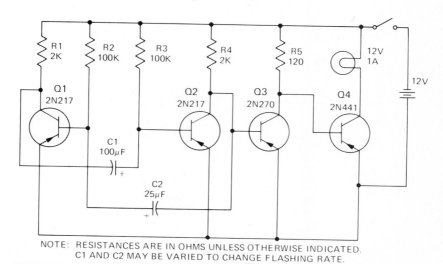

NOTE: RESISTANCES ARE IN OHMS UNLESS OTHERWISE INDICATED.
C1 AND C2 MAY BE VARIED TO CHANGE FLASHING RATE.

Figure 6·10 The completed schematic diagram of a flasher circuit. *(From a circuit in the RCA Transistor Manual.)*

[1] The reader will have to look ahead to Fig. 6·10 to see which resistors have been assigned designations R1 and R4.

ANSI referencing has been used. Transistor identification has been placed close to each envelope with larger-than-average letters. The authors have used this arrangement (putting the designations above and in line with the individual transistors) rather than putting the designations at the top of the diagram, which is also approved practice. Note that the entire drawing appears to be symmetrical and that a fairly uniform density is apparent. There are often notes of a general or specific nature at the bottom of an elementary diagram, and this one is no exception. Instead of locating transistor $Q4$ as shown, we might have drawn a horizontal line from the output of $Q3$ to $Q4$ and thus would have placed $Q4$ "above" the other transistors.

Another popular way to lay out a schematic diagram is to use standard grids. A preliminary sketch can be made on sketching paper having grids, or a grid sheet can be placed under the tracing paper or film on which the drawing is to be made. Figure 6·11 shows a layout of the flip-flop circuit on a $\frac{1}{4}$-in. (6.4-mm) grid. The grid lines have been used not only as guides for the circuit paths but also as guidelines for making the resistor symbols. The $\frac{1}{4}$-in. (6.4-mm) resistors and 1-in. (25.4-mm) transistor envelopes were larger than those that are drawn on most diagrams before they were reduced for printing.

Figure 6·12 shows a null-detector circuit made on a $\frac{1}{10}$-in. (2.54-mm) grid with heavy lines 1 in. (25.4 mm) apart. All signal paths have been drawn right

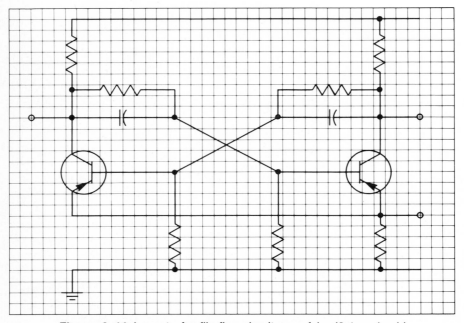

Figure 6·11 Layout of a flip-flop circuit on a $\frac{1}{4}$-in. (6.4-mm) grid.

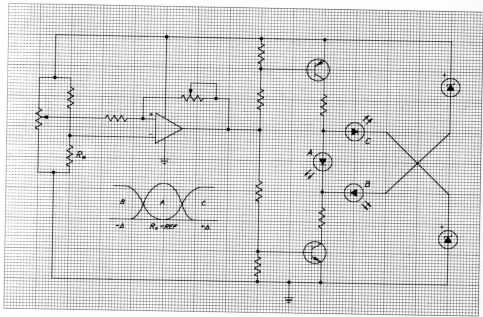

Figure 6·12 Layout of a null-detector circuit on a $\frac{1}{10}$-in. (2.54-mm) grid.

over grid lines, most on the heavy lines. Small grid lines have been used as guidelines for the resistors and to establish the LED envelopes at 0.6 in. (15 mm) diameter. A graph has been added to show that if a resistor is matched against a reference, a diode A will light up when the resistor R_x is exactly or closely matched. If the resistor is less than the reference, diode B comes on, etc. The two transistor envelopes are 0.75 in. (19 mm) in diameter, exactly the size shown in ANSI/IEEE Y32E 1976.* Although most signal paths are drawn horizontally or vertically, in Figs. 6·11 and 6·12 there are lines drawn at other angles. These are two types of circuits, along with multivibrators and bridges, in which it is customary to show certain lines in this manner. In most drawings the paths are horizontal or vertical.

Drawings for manuals and journals are made in ink and often are produced using Leroy equipment.† Working from a rough sketch, the drafter lays out a neat pencil drawing with or without a grid, as above. The final drawing is made on Mylar or similar film, which is placed over the pencil drawing. Then the symbols are made with the template and scriber as shown in Fig. 6·13a.

* This is somewhat larger than in the previous edition of the standard.

† Leroy is a registered trademark of the Keuffel & Esser Company.

The next step is to draw the signal paths between the symbols with a technical fountain pen such as the Rapidograph or Staedtler-Mars "700," as in Fig. 6·13b. Lettering is then put on the drawing. This is done with a Leroy lettering template and scriber, or it is accomplished by using an IBM typewriter and cutting out small rectangles of typed lettering which can be pasted on the drawing at the desired locations. Leroy electrical templates are available with three sizes of symbols and thus provide the layout person with considerable flexibility. Both Leroy lettering templates and typed-lettering "stick-ons" also have much flexibility in the size of letters. Typed letters provide a little more uniformity and a cleaner appearance and save a little time if there is considerable lettering on a drawing.

Still another approach is to use preprinted device symbols in the form of appliqués or stick-ons. Signal paths are made with pressure-sensitive tapes, pencil, or technical fountain pen. As in the method described immediately above, the detailer probably works from a rough sketch. A neat pencil layout of the diagram is then made either on a grid or on plain tracing paper. Symbols may be sketched in, omitted, or partially drawn on this layout if the final layout will be made on another sheet (overlay), which will be placed over the pencil layout.

In the location indicated on the overlay sheet, each pressure-sensitive symbol is oriented in the precise position by permanently and firmly affixing it to the surface of the artwork. A smaller symbol can be separated from the backing material by inserting a knife (with an X-acto No. 11 or No. 16 blade) in between. The symbol is carried to the desired diagram location on the knife, which is also used as a holding and positioning tool. After the symbols are in place, signal paths are put on the diagram. If inked lines are to be used, they can be drawn as shown in Fig. 6·13b with a technical or a ruling pen. Pencil lines will not look as good as ink lines because the printed stick-on symbols will be darker. Lettering is usually put on the drawing at the last. Figure 6·14c shows a sample sheet of pressure-sensitive symbols.

Many students will not have access to appliqués and Leroy equipment. They will probably have to make all lines and symbols with a pencil. The following fundamentals of layout should be observed:

1. Use medium-weight lines for all symbols and signal paths. When emphasis on a component is desired, use a heavier line.

2. Arrange the diagram so that it reads functionally from left to right, if possible.

3. Strive for uniform density so that there are no congested areas and a minimum of large white areas.

4. Draw signal paths vertically or horizontally, except in special cases such as flip-flop circuits.

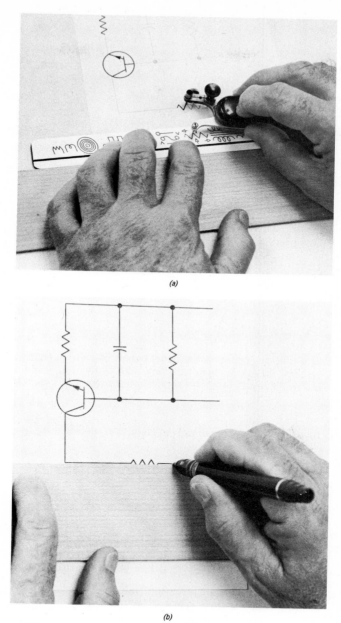

(a)

(b)

Figure 6·13 Using mechanical devices (scribers and pens) to lay out a schematic diagram. *(a):* Using Leroy scriber and template to make symbols. *(b):* Using technical pen to draw signal path between symbols.

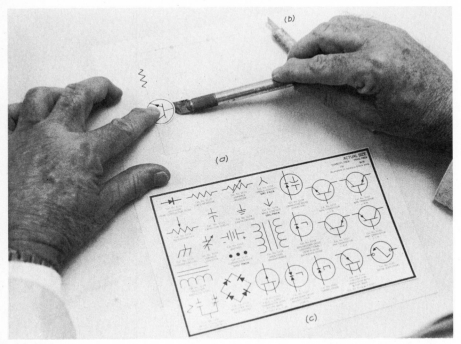

Figure 6·14 The use of appliqués for construction of schematics. *(a)* Applying a symbol. *(b)* A knife with No. 16 X-acto blade. *(c)* Typical symbols that are available as pressure-sensitive adhesives. *(Bishop Graphics, Inc.)*

5. Keep turns or bends to a minimum. Route all connecting lines as directly as possible.

6. Keep the device symbols in proportion to each other. Use only one size of symbol for each device, if possible.

7. Allow adequate space near each symbol for proper identification.

8. Make integrated-circuit and microprocessor outlines large enough for pertinent lettering to be placed within the outline.

9. Center each symbol along its lead line (signal path).

10. Place the complete identification (and values if necessary) beside or within each symbol, whichever is appropriate. Identify each input and output.

11. Check the final diagram for the correctness of symbols, electrical accuracy, and the presence or absence of connecting dots.

Some aids that are not very expensive might facilitate the drawing of schematic diagrams for students. Grid paper, with or without drop-out lines, is one such aid. A template that makes different sizes of circles or that includes electronic device symbols is another aid. The use of templates is discussed in Chap. 1.

6·8 Line arrangements according to voltages

Although the circuit of Figs. 6·8 to 6·10 does not become involved in this problem, some circuit diagrams will have lines representing several different voltages. A good practice is to draw the schematic diagrams so that the highest voltage is on the uppermost line, the next highest is on the line below, and, finally, the most negative voltage is on the lowest line. This makes for easier reading of the elementary diagram in many cases. Typical voltage sequences might be as follows:

235	V
150	V
6.3	V
0	V
−45	V

The top line is what is known as the $B+$ line; the zero line is the $B-$, or common, line and may be joined to the ground or chassis. Transistor circuits, as a rule, do not have as many lines as this, but tube circuits may have this many, depending upon the circuit and types of tubes.

6·9 Examples of elementary diagrams

Figure 6·15 shows the complete schematic diagram of a transistor radio. This particular circuit has a separate mixer and oscillator, two IF stages, a detector and audio amplifier, and a class B push-pull output. The transformer coils are drawn as a series of complete loops. This loop, or helix, is one of two symbols that are approved in the standard. Most drawings now have the simpler symbol, as shown in the power-supply transformer, Fig. 6·16. In Fig. 6·15, $L1$ has a metallic core and $L2$ has an air core. The authors would be tempted to call $L1$ and $L2$ (transformers) $T1$ and $T2$.

Figure 6·16 shows the schematic diagram of a video distribution amplifier that takes the output of a video (TV) camera and provides four independent outputs. The two integrated circuits are an operational amplifier and a current driver in a feedback loop. Inputs and outputs are for coaxial cables. The power supply is placed below the distribution amplifier circuit. This is common practice, although the power circuit could be drawn jointly with the other by joining the 6.8-V lines and (probably) rearranging the whole drawing somewhat. $D3$ and $D4$ are zener diodes used for regulating. A nonstandard symbol in the

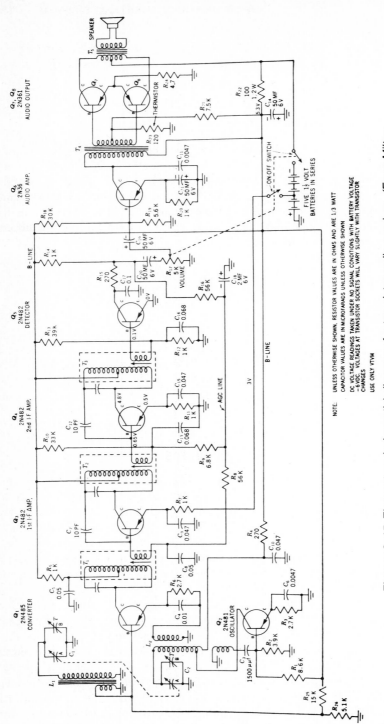

Figure 6·15 Elementary (schematic) diagram of a transistor radio receiver. *(From Milton S. Kiver, "Transistors," 3d ed., McGraw-Hill Book Company, New York, 1962.)*

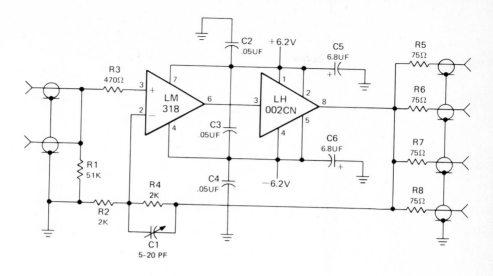

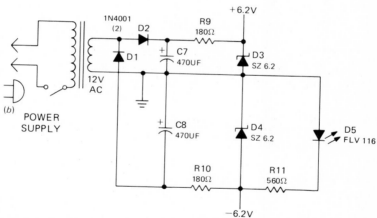

Figure 6·16 Schematic diagram of a video distribution amplifier. *(From "Electronics" magazine, Oct. 13, 1977, p. 121; Copyright © 1977 by McGraw-Hill, Inc.)*

original drawing, shown at *b*, was replaced by the standard connectors at the left of the power supply.

Figure 6·17 is the schematic diagram of a scientific calculator. Its features include the LED display module (top), ICs $Z3$ and $Z4$ that serve as digit drivers, the keyboard module, and ICs $Z1$, $Z2$, and $Z5$. $Z1$ contains the central processing unit. $Z2$ has a scanning capability that handles the keyboard functions. $Z5$

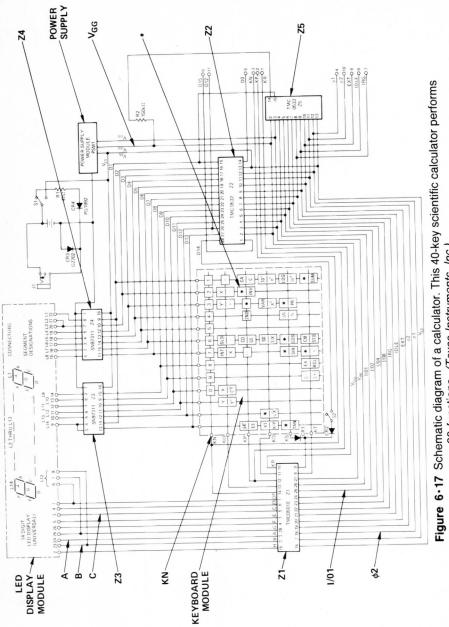

Figure 6·17 Schematic diagram of a calculator. This 40-key scientific calculator performs over 30 functions. *(Texas Instruments, Inc.)*

has the clock function, which enables the data to be entered at the right place at the right time. All three of these integrated-circuit packages have memories.

Connections and conductors are well identified, with lettering either near the pins or approximately halfway between the IC packages. Thus, we have V_{GG} (gate supply); I/O1 (input-out circuit No. 1); $\phi2$ (clock No. 2); A and B (upper and upper-right segments of a digit; A, B, and C would make the number 7 if activated); and KN, an encoding line which is involved in the conversion of decimal numbers to binary numbers.

Keys having the asterisk are not large enough (on the drawing) to accommodate the lettering for the function. Most function keys have two functions. That is, when the function key is pressed after the "2nd" key is depressed, the upper function is performed; and, when the function key is pressed without the 2nd key having been activated, the lower function is performed.

The photograph shown in Fig. 4·26 shows two sides of an older model calculator PCB. Many of the discrete components have been replaced by an integrated-circuit module and the power module shown in Fig. 6·17.

6·10 Tube circuits

Figure 6·18 is the schematic drawing of an audio-amplifier circuit having electron tubes. This older drawing has several features worth mentioning. The tubes, except the rectifier 5Y3-GT, have heaters. The heater circuit XX is not drawn in order to simplify the schematic. Loops (no connections) and dots are used. Finally, the values for each component are placed below the diagram. Such an arrangement may make for slower reading of the diagram but has the advantage of supplying more information about each part than can be conveniently shown with the other system.

6·11 Commonly used units for components

In addition to units such as henrys and ohms, certain prefixes (multipliers) are used when components have extremely large or small values. The most-often-used prefixes are shown in Table 6·2. Suggested units with their prefixes appear in Table 6·3.

Some lettering can be saved by using notes such as the one shown in Fig. 6·10 about resistance values. A similar note might be: *All capacitance values are in microfarads unless otherwise shown.*

Tables 6·2 and 6·3 present what the authors believe to be current practice relative to prefix values for units. We believe this will continue for years to come. However, it should be pointed out that two new standards may bring about some changes. These are ISO 1000, "S1 Units and Recommendation for the Use of their Multiples and Certain Other Units," and ANS Z 210.1–1976 (IEEE Std 268–1976), "Standard Metric Practice." If these standards are adopted by U.S. industry, some of the prefixes will be as we show in Table 6·4.

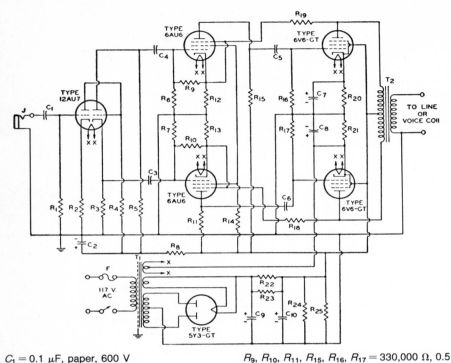

$C_1 = 0.1$ μF, paper, 600 V
$C_2 = 40$ μF, electrolytic, 450 V
C_3, $C_4 = 0.02$ μF, paper, 600 V
C_5, $C_6 = 0.05$ μF, paper, 600 V
C_7, $C_8 = 50$ μF, electrolytic, 50 V
C_9, $C_{10} = 80$ μF, electrolytic, 450 V
F = Fuse, 1 A
$R_1 = 470,000$ Ω, 0.5 W
$R_2 = 6,800$ Ω, 0.5 W
R_3, $R_5 = 39,000$ $\Omega \pm 1\%$ matched, 1 W
$R_4 = 220,000$ Ω, 0.5 W
R_6, R_7, $R_{14} = 1$ mΩ, 0.5 W
$R_8 = 10,000$ Ω, 1 W

R_9, R_{10}, R_{11}, R_{15}, R_{16}, $R_{17} = 330,000$ Ω, 0.5 W
R_{12}, $R_{13} = 1,800$ $\Omega \pm 1\%$ matched, 0.5 W
R_{18}, $R_{19} =$ carbon-film type, 100,000 $\Omega \pm 1\%$, matched, 2 W
R_{20}, $R_{21} = 510$ Ω, 2 W
R_{22}, $R_{23} = 390$ Ω, 2 W
R_{24}, $R_{25} = 150,000$ Ω, 2 W
$T_1 =$ power transformer, 350–0–350 V rms, 125 mA
$T_2 =$ output transformer for matching line or voice coil impedance to 9,000–10,000-Ω plate-to-plate tube load

Figure 6·18 A schematic diagram of a high-fidelity audio amplifier. *(From RCA Receiving Tube Manual.)*

Table 6·2

Prefix	Multiplier	Symbol
Mega	10^6	MEG or M
Kilo	10^3	K or k
Milli	10^{-3}	M or m
Micro	10^{-6}	μ or U
Pico	10^{-12}	P or p

Table 6·3

Range	Units to Be Used	Examples
Up to 999 ohms	Ohms	520 or 210 Ω
1,000–99,999 ohms	Ohms or kilohms	45 kΩ (or K) or 45,000
100,000–999,999 ohms	Kilohms or megohms	500 kΩ or 0.5 MΩ (or MEG)
10^6 or more ohms	Megohms	2.5 MΩ (or MEG)
Up to 9,999 pico-farads (micromicrofarads)	Picofarads	150 pF (or PF) (old $\mu\mu$F)
10,000 or more picofarads	Microfarads	0.05 μF (or UF or MFD)
Up to 0.001 henry	Microhenrys	5 μH (or UH) (= 0.000005 henry)
0.001–0.099 henry	Millihenrys	3 mH (or MH) (= 0.003 henry)
0.1 or more henry	Henrys	20 H

Table 6·4 Suggested Practice, ISO-1000

Prefix	Symbol		Range	Examples
Mega	M		To 999 ohms	210 Ω
Kilo	K		1,000–99,999 ohms	45KΩ or 45,000
Milli	m		100,000– 999,999 ohms	500 KΩ or 0.5 MΩ
Micro	μ			
Pico	p		1 to 9,000 picofarads	150 pF

Another helpful device is that shown in Table 6·5, which shows the highest number of each component, as well as omissions, in a diagram.

The reader may wonder why two resistors were not used. There could be several reasons for such an occurrence. One possibility would be that the original design was altered after testing or use and a particular component was taken out. If, later, another one is added, it is usually given a new number, not the number of the deleted one.

Table 6·5

Highest Reference Designations
C_{23} L_7 R_{43}
NOT USED
R_{22}, R_{23}
C_{16}

6·12 Waveforms

Some common waveform symbols are shown in Fig. 6·19. These are sometimes shown on an elementary diagram at possible test locations. Some common forms that are not shown are sawtooth, trapezoidal, rectangular, exponential, clipped, and triangular.

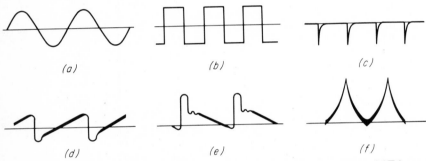

(a)   *(b)*   *(c)*

(d)   *(e)*   *(f)*

Figure 6·19 Symbols of waveforms. *(a)* Sine wave. *(b)* Square wave. *(c)* Trigger. *(d)*, *(e)*, and *(f)* Complex forms.

6·13 Mechanical linkage and other mechanical arrangements

In Fig. 6·15, at the left side, are two capacitors connected by means of dashed lines. The dashed (dotted) lines mean that the two devices are connected mechanically, so that when one is turned, the other is turned to the same position simultaneously. Figure 6·20 shows two other situations in which mechanical connections are shown. Figure 6·20a is part of a diagram that is part pictorial, part schematic, and part connection. The mechanical path may be easily traced. The autopositioner works as follows:

1. The operator turns a remote-control switch to a certain position.
2. The relay (slow-operating in this example) is energized, lifts the pawl out of the sprocket wheel, and also turns on the motor.
3. The motor drives the autopositioner shaft until the position corresponding to the new position of the remote control is reached.

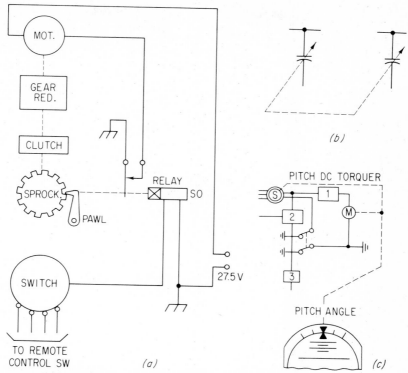

Figure 6·20 Mechanical connections. *(a)* Autopositioner. *(b)* Ganged tuning capacitors. *(c)* Attitude indicator.

4. The relay circuit now opens, the pawl engages the stop wheel, and the energy to the motor is cut off.

Ganged tuning capacitors (Fig. 6·20*b*), like ganged switches, may be shown on an elementary diagram. Figure 6·20*c* shows mechanical connections between the attitude indicator and synchro of a DC-10.

6·14 Separation and interruption

It is often desirable to show a separate package apart from the rest of the circuit or system. For instance, if the same package (or *pack* as it is sometimes called) is to be used several times, it may be drawn as blocks instead of as an elementary diagram. In such a scheme, the elementary diagram would have to be drawn once, but only once. In Fig. 6·21 three ways of drawing a pack schematically, using different terminal or interruption techniques, are shown. Figure 6·21*d* shows how the pack might be drawn diagrammatically as a block. Figure 6·21*a* shows the circuit as removed at terminals. Figure 6·21*b* simply encloses the pack with an optional enclosure line but does nothing about the terminals. (In all four figures, the lines, or terminals, have been identified by

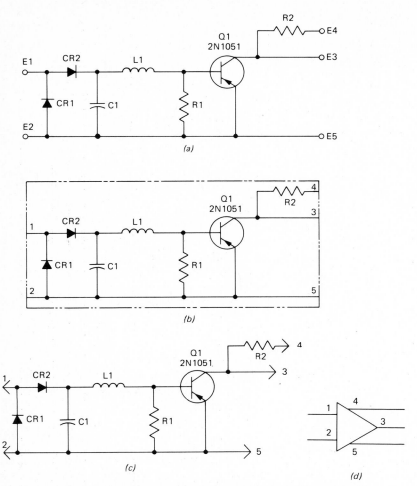

Figure 6·21 Representation of a circuit package separated from the rest of the system. *(From ANSI Y14.15, "Electrical and Electronic Diagrams.")*

means of numbers.) Figure 6·21c shows another method of separating a package. While this is a fairly widely used method, it is slightly confusing. The V-shaped lines at the end of each conductor do not represent any specific kind of connection in this scheme. Yet ANSI/IEEE Y32E uses this symbol to show a male contact. It is so used in other drawings in this book, especially in Chap. 9. The V's, incidentally, are not supposed to be arrowheads. They are correctly drawn at a 90° angle.

Figure 6·22 illustrates other ways of interrupting groups of lines. Lines are generally grouped together in threes. If there are more than three lines to be shown, groups should be separated, as at the right in Fig. 6·22a. Spacing

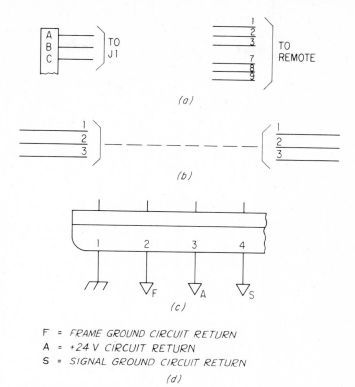

Figure 6·22 Interruption and circuit return. *(From ANSI Y14.15.)*

between lines should be a minimum of $\frac{1}{4}$ in., and between groups $\frac{1}{2}$ in. Brackets may be joined by means of dashed lines, as in Fig. 6·22*b*. Figure 6·22*c* shows hypothetical uses of circuit-return symbols. If the inverted triangle is used, letters and general notes (Fig. 6·22*d*) will probably be necessary. The chassis symbol and the first triangle (with letter F) serve the same purpose, and either one could be used. These symbols are usually oriented as shown here. If it is more convenient, they can be drawn at the ends of lines extending upward or to the right or left. In other words, the entire figure (Fig. 6·22*c*) could be turned at 90 or 180° from its present position. (Letters and numbers might be changed to read in a convenient manner, though.)

6·15 Other examples of schematic diagrams

Figure 6·23 shows the elementary diagram for the preamplifier circuit of the Apollo moon camera. It is enclosed by the optional enclosure lines. The preamplifier is made up of discrete components, primarily to provide low-noise performance. The video signal from the SEC camera is fed into the preamplifier. Its

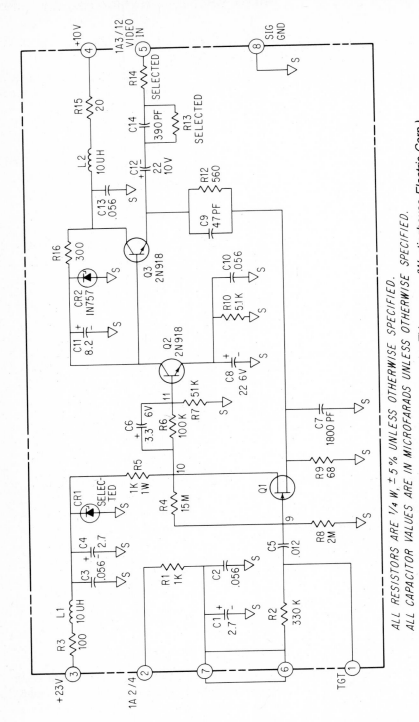

Figure 6·23 Preamplifier circuit of Apollo color TV camera. (*Westinghouse Electric Corp.*)

ALL RESISTORS ARE 1/4 W, ± 5 % UNLESS OTHERWISE SPECIFIED.
ALL CAPACITOR VALUES ARE IN MICROFARADS UNLESS OTHERWISE SPECIFIED.

input is a field-effect transistor (Q_1) stage with a 330-kΩ (330 K in this drawing) load resistor. This is followed by a feedback pair.

Other parts of the camera are a postamplifier, deflection circuits, and a power supply (two, in fact—high and low voltage). In addition to the camera there is a small (85-cc) viewfinder monitor for the astronaut and a transmitter. Figure 5·2 shows the entire system in block form. The no-dot system has been used.[1]

The camera is 17 in. long, including a zoom lens, and weighs 13 lb. It generates a field sequential color signal using a single-image tube and a rotating filter wheel. A ground-station color converter later changes the sequential color signal to a standard NTSC color signal. The lens has a .T number of 5.1 to 51, zoom ratio of 6:1, and focal length of 25 to 150 mm.

Figure 6·24 shows a *systems schematic* drawing (about the left third) produced by a major aircraft manufacturer. This is a *hybrid* diagram that combines information from wiring and elementary and logic diagrams as well as a little pictorial drawing. Other systems schematics may include hydraulic systems and mechanism-type drawings as well as information of the type shown in Fig. 6·24. These hybrid diagrams are of extreme value to persons engaged in maintenance and troubleshooting operations.

The reader will notice a loop-like symbol at the very right edge of lines 49 to 51. This indicates that the three wires are twisted together. Somewhat below this is a similar symbol which indicates that wires 45 and 46 are twisted together and shielded.

SUMMARY

The schematic or elementary diagram shows by means of graphic symbols the functions and connections of a specific circuit arrangement. Symbols for use in such a diagram are extensively covered in ANSI/IEEE Y32E and IEC Publication 117, and the preparation of the diagram itself is covered by ANS Y14.15. There are certain basic arrangements of transistors and tubes which are usually repeated in electronics circuit diagrams. With these and some types of interstage coupling, certain patterns are often discernible—usually early in the game, when one starts to plan an elementary diagram. Designation, or referencing of each component part of a circuit, is important. Certain standard abbreviations and prefixes are used for this purpose. Sufficient space must be made available near each component for referencing.

Conventional treatment is sometimes employed to eliminate the drawing of certain lines in order not to have too cluttered a drawing. On the other hand, additional material, such as general notes and data on waveforms, is often added to a schematic diagram. A symbol is generally spaced midway along the span of circuit path on which it is drawn. Mechanical connections must be shown at times. Certain patterns are usually followed in laying out certain types of circuits in diagrammatic form. Radio circuits, for example,

---

[1] There are connections at points 9, 10, and 11.

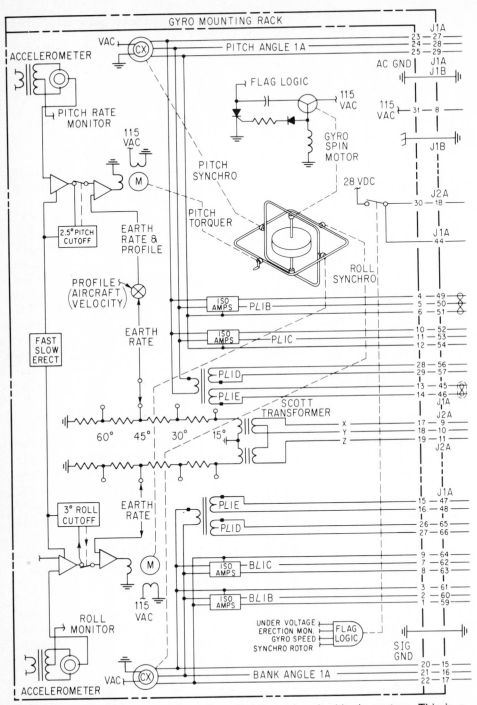

Figure 6·24 Part of a systems schematic of an aircraft altitude system. This is a hybrid drawing. *(Douglas Aircraft Company.)*

usually follow a *cascade* arrangement in which the signal path goes from left to right and the transistors are aligned horizontally. Auxiliary circuits, such as power circuits, are generally placed in the lower part of any schematic diagram. Overall balance and symmetry are other concepts that are observed in diagrammatic layout.

QUESTIONS

6·1 What set of standards governs the preparation of elementary diagrams?

6·2 In what way, or ways, is a schematic diagram different from a connection diagram?

6·3 How would you show the value 100,000 Ω on a circuit drawing? 150,000 pF?

6·4 To what sort of elements do the following terms refer:
   a. 12AV6
   b. Z_2
   c. 50 K
   d. 2N329
   e. CR_2

6·5 What is meant by the term *density* as it applies to elementary diagrams?

6·6 What is meant by the term *symmetry* as it applies to elementary diagrams?

6·7 What do the letters V_{GG} and V_{SS} signify?

6·8 For what purposes are schematic diagrams used?

6·9 Where are auxiliary circuits normally placed on the elementary diagram?

6·10 If a resistor is positioned on a vertical line (signal path), where would you place its identification?

6·11 Where would you place the identification for a vacuum tube in a schematic diagram of a circuit having cascade projection?

6·12 What letters are used to identify the three leads of a **MOSFET**?

6·13 What elements are often found in a hybrid diagram?

6·14 What do the parallel lines adjacent to the outline of a picture-tube symbol represent?

6·15 Define the prefixes micro-, milli-, kilo-, and mega-.

6·16 Is it possible to draw a schematic diagram using only one width of line and in so doing comply with the standards?

6·17 If you are assigned to make a finished drawing of an elementary diagram of a circuit, in what form is the information to which you will refer apt to be?

6·18 What are three basic functions which are performed by transistors?

6·19 What methods are used in coupling different stages together? (Name three.)

6·20 In what way may the number of connecting lines in a diagram be reduced?

6·21 What two shapes are generally used for integrated-circuit modules?

6·22 State briefly, or use sketches to show, how you would separate a package from the rest of a circuit.

6·23 What is the difference between a picofarad and a micromicrofarad?

PROBLEMS

6·1 Draw the schematic diagram of the all-channel CB monitor/receiver shown in Fig. 6·25. Use standard identification for each component. We have shown capacitor C_1 (10 pF) at the upper left. Each capacitor and resistor should have a similar reference designation.

    Make any changes or improvements in the symbols or designations/values of components that you think would be appropriate. Add a suitable title and the following note: Unless otherwise specified all resistors are $\pm$ 5%, $\frac{1}{4}$ W. There is no connection at pin #6 of the amplifier LM 380. If well planned, this diagram, with a suitable title, will fit on an $8\frac{1}{2} \times 11$ sheet.

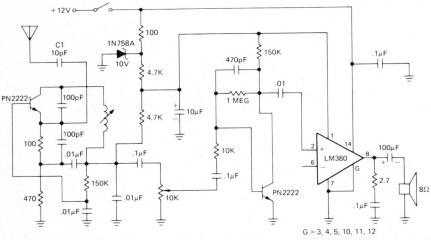

Figure 6·25 (Prob. 6·1) Elementary diagram of CB monitor/receiver. *(Kantronics, Inc.)*

6·2 Draw the schematic diagram of a digital clock that is shown in Fig. 6·26. This will fit on an 11×17 sheet. Complete the drawing by showing resistors R_4 through R_9 and drawing the CRB1 full-wave bridge where indicated. Use ANSI practice in identifying the following components:

C1 220 UF	S1, S2 .5 AMP
R1 18K	Q1 − Q8, 2N2222
R2 − R10 10K	Q9 − Q16 2N3866
	(Darlington)
CR5 1N914	CRB1 WE460J

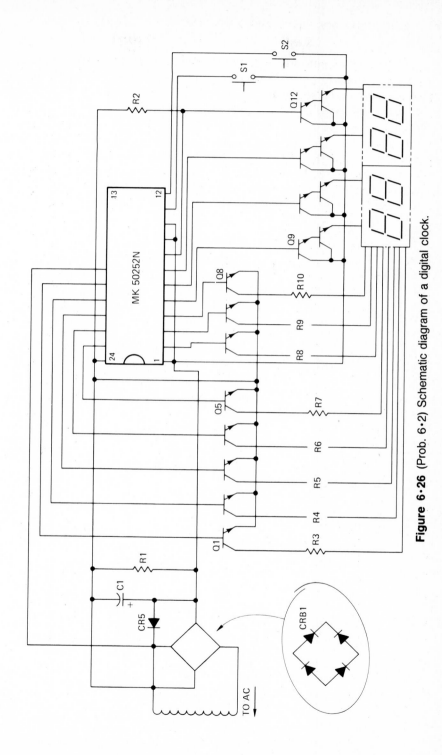

Figure 6·26 (Prob. 6·2) Schematic diagram of a digital clock.

The corner pins are numbered on the Mostek clock chip. There are no connections to pins 13 through 17. The coil symbol at the left is part of transformer T_1, which has an iron core. The leads from the left winding of the transformer go to the 115-V ac supply.

6·3 Complete the elementary diagram of Fig. 6·27 by insertion of the correct symbols as indicated by standard notation. Add the identification of each component and the capacity if given below. Squares Z_1, Z_2, etc., are integrated circuits drawn as squares or rectangles with identification shown within the square. Video input is at 12.

Z_1	SE501G	C_{10}	15 pF
Z_2	MD2905F	R_1	20 kΩ
Z_3	SE501G	R_2	510
Z_4	SE501G	R_5	51 kΩ
CR_1	IN914	R_6	10
CR_2	IN914	R_7	470
S	ground	C_{11}	6800 pF
C_1	3.3 μF	R_4	3.3 kΩ
C_2	3.3 μF	L_1	10 μH
C_5	22 μF	L_2	10 μH
C_6	0.056 μF	C_{14}	3.3 μF
C_7	22 μF		
C_8	056 μF		
C_9	15 pF		

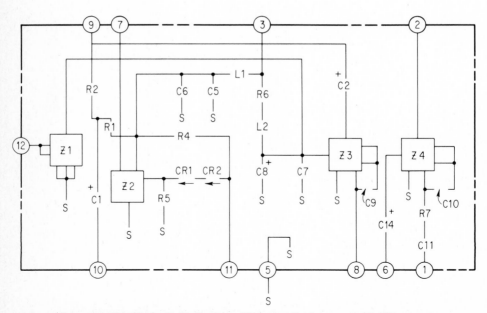

Figure 6·27 (Prob. 6·3) Postamplifier circuit of moon-walk TV camera.

This drawing will be slightly crowded on $8\frac{1}{2} \times 11$ paper, and will have more than enough room on 11×17 paper.

6·4 Complete the elementary diagram of the four-function calculator shown in Fig. 6·28. Major elements are the terminal (at top) that connects to the LED strip (not shown), the calculator chip TMC 972, and the keyboard module. The unfinished leads should be completed by drawing lines horizontally and/or vertically to the connector indicated. Use 11×17 paper. Use the standard battery symbol. Show stub lines at NCs (no connections).

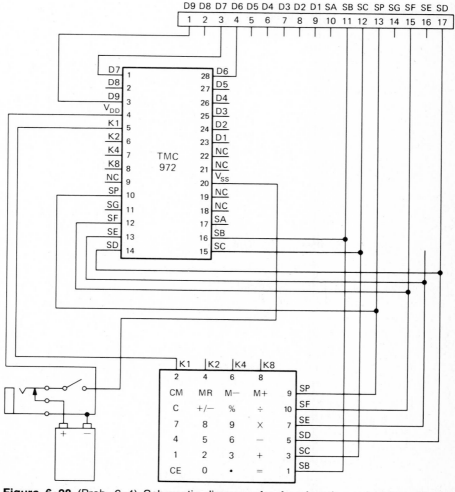

Figure 6·28 (Prob. 6·4) Schematic diagram of a four-function calculator. *(Texas Instruments, Inc.)*

6·5 Complete the elementary diagram of the FM tuner circuit shown in Fig. 6·29. C_1 and C_6 are ganged, 10-365 pF; C_3 and C_4 are 0.01 μF; all other capacitors are 10 pF, except C_8 and C_{10} which are 250 μF. R_1 is 180 Ω; R_2, 5 kΩ; R_3, 33 kΩ; R_4 and R_6, 1 kΩ; R_5, 10 kΩ; R_7, 150 kΩ; R_8, 1.5 kΩ; R_9, 470 kΩ; R_{10}, 7000 Ω. Use 11 $\times$ 17 or 12 $\times$ 18 paper.

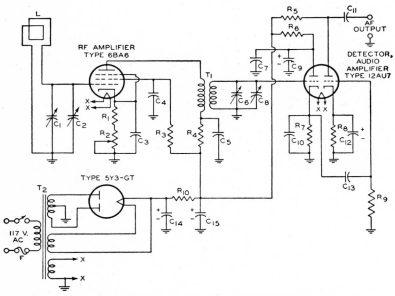

Figure 6·29 (Prob. 6·5) Schematic diagram of FM tuner circuit.

6·6 Redraw the high-fidelity amplifier shown in Fig. 6·18 to about $2\frac{1}{2}$ or 3 times the size that it appears in the book. Instead of the component identification system shown, use the ANSI system of identification shown in this chapter. Use 11 $\times$ 17 or 12 $\times$ 18 paper.

6·7 Redraw the schematic diagram shown in Fig. 6·30 to approximately twice the size it is shown in the book. Use standard identification for resistors, capacitors, etc. Add the following information on or near ICs 1 and 2: Pin 1, Trigger; 2, V_{REF}; 3, R/C; 4, GND; 5, V+; 6, C; 7, E; 8, LOGIC. On IC 3: Pin 1, CL; 2, D; 3, CLOCK; 4, GR; 5, Q; 6, $\overline{Q}$. Use $8\frac{1}{2} \times 11$ paper.

6·8 Figure 6·31 is a freehand sketch of an isolation circuit that is not complete. Make an instrument drawing of this circuit that is complete, uses correct symbols in accordance with ANSI/IEEE Y32E, and has maximum identification of components. Q_1 and Q_2 are optical couplers, FCD 810 or equivalent. A_1 and A_2 are operational amplifiers. A_2 is separated into two parts,

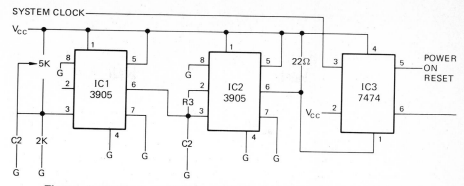

Figure 6·30 (Prob. 6·7) Schematic diagram of power-on reset circuit.

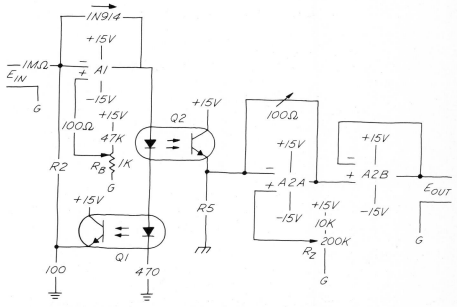

Figure 6·31 (Prob. 6·8) An isolation circuit with optical couplers.

A_{2A} and A_{2B}. R_B is a bias resistor, and R_Z is a zero-adjust potentiometer. R_2 and R_5 are 100 kΩ, each. Make any corrections which appear to be appropriate. Use a standard termination scheme for E_{IN} and E_{OUT}. This problem will fit on $8\frac{1}{2} \times 11$ paper; 11×17 paper would also be OK.

6·9 Make a drawing of the clock-radio circuit shown in Fig. 6·32. Correct the symbols where obsolete or incorrect ones appear. Add identification for semiconductors as follows: Q_1, converter, 95101; Q_2, first IF, 95103; Q_3, second IF, 95102; CR_1, detector, IN295; Q_4, audio driver, 95201; Q_5 and Q_6, audio outputs, 95220; CR_2, rectifier, IN290. All resistors are $\frac{1}{2}$ W. The no-dot system has been used except at connections J_1 and J_2. Resistance values are in ohms unless otherwise noted, and all capacitance values are in microfarads. Use 11×17 or 12×18 paper.

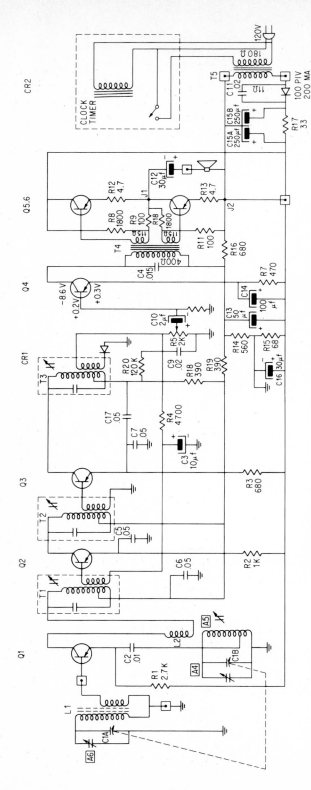

Figure 6·32 (Prob. 6·9) Schematic diagram of transistor clock radio.

6·10 Figure 6·33 is a sketch of a volt-ohmmeter circuit with many of its symbols (resistors and capacitors) missing. Make a complete elementary drawing of this circuit. Values are as follows:

C_1	0.1 μF	$R_{13,\,18}$	1 MΩ
C_2	0.33 μF	R_{14}	10 kΩ
C_3	10 μF	R_{15}	1000 Ω
C_4	0.01 μF	R_{16}	10 Ω
R_1	5 MΩ	R_{17}	330 Ω
R_2	800 kΩ	R_{19}	15 kΩ
R_3	1.36 MΩ	R_{20}	15 kΩ
R_4	250 kΩ	R_{21}	7.5 kΩ
R_5	67.8 kΩ	$R_{22,\,25}$	1.5 kΩ
R_6	36.1 Ω	R_{23}	470 Ω
R_7	3.7 MΩ	R_{24}	12.5 kΩ
R_8	1 MΩ	R_{26}	12 kΩ
R_9	0.2 MΩ	R_{27}	47 kΩ
R_{10}	37.5 kΩ	R_{28}	130 kΩ
R_{11}	12.5 kΩ	$R_{29,\,30}$	68 kΩ
R_{12}	10 MΩ		

Use 11 × 17 or 12 × 18 paper.

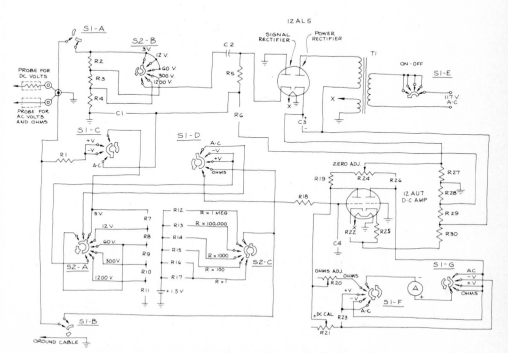

Figure 6·33 (Prob. 6·10) Sketch of volt-ohmmeter circuit.

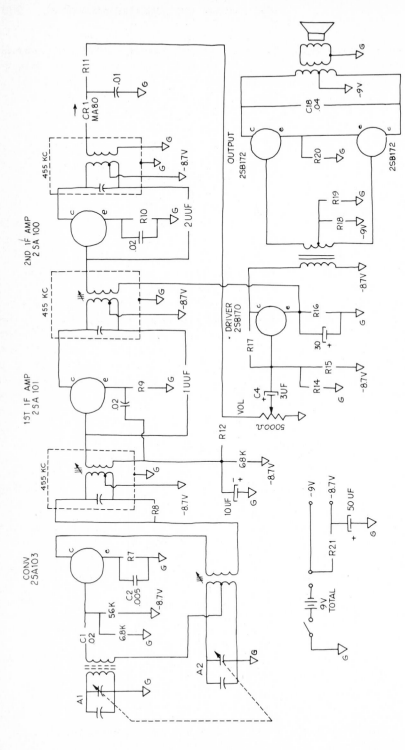

Figure 6·34 (Prob. 6·11) Incomplete sketch of a six-transistor AM receiver.

6·11 Figure 6·34 is a sketch of a six-transistor radio-receiver circuit. Many symbols are missing, but clues to their identity are given as reference designations or values. All transistors are PNP; their collector and emitter leads are identified. Capacitors at A_1 and A_2 are variable; no values are available. Capacitors in 455 kHz (KC is an older term) packages also have no values. Other capacitances are given as microfarads (μF) unless otherwise shown. Resistance values not already shown are as follows:

R_7	2.2 kΩ	R_{12}	3900 Ω	R_{19}	680 Ω
R_8, R_{16}	100 kΩ	R_{14}	5600 Ω	R_{20}	5 Ω
R_9, R_{10}	1 kΩ	R_{15}, R_{18}	33 kΩ	R_{21}	100 Ω
R_{11}	470 Ω	R_{17}	220 kΩ		

Some symbols are incorrectly shown. Reference designations may not be in the best sequence. Correct and complete the elementary diagram of this circuit, using the standard symbols and reference-designation procedure. Use 12 × 18 paper or larger.

6·12 Figure 6·35 is the incomplete schematic diagram of a circuit that indicates at a remote location what the status of a telephone line is. LED T1L209 is dark when the line is not in use, on when the line is in use, and flashes when the phone is ringing. Complete the schematic by showing standard symbols for discrete elements and distinctive shapes for the logic elements. Elements should be properly identified. MCT-2 is an optical coupler. This is a tight fit on $8\frac{1}{2}$ × 11 paper.

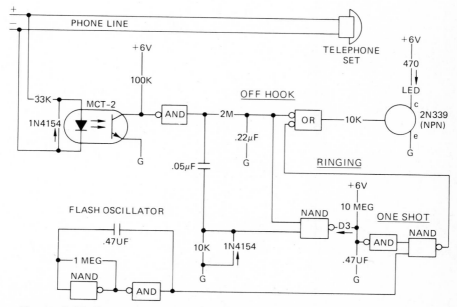

Figure 6·35 (Prob. 6·12) Schematic diagram of telephone-line status circuit.

6·13 Figure 6·36 is a sketch of the logic diagram and elementary diagram of a flip-flop circuit. Make a complete instrument drawing of this sketch, improving symbology or layout wherever you think it is advisable. Add the following information: total capacitance, 115 pF; total resistance, 70 kΩ; tunnels (crossover paths), 14; Vcc, Pin #3; Gnd, Pin #8. Use 11 × 17 or 12 × 18 paper.

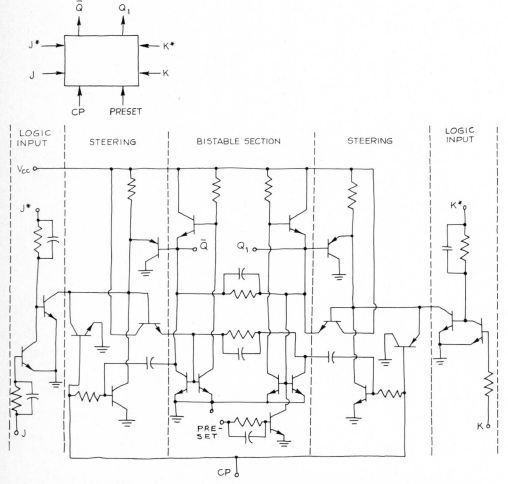

Figure 6·36 (Prob. 6·13) A logic diagram and elementary diagram of an integrated semiconductor flip-flop circuit. *(Circuit from Texas Instruments, Inc.)*

6·14 Figure 6·37 shows an incomplete sketch of a satellite beacon oscillator. Many symbols are missing or incorrect, but clues are present as to their correct identities and/or values. All transistors are PNP; their emitter and collector leads are identified. Capacitors are identified as C_1, C_2, or with symbols, some obsolete or nonstandard. Complete the elementary diagram insofar as possible using standard symbols and reference designations.

Some values are: R_1, 1780 Ω; R_2, 5210 Ω; R_5, 178 Ω; C_1, C_4, C_9, and C_{12}, 1 μF; C_3, 0.1 μF; C_5, C_8, and C_{11}, 0.01 μF; and C_6, 90 μF; and C_7, C_{10}, and C_{13}, 1000 μF. L_1, L_2, and L_3 are F54796. TR_1 is F54871; TR_2 is F54870; TR_3 is F54868; and CR_1 is F54857.

Use 11 × 17 or 12 × 18 paper.

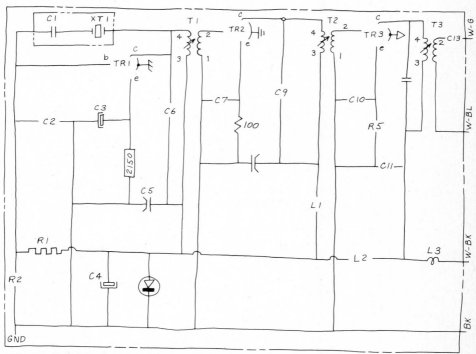

Figure 6·37 (Prob. 6·14) Elementary diagram of an oscillator for a communications satellite. Rough sketch.

6·15 Figure 6·38 has a pictorial view of an IC comparator circuit used as a clipper in *a*. An improvement to that circuit including a Schottky diode is shown in *b*. Make a complete schematic diagram of the improved circuit, including waveforms. Use $8\frac{1}{2} \times 11$ paper.

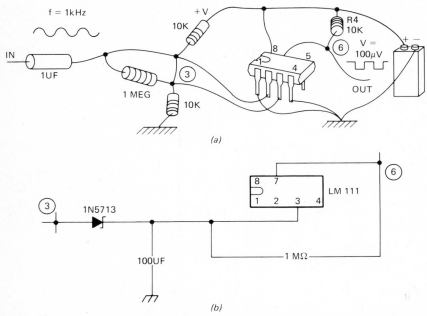

Figure 6·38 (Prob. 6·15) IC comparator used as a clipper. *(a)* Pictorial diagram. *(b)* Partial elementary diagram.

Chapter 7
Microelectronics and microprocessors

One of the most fantastic and far-reaching developments of the twentieth century is that of microelectronics. It began in 1948 with the transistor and continues to grow and evolve at a phenomenal pace. It has made a new industry and many new manufacturing companies, and has brought about a new technology in which graphics plays an important part. In this chapter the authors will show some of the drawing that is involved and some of the background that surrounds it.

7·1 Types of integrated circuits

Figure 7·2 lists the types of integrated circuits that have been developed to this point in time. Typical of many of these circuits is the wafer, shown in Fig. 7·1, which includes many dice, each of which bears an individual integrated circuit (IC). If we exclude the hybrid ICs for the moment, we notice that there are two fundamental processes, bipolar and MOS (metal-oxide semiconductor). The glossary that follows briefly describes the differences and variations among the processes shown in Fig. 7·2.

Bipolar: One of two fundamental processes for fabricating ICs and currently the more popular. A bipolar IC is made up of layers of silicon with differing electrical characteristics. Current flows between the layers when a voltage is applied to the "junction," or boundary between the layers.
TTL: Transistor-transistor logic, by far the most successful bipolar IC logic. It gets its name, as do other digital IC product families, from the way the components are combined to form the logic elements. Digital ICs solve problems by manipulating electrical signals that represent bits of information. Basic TTL is a mature product, but faster and lower-power versions are expected to extend the life of TTL into the 1980s.
ECL: Emitter-coupled logic, a bipolar digital IC family that uses a more complex design than TTL to speed up IC operations. It is costly, power hungry, and difficult to use, but it could become important in the next generation of large computers because it is four times faster than TTL.

Figure 7·1 Partial view of a wafer on which many LSI chips have been fabricated. *(Intel Corp.)*

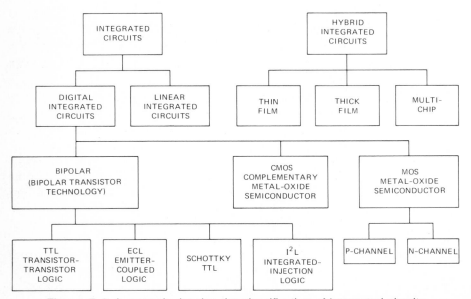

Figure 7·2 A network showing the classification of integrated circuits.

I²L: Integrated-injection logic, a high-density bipolar logic design. It holds promise of producing ICs with a circuit density approaching MOS, the speed of TTL, and the low power requirements of CMOS.

Linear IC: Like a digital IC, a linear circuit combines bipolar transistors and other devices on a single chip. But instead of manipulating bits of data that represent numbers, as a digital device does, it operates as an analog: amplifying, shaping, or comparing electrical signals. Linear ICs are widely used in instruments and communications equipment.

MOS: Metal oxide semiconductor is the second fundamental process for fabricating ICs and the fastest growing. The active area of a MOS chip is at the surface, where a gate electrode applies a voltage to a thin layer under it to create a temporary "channel" through which current can flow.

PMOS: The oldest MOS circuit technology uses a channel of P-type material, where the flow of current is made up of positive charges. PMOS now accounts for most MOS volume. Because the action in a MOS circuit is near the surface, contamination that would be minor in a bipolar device caused serious problems in the early days of MOS.

NMOS: N-channel MOS, where the flow of current is made up of negative charges, has moved into production. It is two to three times faster than PMOS circuits, but its manufacture is harder to control than PMOS.

CMOS: Complementary MOS combines P-channel and N-channel transistors to create an IC that is as fast as a NMOS circuit, but consumes very little power. Although it is more expensive to manufacture than other MOS circuits, CMOS is rapidly catching on for low-power applications, such as digital watches.

SOS: The silicon-on-sapphire process is still being developed, but its potential to achieve bipolar speeds with MOS densities may soon move SOS circuits into production even though sapphire now costs much more than silicon. Sapphire increases the operating speeds because it insulates the MOS transistors from one another.

OTHER MOS: New variations of the MOS process are coming from the R&D labs almost every year. H-MOS (high-performance MOS) is one example. V-MOS (transistors built on the slanting surface of a V groove) is another.

7·2 Transistors, the basic building block

Although this is not a book on electronic theory, Fig. 7·3 is a diagram of several transistors or transistor-like devices. The original bipolar PNP or NPN transistor has acquired some companions, but they are basically made of the same material, silicon being the most widely used. To enable the crystal to conduct electric current, small quantities of impurities are mixed with the pure silicon. If phosphorus is added, N type is the result. If boron is added, P type is produced.

By itself neither N- nor P-type material can accomplish a junction. But, when put together in various combinations and configurations, they provide a number of practical working semiconductor devices. It is the behavior of elec-

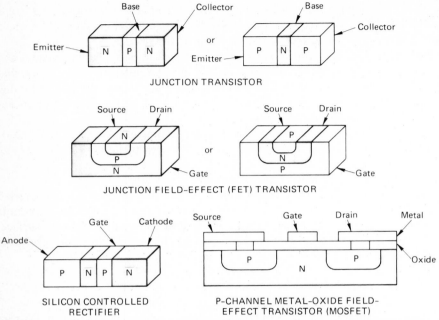

Figure 7·3 Some basic semiconductors. From top to bottom: Junction transistor, junction field-effect transistor, silicon controlled rectifier (SCR), and MOS field-effect transistor.

trons and holes in the vicinity of the PN junction that gives bipolar junction transistors their unusual properties. MOS transistors (MOSFET), however, have the active area at their surface. The silicon controlled rectifier (SCR) is widely used in the electric-power and industrial fields and is discussed at more length in the next two chapters.

Transistors are still widely available and used as discrete components. Along with diodes, capacitors, and resistors, they can be made to be part of an integrated circuit. A summary of semiconductor devices follows.

1. Thyristors
   Silicon controlled rectifier (Electrically controlled switch for dc loads.)
   Triac (Electrically controlled switch for ac loads.)

2. Diodes (Exhibit very high resistance to current flow in one direction and little resistance in the other.)

3. Discrete transistors (Three-terminal devices used for amplifying or switching electrical signals.)
   Bipolar
   FET (field-effect transistor)
   Unijunction

4. Charge-coupled devices (CCDs) (Storage devices in which packets of electric charge are moved across the surface of a semiconductor by electrical signals.)

5. Photo-semiconductors (Emit or absorb electromagnetic radiation, usually in the visible or infrared band.)
Phototransistor
LED (light-emitting diode)
Photothyristor

6. Integrated circuits (To be defined and discussed later.)

7. Hybrid integrated circuits (To be defined and discussed later.)

7·3 An integrated-circuit processing glossary[1]

In describing the manufacture of ICs, we believe that presentation of the following terms will make it easier for the reader to understand the processes.

Alignment: A technique in the fabrication process by which a series of six to eight masks are successively registered to build up the various layers of a monolithic device. Each mask pattern must be accurately referenced to or aligned with all preceding mask patterns.

Die: A portion of a wafer bearing an individual IC, which is eventually cut or broken from the wafer.

Diffusion: A high-temperature process involving the movement of controlled densities of N-type or P-type impurity atoms into a solid silicon slice in order to change its electrical properties.

Emulsion: A suspension of finely divided photosensitive chemicals in a viscous medium, used in semiconductor processing for coating glass masks.

Etching: A process using either acids or a gas plasma to remove unwanted material from the surface of a wafer.

Mask: A chrome or glass plate having the transparent circuit patterns of a single layer of a wafer. Masks are used in defining patterns on the surface of a resist-covered wafer.

Master mask: A chrome mask of a complete wafer's multiple images. It is used either in projection printing on a wafer or to contact-print additional masks.

Photoresist: A material that allows selective etching of a wafer when the wafer is photographically exposed. With a *negative resist,* the resist film beneath the clear area of a photomask undergoes physical and chemical changes that render it insoluble in a developing solution. In a *positive resist,* the same areas after exposure are soluble in the developing solution, so they disappear, permitting development of the exposed pattern underneath.

Plasma etching: An etching process using a cloud of ionized gas as the etchant.

---

[1] Reprinted from *Electronics,* July 21, 1977: copyright © McGraw-Hill, Inc., 1977.

Reticle: A glass-emulsion or chrome plate having an enlarged image of a single IC pattern. The reticle is usually stepped and repeated across a chrome plate to form the master mask.

Step and repeat: A method of positioning multiples of the same pattern on a mask or wafer.

Stripping: A process using either acids or plasma to remove the resist coating of a wafer after the exposure, development, and etching steps.

Yield: The number of usable IC dice coming off a production line divided by the total number of dice going in. Yield tends to be reduced at every step in the manufacturing process by wafer breakage, contamination, mask defects, and processing variations.

7·4 Integrated-circuit fabrication process

Figure 7·4 shows the major milestones of the silicon IC fabrication process. Our description will begin at step 2, in which the crystal has been sawed into thin wafers which have been lapped and polished to a mirrorlike finish with no scratches or defects. Next, a layer of silicon dioxide (SiO_2) is grown on the surface to protect it from contamination and prevent the diffusion of impurities in some of the processes that follow.

The next step is to provide "windows" through which impurities can be deposited on the surface. Special photolithographic techniques are used to remove

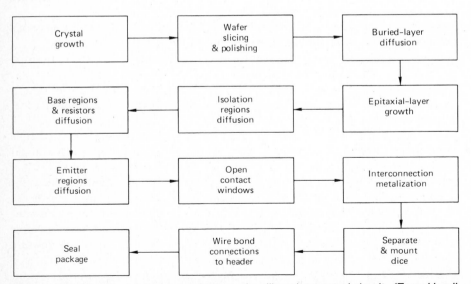

Figure 7·4 The steps in the manufacture of a silicon integrated circuit. *(From Hamilton, D. J., and W. G. Howard, "Basic Integrated Circuit Engineering," McGraw-Hill Book Company, New York, 1975.)*

the oxide in certain selected regions. When the wafer is placed in a diffusion furnace at 1000°C and a gas containing the required impurity is passed over the surface, the impurities are deposited where there are windows and then are "diffused" into the silicon. This *buried layer* reduces the series collector resistance of transistors. Table 7·1 shows the masks that are used in the basic fabrication techniques. The windows mentioned above have been made using mask 1.

After the buried-layer diffusion is complete, the SiO_2 is removed from the wafer surface and a new layer of N- or P-type silicon is grown on the surface by means of epitaxial growth. In this step the wafer is exposed to an atmosphere containing a silicon compound at high temperature, and a new layer of silicon atoms is deposited in accordance with the crystal structure of the substrate. This *epitaxial layer* is also doped by including small amounts of the desired impurity in the gas stream during its growth.

As shown in Table 7·1 and Fig. 7·4, successive oxide growths, photolithography steps, and diffusions are undertaken to provide base and emitter regions, isolation walls, etc. The last steps include making windows for interconnections among the many devices now present on each circuit, followed by the metallization process. In this process a metal, often aluminum, is evaporated over the surface of the wafer; then unwanted portions are etched away and the interconnection contacts are left.

Finally the wafer is broken up into small dice, or chips, each having a single circuit or assembly, such as a microprocessor. Each chip is then mounted on a header, such as a dual-in-line mounting, and the aluminum contacts are connected to the header leads (pins) by wire or one of the newer processes, such as film striping.

Wafers (and the ICs themselves) have been increasing in size. Typical diameters that can be processed by step-and-repeat photography are 2, 3, and 4 in. For larger sizes, the electron beam is being used for mask making and exposure of patterns on wafers (Fig. 7·5).

7·5 Photolithography

This process accomplishes two important functions: (1) reduction of the original mask drawing to the final mask size and (2) the photoresist process itself. The process of reducing the mask size also includes producing multiple images of the layout.[1]

The photoresist process involves making windows in the SiO_2 layer of the wafer. The photoresist is a light-sensitive coating that is placed over the oxide. The glass mask with the desired portion of the circuit pattern is placed over the photoresist. After exposure to ultraviolet light, the unexposed coating is dissolved; that which remains is chemically resistant to the solution or gas that is used to etch away the oxide. Later the remaining photoresist can be

---

[1] Many identical circuit chips are made from one wafer (Fig. 7·1).

Table 7·1 The Basic Fabrication Sequence and Masks Required

	Mask 1	Mask 2	Mask 3	Mask 4	Mask 5	Mask 6	Mask 7
Process	Select buried-layer locations	Select isolation areas	Select base region	Select emitter region	Select capacitor locations	Select openings for ohmic contacts	Select areas where metal is removed
	Buried-layer diffusion	Isolation diffusion	Base diffusion	Emitter diffusion	Capacitor oxidation	Metalli-zation	Gold backing if required
	Epitaxial layer and diffusion						

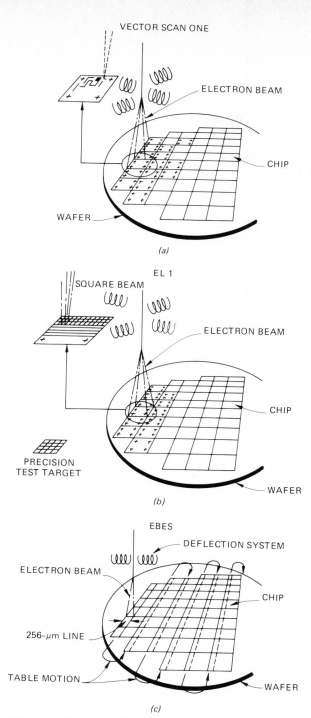

VECTOR SCAN ONE

ELECTRON BEAM

CHIP

WAFER

(a)

EL 1

SQUARE BEAM

ELECTRON BEAM

CHIP

PRECISION
TEST TARGET

WAFER

(b)

EBES

DEFLECTION SYSTEM

ELECTRON BEAM

CHIP

256–μm LINE

TABLE MOTION

WAFER

(c)

Figure 7·5 High-resolution pattern registration on a wafer. *(a)* and *(b)* Step-and-repeat scanning. *(c)* Continuous-moving table. *(Reprinted from "Electronics," May 12, 1977; Copyright © McGraw-Hill, Inc., 1977.)*

removed, so that additional masking and etching can be performed. The process is illustrated in Fig. 7·6.

Figure 7·7 includes, greatly enlarged, three steps of the diffusion process as they would appear if one were looking at the surface of an IC die. (In other circuits, the shapes of these components may be different from those shown in Fig. 7·7.)

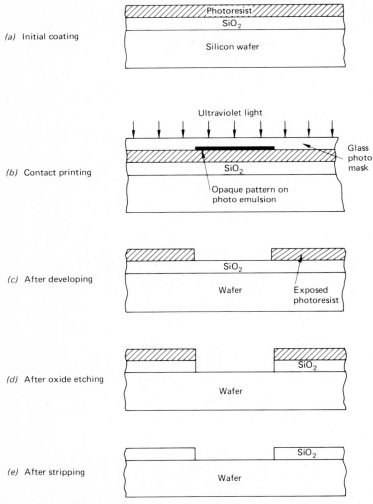

(a) Initial coating

(b) Contact printing

(c) After developing

(d) After oxide etching

(e) After stripping

Figure 7·6 Steps in the photoresist process. *(From Hamilton, D. J., and W. G. Howard, "Basic Integrated Circuit Engineering," McGraw-Hill Book Company, New York, 1975.)*

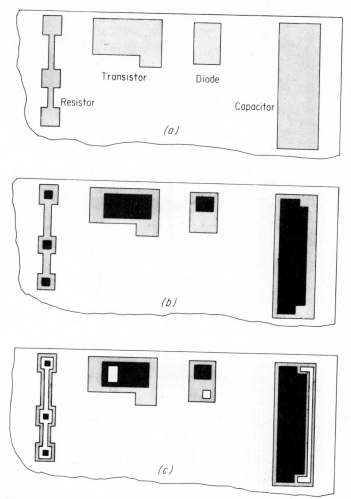

Resistor   Transistor   Diode   Capacitor

(a)

(b)

(c)

Figure 7·7 Three steps in the manufacturing process of inte-grated-circuit semiconductors.

7·6 Drawing and artwork for integrated-circuit design

Integrated circuits can be classified by the number of logic elements or "gates" they contain. Small-scale integration (SSI) refers to chips that have 100 or fewer gates. Medium-scale integration (MSI) refers to chips containing 100 to 1000 logic gates, while large-scale integration (LSI) refers to chips with more than 1000 elements. Figure 7·8 shows the first mask of a series for an older small-scale integrated circuit. As shown in Fig. 7·9, this chip has 31 elements. Figure

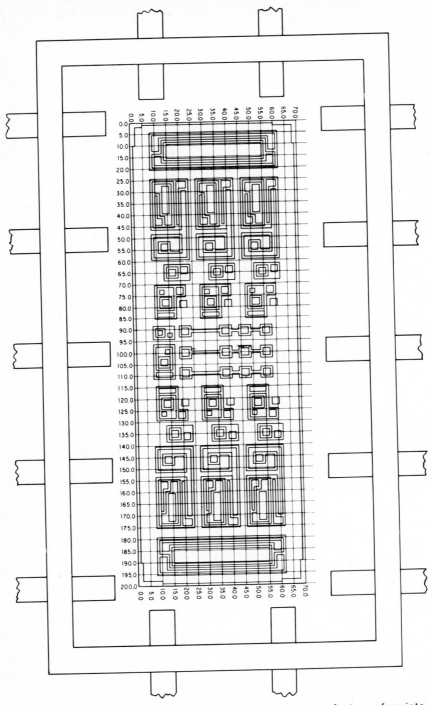

Figure 7·8 The first of a series of drawings for the manufacture of an integrated circuit. *(Texas Instruments, Inc.)*

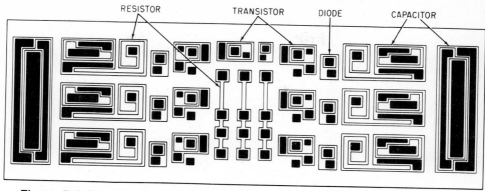

Figure 7·9 The integrated circuit (see Fig. 7·8) will appear about like this after several diffusion processes. *(Texas Instruments, Inc.)*

7·10 shows a series of masks for a latch circuit which is part of a microprocessor chip. The schematic diagram of this circuit is shown in Fig. 7·11, and the photograph of the LSI microprocessor chip that contains the latch is shown in Fig. 7·12.

The original drawing for the mask shown in Fig. 7·8 was drawn to a scale 150 times the actual size. Notice the grid system that is used to facilitate accuracy. Spacing between components is actually as close as 0.0005 in. at times; therefore, alignment errors cannot exceed a few *ten-thousandths* of an inch. Such accuracy is usually obtained only with computer-generated drawings. Figure 7·13 shows some of the equipment that is utilized in the fabrication of integrated circuits, including a computer-operated automatic drafting machine. The actual size of the SSI chip (Figs. 7·8 and 7·9) is about 0.125 (3.2) × 0.250 in. (6.4 mm). It is in the form of a flatpack package with 14 external leads.

Original drawings for LSI chips such as the microprocessor are often drawn 500 times the actual size. The TMS 9900 microprocessor was produced using N-channel silicon-gate MOS technology. The MOS devices can be made with very small geometry, permitting high densities. For example, the memory space of the 9900 is 65,536 bytes (where one byte = eight binary units). Only a small part of the drawing for the entire unit is represented by the latch artwork (Fig. 7·10). The microprocessor is housed in a ceramic dual-in-line package 0.90 (23) × 3.20 in. (81 mm) and having 64 pins.

Artwork is made from each drawing on a plastic material called Rubylith. Cuts are made in this material—Mylar with a red plastic coating—and the coating is peeled off in the desired places. When the Rubylith artwork is photographed, the red areas will appear opaque to photographic film. The photograph looks pretty much like a negative.

In order to check the accuracy of medium- and large-scale IC drawings, debugging plots composed of all masks are made. Carefully following registra-

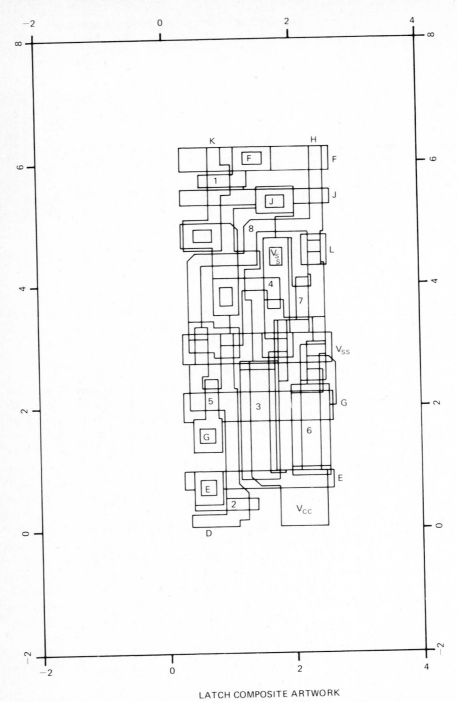

LATCH COMPOSITE ARTWORK

Figure 7·10 The composite artwork for a latch circuit, part of a microprocessor. *(Texas Instruments, Inc.)*

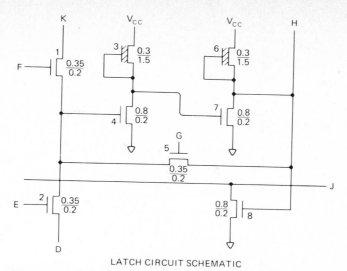

LATCH CIRCUIT SCHEMATIC

Figure 7·11 The schematic diagram for the latch circuit which is part of the microprocessor shown in the next figure.

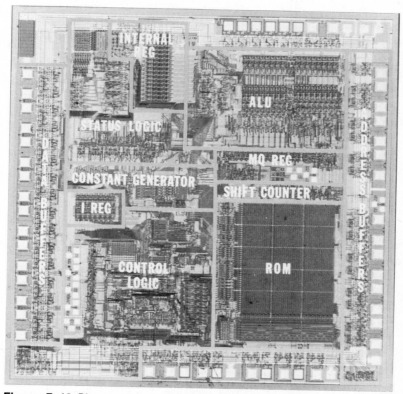

Figure 7·12 Photograph of the TMS 9900 microprocessor greatly enlarged. Latch circuits, which are really flip-flop units, are part of the register (reg) circuits and are usually found in input-output areas of microprocessors. *(Texas Instruments, Inc.)*

223

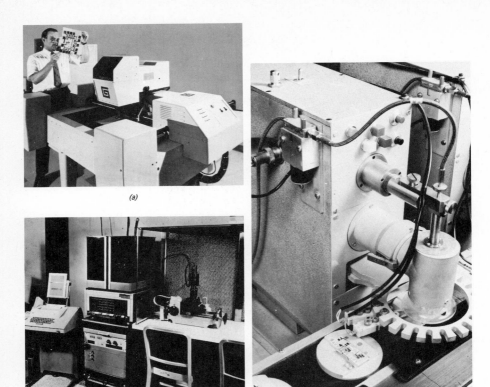

(a)

(c)   (b)

Figure 7·13 Equipment used in production of integrated circuits. *(a) An automatic drafting machine. (Gerber Scientific Instrument Co.) (b) An abrasive trimming machine for trimming thin-film resistors. (Penwalt S.S. White Industrial Products) (c) Computer-controlled step-and-repeat cameras. (Copyright © 1968, Bell Telephone Laboratories, Inc. Used by permission.)*

tion, as in Fig. 7·14, the masks are superimposed on one another. Unwanted overlapping or less-than-minimum clearances can be observed. The heavy dark horizontal line at center right might be such a deficiency.

Finally, to connect the elements of the circuit to their respective leads or other components (or both, in some cases) an aluminum pattern must be placed over the surface and in contact with the appropriate component. After openings for ohmic contacts have been made (Mask 6 in Table 7·1 and step 8 in Fig. 7·4), the metal is evaporated over the entire surface of the wafer (step 9, Mask 7), where it makes good contact with the silicon. The pattern could look like Fig. 7·15. Unwanted metal is etched away, and the chip is ready to be mounted.

7·7 Packaging of integrated circuits

Until 1976 or so, two types of packages were being used for most integrated circuits, flatpacks and dual-in-line packages (DIPs), ceramic or plastic, although the transistor TO "can" provided a third alternative (Fig. 7·16). Growing pin

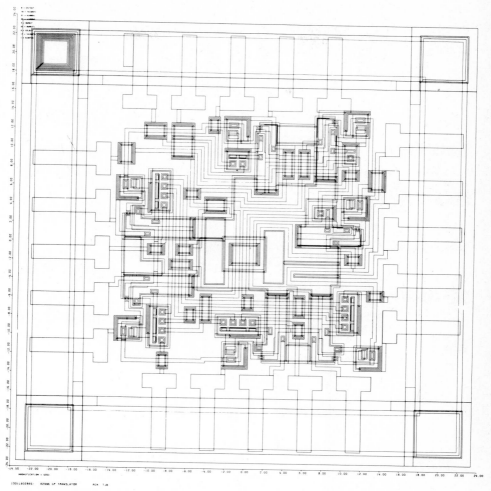

Figure 7·14 Series of drawings made by a computer. These "debugging plots" are composite views of all nine masks, which have been superimposed. *(Copyright © 1969, Bell Telephone Laboratories, Inc. Used by permission.)*

counts have been forcing some changes, particularly in LSI. It has become necessary to find the right package for both the system and the chip.

One 3.20 in. (81 mm) long DIP has 64 pins (the one in Fig. 7 · 26 is slightly smaller). The DIP is reliable and easy to assemble and remove from circuit boards. But, since 3-in. bodies take up a lot of board space, some designers have turned to more economically sized packages. A redesigned square flat-pack with an equal number of leads from all four sides (up to 64 or 80 total) has been gaining acceptance for several devices, including digital watches

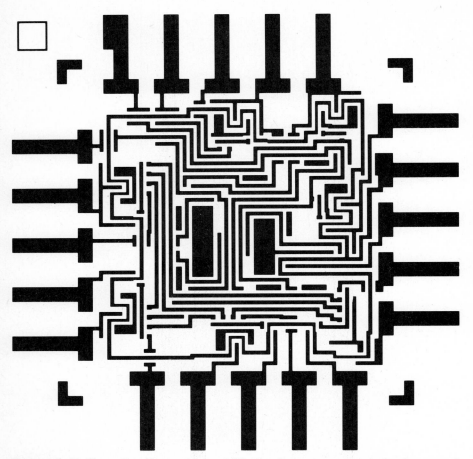

Figure 7·15 The pattern for contact metallization (interconnections) of an integrated circuit. *(Copyright © 1969, Bell Telephone Laboratories, Inc. Used by permission.)*

and calculators. Multilayered (cofired) DIPs have been in production, including one that has four rows of pins and is called a QUIP.

Another approach is to use a film carrier, usually for several LSI units. The carrier, which comes in different shapes and sizes, is a multilayer package on the bottom of which is a pattern of "bumps" on 40- or 50-mil centers.[1] There is a metal frame and base pad, making it possible to connect the internal IC pad to the external bumps through grooves in the sides. Another type of carrier uses film, including 35-mm film, which is quite compatible with the automation process. Figure 7·16 shows three of the older types of packages,

---

[1] A mil is 0.001 in. (0.025 mm). Most DIP and printed-circuit-board construction is based on 100-mil (2.54-mm) construction. (100 mils is also = 0.10 in.)

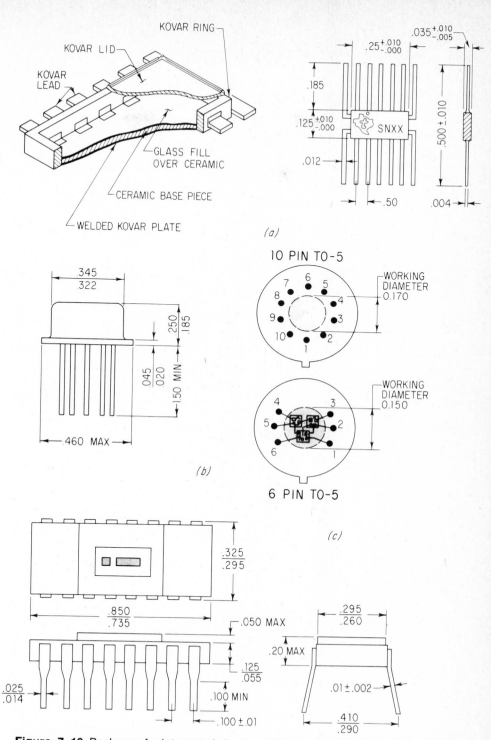

Figure 7·16 Packages for integrated circuits. *(a)* Flatpack. *(Texas Instruments, Inc.)* *(b)* TO-5 packages. Three chips are shown in the lower right-hand figure. *(c)* Dual-in-line package (DIP), ceramic. *(Intel Corp.)*

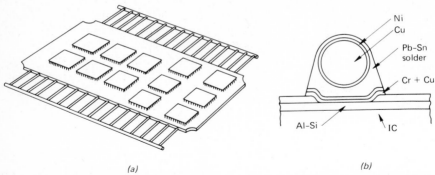

(a)

(b)

Figure 7·17 Chip carrier and solder bump. *(a)* A carrier used for assembling and holding several MSI or LSI chips. *(b)* Greatly enlarged section of a solder bump.

all of which are still used. Figure 7·17 shows a chip carrier at *a* and the cross section of a solder bump at *b*.

Small copper spheres are placed on the contact points of the circuit while it is still on the wafer. The wafer is then heated until solder melts over the spheres and forms a solder-coated bump on the contact pad as shown in Fig. 7·17*b*. When an individual IC is placed in a header and the assembly is heated,

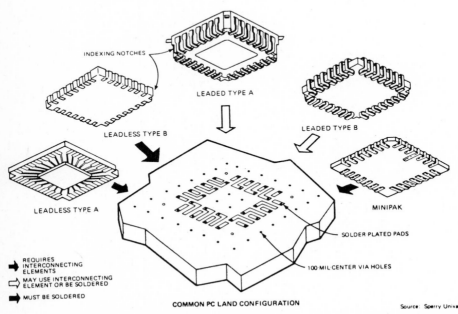

Figure 7·18 LSI package standard for devices with connecting leads on 50-mil (0.127-mm) centers. Such carriers would interconnect with common printed circuit boards having 100-mil (2.54-mm) hole spacing. *(Reprinted from "Electronics," March 17, 1977; Copyright © 1977 by McGraw-Hill, Inc.)*

the IC becomes connected to the header through the bumps. Some variations of this appear in Fig. 7·18, where devices with 50-mil (1.27-mm) distances between pins can be placed on a board with the common 100-mil (2.54-mm) PC-board hole spacing. Thus, LSI chips with leads only 50 mils apart can be through-connected to boards that have holes 100 mils apart. The EIA (Electronics Industries Association) hopes to achieve a nationwide standardization of LSI packaging and connecting.

7·8 Film (hybrid) circuits

A hybrid microcircuit is usually a mixture of etched or silk-screened circuit elements and individual (discrete) elements or standard IC chips. Manufacturing tolerances of resistors and capacitors are smaller in film techniques than in the diffusion process. Therefore, hybrid circuits combining silicon chips that have *active* devices (transistors and diodes) with film circuits composed of *passive* devices (resistors and capacitors), are sometimes the most appropriate design for a particular purpose. Hybridization may also minimize weight, space, and interconnections, as well as provide a larger range of values for some elements. (See Table 7·2.)

Table 7·2 Data on Film and Diffused Circuit Elements

	Thick film	Thin Film	IC (Diffused Silicon)
Manufacturing Tolerance			
Resistors	0.50 to 10%	0.01 to 5%	5 to 20%
Capacitors	6%	3%	15 to 50%
Resistance Values	To 100 MΩ	10 Ω to 10 MΩ	100 Ω to 50 kΩ
Thickness	0.013 to 0.05 mm	$\approx 0.00x$ mm	$\approx 0.00x$ mm
Resistive Material	Lithium alloy	Tantulum alloy	Silicon, SiO_2

Figure 7·19 shows the same circuit manufactured as a thick-film device and as a thin-film device. *Thick-film* circuits are made by firing a paste onto an insulating substrate of a glossy or ceramic material. Formerly, expensive noble metals were used, but alloys of lithium, nickel, and copper have recently been used successfully. The deposited material can be from $\frac{1}{2}$ (0.0127) to 2 mils (0.05 mm) in thickness. They are trimmed to the desired accuracy (0.5 to 2 percent) with lasers. (Two thin light grey lines appear on the right side of each dark-area resistor in the photograph of Fig. 7·19, where trimming has been performed.)

Thin-film circuits are different in that the resistor and conductive films are sputtered or vacuum-deposited onto the substrate. The film can be deposited as a pattern or as a film for subsequent photoprocessing and etching. Film thickness is a few micrometers, one micrometer being equal to 0.001 mm. Elements are trimmed to the desired size by such processes as abrasive trimming. As in thick-film technology, thin-film techniques are used to manufacture re-

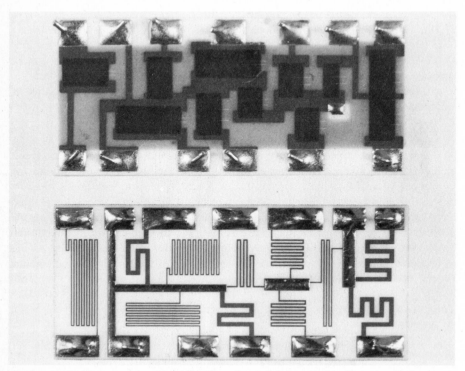

Figure 7·19 Photograph of a thick-film circuit (top) and thin-film circuit (bottom). These devices, each about 1 in. in length, are the same circuit shown in Fig. 7·20.

sistors, capacitors, and conductors. They have better resistor stability, resolution, and circuit-element density than thick-film circuits. However, they are more expensive and cannot produce as large resistance values. (See Table 7·2.)

Figure 7·20 shows a schematic diagram of the resistor network that appears in the photograph (Fig. 7·19). The lower drawing (Fig. 7·20b, which will be referred to in a later drawing) shows the outline of the substrate and its edge tolerances. Figure 7·21 includes the plan view of the thick-film circuit, a table, and some notes. The drawing shows the contacts along the top and bottom, and the resistor and conductor areas as hidden lines. A coating to protect against unwanted contact with foreign objects is placed over the resistors and is shown by solid lines which outline its perimeter. A table is included that gives the desired resistance value, resistivity in ohms per square, and desired tolerances. The resistor sheet resistances range from 1 to 100 kΩ per square depending on the film thickness. Tolerances are given in two categories, end of life (E.O.L.) and initial trim. Thick-film resistances have a tendency to drift after they are put into use; hence there are two maximum allowable tolerances, the smaller

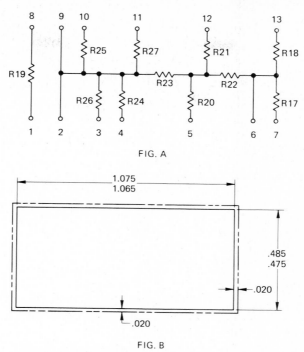

FIG. A

FIG. B

Figure 7·20 The circuit that has been produced in both thin-film or thick-film form. *(a)* Schematic diagram. *(b)* Outline drawing of the substrate.

of which is for the initial trim. Over the life of the circuit (perhaps 20 years), the values are expected to change no more than the percentages listed. This drawing was drawn 10 times actual size.

Figure 7·22 includes the plan view of the thin-film circuit, drawn 10 times desired actual size, notes, and a table. If the resistors have a sheet resistance of, say, 75 Ω per square, the length of a resistor can be determined by an equation such as $L = (R/75)(W)$. Resistor R_{27}, with a width of 0.015 in. (0.38 mm), would have a length of $(2000/75) (0.015) = 0.40$ in. (10 mm). The designer has to select a route that will provide a 0.015-in.- (0.38-mm-) wide line that is 0.40 in. (10 mm) long. This has been done for each of the resistors in Fig. 7·22, and results in the patterns that are seen in the drawing and photograph. Note that the higher-value resistors have the narrower lines, 4 mils (0.1 mm). Line spacing has been kept equal to line width. In order to obtain accurate drawings and production models of the circuit, lines of this size are usually drawn by a computer or automatic drafting machine. Notes 3 and 4 of the drawing refer to Fig. 7·20*b*.

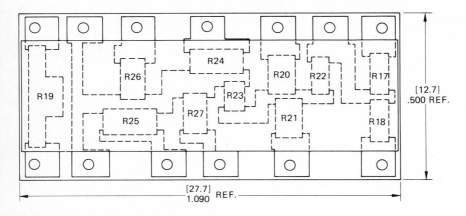

RESISTOR	VALUE, Ω	RESISTOR INK	TOLERANCE	
			E.O.L.	INITIAL TRIM
R17	2.59K	1KΩ/□	±1.0%	±0.5%
R18	2.59K	1K	±1.0%	±0.5%
R19	60.4K	10K	±2.5%	±1.0%
R20	15.4K	10K	±1.0%	±0.5%
R21	15.4K	10K	±1.0%	±0.5%
R22	24.3K	10K	±3.0%	±1.5%
R23	13.3K	10K	±1.0%	±0.5%
R24	40.2K	100K	±1.0%	±0.5%
R25	43.2K	100K	±1.0%	±0.5%
R26	2000	1K	±1.0%	±0.5%
R27	2000	1K	±1.0%	±0.5%

NOTES:
1. PROCESS PER C 6920.
2. LASER TRIM PER C 6950.

Figure 7·21 Drawing and notes for a thick-film circuit. This is the circuit that is shown in the upper part of Fig. 7·20.

Hybrid circuits having different combinations of elements are in use. The circuit in Fig. 7·23 has discrete devices such as transistors mounted on a thin-film network. Similar arrangements using silicon IC chips have been produced in the circuit shown in Fig. 7·24. As the exploded-view drawing shows, active components are formed within a silicon block and the passive-component pattern is deposited by film techniques on top of a passivating SiO_2 layer, which covers the "active" circuit. A photograph of such a circuit appears in Fig. 7·25.

7·9 Microprocessors and microcomputers

The development and maturity of LSI technology made possible the design and quantity production of microprocessors. A *microprocessor* provides on one chip the CPU, or central processing function, of a digital computer. (See glossary for microprocessor terms.) One microprocessor chip might do the work of 80 standard integrated circuits. A refinement of the microprocessor has resulted in the development of a chip called a *microcomputer* which has most of the

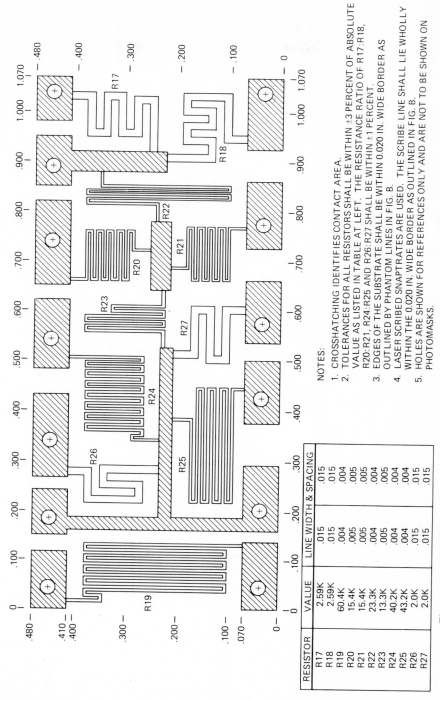

RESISTOR	VALUE	LINE WIDTH & SPACING
R17	2.59K	.015	.015
R18	2.59K	.015	.015
R19	60.4K	.004	.004
R20	15.4K	.005	.005
R21	15.4K	.005	.005
R22	23.3K	.004	.004
R23	13.3K	.005	.005
R24	40.2K	.004	.004
R25	43.2K	.004	.004
R26	2.0K	.015	.015
R27	2.0K	.015	.015

NOTES:

1. CROSSHATCHING IDENTIFIES CONTACT AREA.
2. TOLERANCES FOR ALL RESISTORS SHALL BE WITHIN ±3 PERCENT OF ABSOLUTE VALUE AS LISTED IN TABLE AT LEFT. THE RESISTANCE RATIO OF R17:R18, R20:R21, R24:R25 AND R26:R27 SHALL BE WITHIN ±1 PERCENT.
3. EDGES OF THE SUBSTRATE SHALL BE WITHIN 0.020 IN. WIDE BORDER AS OUTLINED BY PHANTOM LINES IN FIG. B.
4. LASER SCRIBED SNAPTRATES ARE USED. THE SCRIBE LINE SHALL LIE WHOLLY WITHIN THE 0.020 IN. WIDE BORDER AS OUTLINED IN FIG. B.
5. HOLES ARE SHOWN FOR REFERENCES ONLY AND ARE NOT TO BE SHOWN ON PHOTOMASKS.

Figure 7·22 Drawing and notes for a thin-film circuit. This circuit is shown in the lower part of Fig. 7·20.

Figure 7·23 A hybrid circuit in which discrete devices such as transistors and diodes have been mounted on a thin-film network. *(Motorola Semiconductor Products Division.)*

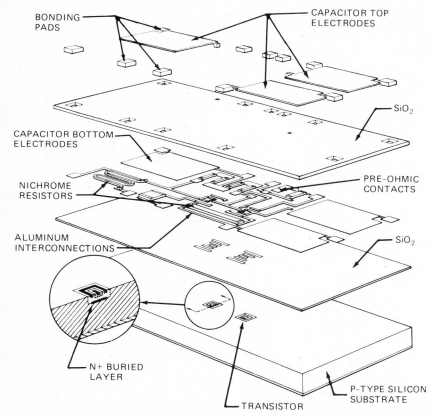

BONDING PADS

CAPACITOR TOP ELECTRODES

SiO₂

CAPACITOR BOTTOM ELECTRODES

PRE-OHMIC CONTACTS

NICHROME RESISTORS

ALUMINUM INTERCONNECTIONS

SiO₂

N+ BURIED LAYER

TRANSISTOR

P-TYPE SILICON SUBSTRATE

Figure 7·24 Another type of hybrid circuit. Thin-film patterns of passive elements have been laid down on top of the insulated SiO₂ surface of a semiconductor active network. *(Motorola Semiconductor Products Division.)*

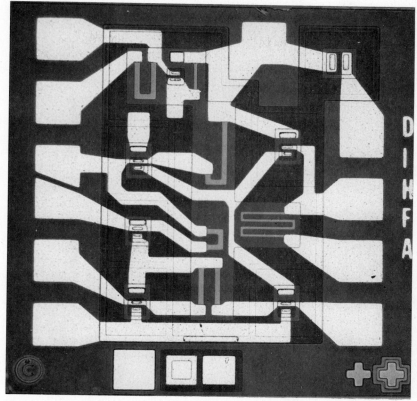

Figure 7·25 In this hybrid circuit contacts and connections are nearly white in color. Thin-film elements (light grey) can be seen superimposed on the active elements. *(Fairchild Semiconductor Division.)*

functions of a digital computer (see Fig. 7·26). Actually the two terms are rather loosely applied, and there will probably be some confusion about these titles for years to come.

The famous 8080 microprocessor (manufactured by Intel Corp.) has an arithmetic logic unit (ALU), accumulator, registers (three types), and control logic and can address up to 512 input-output ports. Memory is minimal. The TMS 9900 microprocessor (Fig. 7·12) has an ALU, three registers, control logic, ROM, and input-output (I/O) control lines. It is usually augmented with other chips, including RAM, as is shown in Fig. 7·28. The 8748 microcomputer has a CPU, a clock circuit that times and coordinates data flow, ROM and RAM, I/O circuits, and expansion circuitry which enables the user to extend the capabilities of the chip. As powerful as this chip is, it is usually mounted on a board with still other components.

Figure 7·26 The 8748 microprocessor, or microcomputer, package. This dual-in-line package is 3 in. long. It can be reprogrammed by focusing an ultraviolet light on the open part in the center. *(Intel Corp.)*

The functional diagram of the 8080A microprocessor is shown in Fig. 7·29, and the architecture of a single-chip microcomputer is shown in Fig. 7·30. The latter includes a CPU, a clock circuit and other auxiliary circuits, a memory section, expansion circuits, and input-output ports. Its memory section includes both RAM and ROM types. The expansion circuitry makes it possible to extend the capabilities of the chip.

It is very common to find combinations of microprocessors or microcomputer chips and other devices mounted on a printed-circuit board. These single-board computers can perform many simultaneous or sequential operations for computation or for process control. Figure 7·31 is the photograph of such a board. Such arrangements provide essentially the same functions as a minicomputer. This is a low-cost (hundreds of dollars) way to control many parts of a system or process, and to provide convenient monitoring (video tube, LED), input (teletype, sensors), and ease of programming (FORTRAN and BASIC languages and floating decimal point). A functional flow diagram of the SBC 80/10 is shown in Fig. 7·32.

Figure 7·27 The chip of a multifunction microprocessor (microcomputer), greatly enlarged. Most easily recognized functions are the read-only memory at the bottom and input-output ports around the edge. *(Intel Corp.)*

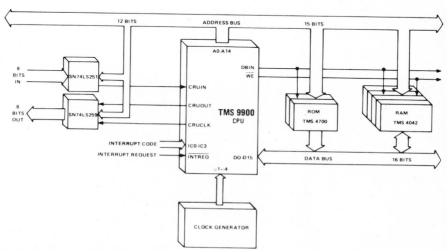

Figure 7·28 The minimum TMS 9900 system includes eight bits of input and output interface, clock generator, ROM (1024 × 16) and RAM (256 × 16) blocks, an address bus, and a data bus. *(Texas Instruments, Inc.)*

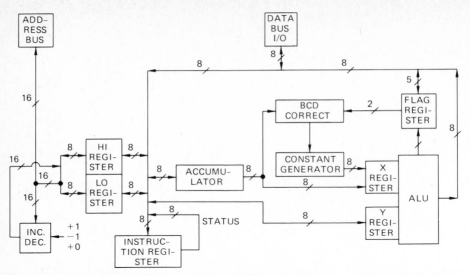

Figure 7·29 Functional diagram of the 8080A microprocessor. Separate 8- and 16-bit address busses allow direct addressing of 65,536 bytes of memory. Numbers indicate the number of data bits at the points indicated. Block at lower left represents the incrementer-decrementer unit.

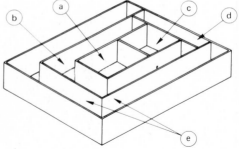

Figure 7·30 Pictorial schematic of a microcomputer chip. *(a)* Central processing unit. *(b)* Clock circuit and other auxiliary circuits. *(c)* Memory section. *(d)* Expansion circuitry. *(e)* Input-output ports.

Figure 7·31 Photograph of the SBC 80/10 single-board computer. It contains an 8080 CPU, 1K bytes of RAM, 4K bytes of EPROM or ROM, and 48 programmable I/O lines. *(Intel Corp.)*

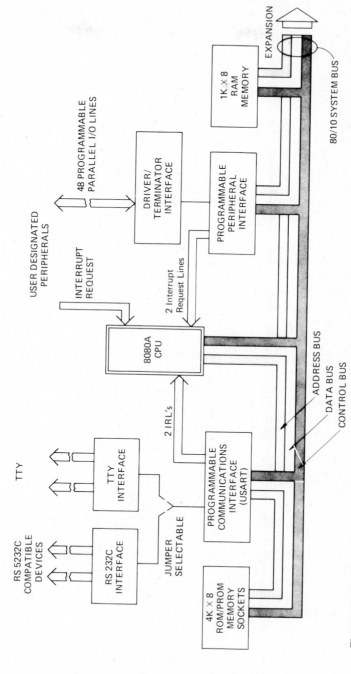

Figure 7·32 Flow diagram of the SBC 80/10 single-board computer. 4K × 8 means 4000 bytes of memory. *(Intel Corp.)*

Microprocessor Glossary

CPU: Central processing unit, the heart of a microcomputer. It contains the arithmetic logic unit, control circuitry, and registers for the temporary storage of data.

ALU: Arithmetic logic unit, that part of the chip that performs all the arithmetic using binary techniques. Most ALUs perform boolean logic operations and have shift capabilities.

Input-Output (I/O) Circuits: These are "ports" to keyboards, printers, cathode-ray-tube or LED displays, sensors in industrial and automotive control systems, and other possible devices.

Register: A temporary storage unit. Program counters and instruction registers have dedicated uses. Accumulators have more general purposes.

RAM: Random-access memory, a volatile memory requiring electric power to retain data. Data can be stored for manipulation by other parts of the processor or computer.

ROM: Read-only memory, a nonvolatile memory in which data are stored permanently; no electric power is required to retain the data.

PROM: Programmable read-only memory.

EPROM: Erasable programmable read-only memory, in which data can be erased by ultraviolet light.

EAROM: Electrically alterable (erasable and reprogrammable) read-only memory.

EEPROM: Electrically erasable programmable read-only memory.

Bits and Bytes: A bit is a binary digit, a 0 or 1. A byte is an 8-bit "word" or unit. A kilobyte represents 1000 bytes or 8000 bits. Microprocessors that handle 8-bit words (bytes) are suitable for complex tasks such as educational, automotive, or home control systems. Devices that handle 16-bit words are designed for even more sophisticated systems, such as navigation or data processing.

7·10 Applications of microprocessors

It is impossible to give the reader a comprehensive idea of the many present and future uses of microprocessors. Rather, we can present only a few examples in this and the next two chapters. From a control standpoint these low-cost microcircuits are making switches, gears, levers, springs, and relays obsolete in many applications.

Automotive control is one area that readers will appreciate. The Delco-Remy Misar system includes four inputs: crankshaft position, coolant temperature, reference timing, and manifold vacuum. Basically, this system "looks up" the optimum spark advance in a memory map that corresponds to the driving conditions and passes this information to the distributor. Figure 7·33 shows this system in diagrammatic form. Another approach, shown in Fig. 7·34, puts both spark advance and exhaust gas recirculation under the control of a microprocessor.

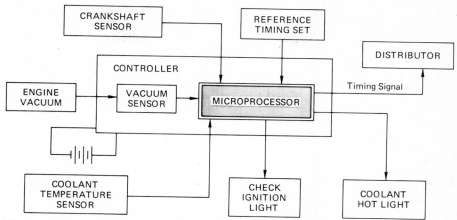

Figure 7·33 Spark-advance system using a microprocessor. Four inputs are funneled into the controller, which is preprogrammed with a "map" of spark timings appropriate to different engine conditions. *(Reprinted from "Electronics," Sept. 29, 1977; Copyright © 1977 by McGraw-Hill, Inc.)*

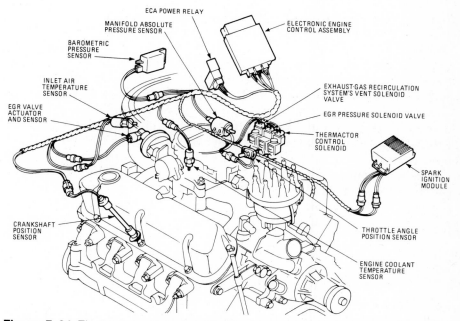

Figure 7·34 Electronic engine control by Ford puts both spark timing and exhaust-gas recirculation under the control of a 12-bit microprocessor. It uses seven sensors, and its output goes to two actuators, the spark-ignition module and the EGR valve actuator. *(Reprinted from "Electronics," September 29, 1977; Copyright © 1977 by McGraw-Hill, Inc.)*

Other applications are in the area of industrial controls. One such arrangement checks the dimensions of sheet-metal parts, reads a binary digital signal determining whether each part is within the allowable tolerance or not, and provides whatever statistical analysis is required. Another, incorporated in a cash register, drives a seven-segment digital readout, prints a receipt with the amount of the purchase on it, and makes change—all at practically the same instant.

Programming a microprocessor systen may require quite a bit of effort. In fact, because of the low cost of these circuits, the programming may cost more than the initial cost of the hardware. Program writing is often done in machine language or mnemonic (assembly) language. To use either, the user must develop an understanding of the language and how the processor works and skill in using the language and processor. Typical mnemonic instructions might be:

Instruction	Description
INXB	Add 1 to BC register pair
JNC	Jump if C flag is false
DCXB	Subtract 1 from BC register pair
OUT	Output data from accumulator
STA	Store A value at direct address

These are a few of more than a hundred such instructions for a particular microprocessor system. A routine for a programmable controller using this type of language is shown in Fig. 8·26. By spending more money for hardware or software, it is often possible for a user to have a system that utilizes a language that is easier to learn and work with, such as FORTRAN and BASIC. In BASIC, for example, 2 times 3 is simply written or typed 2*3. PRINT 2*3 produces the answer, 6, on a typewritten sheet or video screen.

Personal computers represent a logical development in the use of microprocessors. Because of their low cost, compactness, and versatility, they are well suited for use around the home. They have been used to control lighting and sound throughout a house, for video-type games, to plan menus for those on a diet, to turn on the coffee and furnace in the morning, to lock the front door at night, to keep track of investments, to keep telephone numbers and Christmas card lists up to date, and to water the yard when it reaches a certain aridity. Two models of these home computers appear in Fig. 7·35. The smaller computer incorporates a hexadecimal keyboard and nine-digit LED display. It is suitable for controlling lights, heat, appliances, etc., throughout a house. The larger computer system has the capability of supplying more than 100 accessory products, such as disk drives, high-speed printers, and audio cassette-based systems, and of performing scientific computations. Software for FORTRAN and BASIC languages is available.

(a)

(b)

Figure 7·35 Two models of personal computers. *(a)* 8048 single-board control computer. Inset shows the computer without the cover. *(b)* The personal computer system PCS 80/30 is a high-performance computing system. *(IMSAI Corporation.)*

SUMMARY

The development of integrated-circuit manufacturing has changed the face and substance of the electrical and electronics industry as well as the lives of millions of users of these products in the United States and throughout the world. The diffused silicon process—and later metal-oxide technology—made possible medium-scale integration (100 to 1000 transistor functions) and large-scale integration (more than 1000 functions).

Manufacture of some circuits, namely thin-film and thick-film, is more analogous to the printing process than to the diffusion process. They also provide more accurate values of such passive devices as resistors. Accuracy of manufacture requires extremely high precision in drawings for integrated circuits, with the result that many are generated by the computer. These drawings are usually executed many times the size of the circuit they help produce.

LSI technology has made possible such consumer products as digital watches, pocket calculators, microprocessors, and microcomputers. Microprocessors are usually employed in dedicated situations, whereas microcomputers can be programmed for different uses. Industrial applications will be presented in the next chapter.

QUESTIONS

7·1 What are the advantages of thin-film circuits over integrated silicon circuits?

7·2 How much thicker is a typical thick-film resistor than a thin-film resistor?

7·3 What drawings would be required for the manufacture of a thin-film circuit? What else would probably be included in the drawings?

7·4 What are four specific problems brought on by the problem of great numbers of components?

7·5 List four scales (other than full-size) to which drawings for miniature or microminiature circuits have been drawn.

7·6 Why is the automated drafting machine sometimes used for drawings of integrated circuits?

7·7 What are two functions of the photolithographic process?

7·8 What are some typical widths of thin-film resistors?

7·9 How do you determine the differences between SSI, MSI, and LSI?

7·10 What is your concept of a matrix?

7·11 What material or materials are generally used as substrate(s) for integrated circuits?

7·12 What is an effective masking agent for the above? Name an acceptor impurity. A donor impurity.

7·13 How is it possible to get a series of 10 or more circuits from one integrated semiconductor pattern that has capacitors, diodes, resistors, and transistors?

7·14 What are five drawings, of a series, required for production of a completed integrated- (semiconductor-) circuit package? What scales would be appropriate for these drawings?

7·15 What is your concept of hybridization insofar as this chapter is concerned?

7·16 What trends do you identify in electrical drawing for miniature circuits and devices in the next decade?

7·17 What are the essential components (sections, circuitry, etc.) of a microprocessor?

7·18 What is the difference between a RAM and a ROM? Between a PROM and EAROM?

7·19 Why does programming a microprocessor sometimes cost more than the initial cost of the processor itself?

7·20 What kinds of busses are likely to be on a single-board computer? Name at least one type of component that would be connected to each type of bus.

7·21 List six functions around a large office that might be controlled by a microcomputer.

PROBLEMS

7·1 MOS technology is somewhat different than the bipolar method for the manufacture of integrated circuits. The following steps are employed: (1) The initial slice is N-type oxidized silicon. (2) The first photoresist process cuts a window in the oxide. (3) The surface in the window is reoxidized (thin layer). (4) Windows for the source and drain are cut by the second photoresist process. (5) Boron is diffused "in" to form the source and drain. (6) The oxide in the main window is stripped off. (7) Pure oxide is formed for the gate region. (8) Windows for the source and drain are cut by the third photoresist process. (9) Aluminum contacts are deposited and then defined by the fourth photoresist. Make a flow diagram of a MOSFET fabrication process similar to that shown in Fig. 7·4.

7·2 The integrated-circuit bar diagram shown in Fig. 7·36 can be connected to form a common-emitter dc amplifier, shown in Fig. 6·4b. Draw the mask that would be required for metallization that would produce the amplifier circuit. Your instructor may require you to draw the complete IC also. Suggested scale: 1 in. = 10 units (mils), shown around the edges. Let the conductor paths be 1.5 units wide. The capacitors have the same value. Resistor R_1 is U-shaped, R_2 is S-shaped (bottom left), and R_3 is L-shaped (bottom right). R_3, R_4, and R_5 are I-shaped. (Use $8\frac{1}{2} \times 11$ paper or Mylar.)

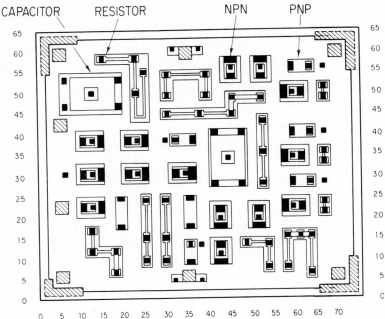

Figure 7·36 (Prob. 7·2) Bar diagram of a bipolar small-scale integrated circuit. Typical devices have been labeled. Units around edge are mils (0.025 mm).

7·3 Figure 7·37 shows the schematic diagram of a thin-film resistor and a scale drawing that is complete except for the resistors. It is desired to fabricate resistors 0.02 in. (0.51 mm) wide. Compute the length of the resistors using the equation $L = [(R - 120)/40] W$.

Make a table for the resistors, including value in ohms, width, and length for each. Make a scale drawing of the complete thin-film layout using these widths and lengths. A suggested scale is 1 in. = 0.20 in. Including the 0.2 grid system as shown should enable you to make a reasonably accurate scale drawing. Use $8\frac{1}{2} \times 11$ paper for the layout only, 11×17 paper for layout, table, and schematic.

7·4 Figure 7·38 shows the schematic and scale layouts of a circuit that is to be manufactured by thin-film technology. Using the table of squares and/or the equation $L = [(M - 120)/40] W$, calculate the length of R_1. Then make a 5:1 scale layout of the complete circuit. (Drawing the schematic diagram and including the table is optional.) Your instructor will specify what is to be included in the drawing. Size: $8\frac{1}{2} \times 11$ for layout only, 11×17 if schematic and tables are included.

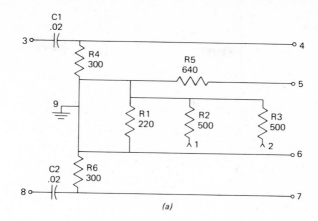

(a)

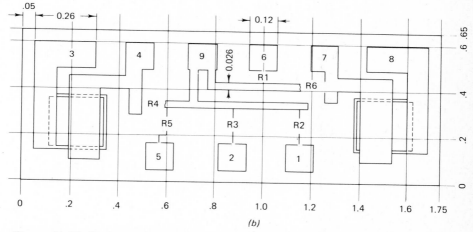

(b)

Figure 7·37 *(Prob. 7·3)* A thin-film circuit drawing. *(a)* Complete elementary diagram. *(b)* Scale drawing, incomplete. Drawing scale: 1″ = 0.200 units (5:1).

7·5 Figure 7·28 shows the block diagram of the minimum TMS 9900 system. It is desired to augment this system with an interrupt interface (near the 8-bit input and output), a DMA (direct-memory-access) interface (using the same address and data busses that are used for ROM and RAM), and a control bus which connects ROM, RAM, and DMA to DBIN and $\overline{\text{WE}}$ on the CPU. Draw the complete diagram, and label all blocks and busses.

7·6 Figure 7·32 shows the flow diagram of the SBC-80/10 single-board computer. It is desirable to add two 8900 programmable timers and a 8185 RAM/I/O Timer to the 80/10 system. The 8185 will be connected to RAM. The programmable timers will be connected to the system bus

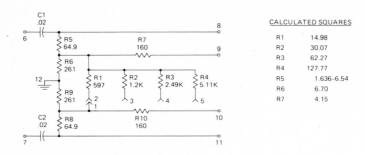

CALCULATED SQUARES

R1	14.98
R2	30.07
R3	62.27
R4	127.77
R5	1.636-6.54
R6	6.70
R7	4.15

CAPACITORS

	LENGTH	WIDTH	VAL.	EFF. AREA
C1	0.246	0.130	.02 µF	0.032 □″
C2	0.246	0.130	.02 µF	0.032 □″

$$\text{LENGTH} = \left(\frac{M-120}{40}\right)(\text{WIDTH})$$

RESISTORS

	LENGTH	WIDTH	OHMS(M)	M-120
R1		0.010	597	477
R2	0.270		1200	1080
R3	0.592		2490	2370
R4	1.247		5110	4990
R5a	0.0349		259.6	139.6
R5b	.0349		259.6	139.6
R5c	.0349		259.6	139.6
R5d	.0349		259.6	139.6
R6	.035		261	141
R7	.010		160	40
R9	.035		261	141
R8a	.0349		259.6	139.6
R8b	.0349		259.6	139.6
R8c	.0349		259.6	139.6
R8d	.0349		259.6	139.6
R10	.010		160	40

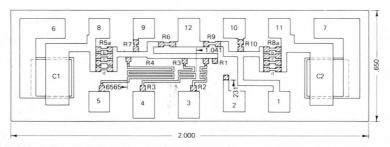

Figure 7·38 (Prob. 7·4) A schematic diagram, scale layout, and calculation tables for a thin-film circuit. Suggested scale for the layout: 1″ = 0.20 units, or 5:1.

and will have a data link to the parallel I/O lines. Draw the complete functional flow diagram of this augmented system. Use 11 × 17 paper.

7·7 Figure 7·39 shows a rudimentary drawing of the Ford dual-displacement engine. Sensors measure temperature, transmission gear, throttle position, and engine load and speed. At a certain condition the control module shuts down three of the cylinders. Study Fig. 7·34. Then draw the outline-

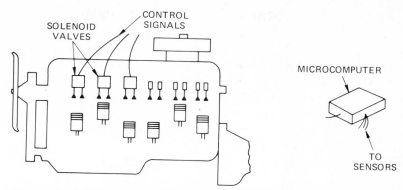

Figure 7·39 (Prob. 7·7) Rudimentary outline of six-cylinder dual-displacement motor and microprocessor.

schematic drawing of this system showing the engine, sensors, microcomputer, control signals, and connecting lines. (Use $8\frac{1}{2} \times 11$ paper.)

7·8 Figure 7·40 shows a gaging system which is used to determine whether or not a part is within acceptable tolerance. Item 1 is a junction box for a line leading from each probe. Item 2 is a manifold which also has a line from each probe. The manifold and junction box each feed into an 80/20 microcomputer board. The board has outputs to video display (3) and printer (4). Show a schematic flow diagram of this complete system. (Use $8\frac{1}{2} \times 11$ paper.)

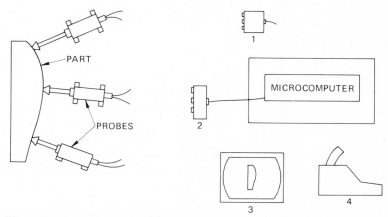

Figure 7·40 (Prob. 7·8) A manufactured-part gaging system in which a single-board microcomputer is employed.

7·9 Figure 7·41 includes the incomplete flow diagram of a cash-register system. Make a flow diagram of the completed diagram in which item 1 is the 8085 CPU, 2 is the 8155 RAM/I/O/Timer, 3 a receipt/ticket printer, 4 an 8355 RAM, and 5 a change dispenser. (Use $8\frac{1}{2} \times 11$ paper.)

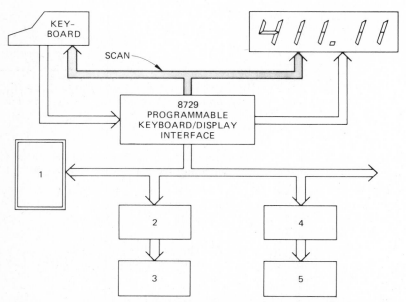

Figure 7·41 (Prob. 7·9) Incomplete diagram of a microprocessor-based cash-register system. Shaded busline is for internal scanning. Unshaded busses are for control and data transmission.

7·10 The flow diagram of a typical microcomputer is presented in Fig 7·42. Draw the flow diagram with the following additions: At (1) add the Keyboard Display block with flows to Keyboard Switches & Sensors and the Numeric/Alpha Display blocks as shown in the upper part of the figure. At (2) add the Standard Interface block with flows to PROMS, etc., as shown in the upper part of the drawing. Add light shading to the CPU and data bus. (Size: 11 × 17.)

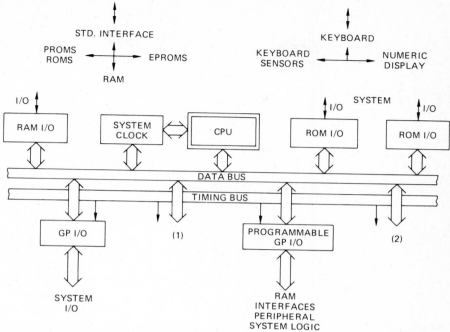

Figure 7·42 (Prob. 7·10) Block diagram of a typical microcomputer. Additional blocks are to be added at (1) and (2).

Chapter 8
Industrial controls

8·1 Introduction

In recent years, there has been a steady evolution toward the use of solid-state components in industrial control. This does not mean that all electromechanical devices will be replaced, but they will share a smaller part of the industrial control spectrum in the future.

This chapter covers the drawings encountered in many types of industrial control situations, from basic motor control to computer control. Almost all industrial control drawings follow some drawing standard or recommended practice, and these standards will also be described.

In general, as mentioned previously in the text, there are several types of drawings that will be associated with all types of control. They are: (1) the elementary diagram or schematic, (2) the wiring diagram, (3) the layout, (4) the block plan, (5) the parts list, and (6) the assembly drawings. However, some types of industrial control have associated specialized drawings, which will also be covered. An interesting phenomenon takes place in industrial control drawings; namely, as the control circuits become more sophisticated and complex, the drawings tend to become simpler. This will become obvious as the chapter proceeds.

A word of caution should be inserted here. Although this chapter presents industrial control drawings in discrete categories (e.g., electromechanical, solid-state logic, etc.), in reality, these types of controls are sometimes used together. This can, in some instances, present nightmares to the drafter putting them down on paper. Basic concepts will show how to ease this interfacing chore.

8·2 Basic motor control

Most designers and drafters doing industrial control work start by "cutting their teeth" on some type of motor control circuit. Before delving into the types of drawings used in motor control circuit design, it is a good idea to first understand the functions of motor control and some of the components associated with motor controllers. There are certain definite functions that are performed or governed by motor control. These are: (1) starting, (2) stopping, (3) running, (4) speed regulation, and (5) protection. Starting technically refers to rotating the motor from zero speed to maximum (breakdown) torque speed.

Running refers to propelling the motor from maximum torque speed to load speed (usually faster). These two functions are accomplished by one device in most motor control circuits. Speed regulation is done by some device such as a rheostat in a motor control circuit. Motor control circuits provide protection for both the power source and the motor. An example of one of the motor control functions is "reduced voltage starting," which reduces the current inrush during the starting period of a motor or machine. This eliminates or minimizes the shock of a quick start on the driven machine, the reduction of voltage in the line or system to which the machine is connected, and the dropping out of synchronism of synchronous motors on the line. Reduced-voltage starters, which provide smooth accelerated starting without a serious drop in line voltage, are available.

A large motor in a New England plant requires 56 s starting time under normal load. An oil-well pump in Texas will suffer serious damage if its rotor locks and the motor is not tripped from the line in 20 s. A conveyor drive motor in a Florida potash plant can withstand 25 percent overload for 30 min, but a compressor motor in Missouri may burn up in 3 min at 25 percent overload. Each of these motors must be protected from abnormal overload currents, which creep in from time to time.

An electric motor controller is a device, or group of devices, which controls the electric power delivered to the motor. Motor controls range in complexity from a simple manual motor starter to an elaborate motor control center. Some basic control devices which may be built into motor controllers, as well as other industrial controls, are described in the following paragraphs.

A circuit breaker has the function of interrupting the flow of power in a circuit under normal and abnormal conditions. Although it might be used as a power (disconnect) switch, its primary function is to protect the line against abnormally high currents. Hence, it is designed for infrequent operation only. Just as frequently, a line switch and a set of fuses are used in place of a circuit breaker. The switch is used as the disconnect, while the fuses protect the line against faults.

A contactor is a device for repeatedly establishing and interrupting an electric power circuit. In essence, it is a specialized type of relay. Usually operated magnetically, it can be operated in line or can be remotely governed by pilot devices or relays. The overload relay is used to interrupt maximum overload current or to remove the power supply from a starter and motor under a normal overload. Pushbutton switches are used to energize or deenergize the motor controller, and thus start and stop the motor, respectively. Transformers are used to reduce the line (power-source) voltage to the control-circuit voltage. Sometimes, transformers are not used, but most designers prefer to have their control voltages at 120 V ac or less for safety reasons. Many motor controllers have indicating, or pilot, lights on them to show whether the motor is running or stopped.

The types of drawing encountered when doing motor control work are the elementary diagram, the wiring diagram, the layout, the motor-control-center layout, and the motor-control-center schedule. Figure 8·1 shows two types of motor controllers. Figure 8·2 shows the elementary diagram of a motor controller. This diagram may appear a little strange to the reader because of the unusual symbols. Unusual symbols can appear because of two factors: (1) different devices are often used in industrial controls than in computer and communication equipment, and (2) different, or alternative, symbols are often used for components that are commonly used. A number of alternative symbols are used in diagrams of this type. These include the contacts and heaters. The unit shown in this diagram is a combination motor starter. Combination means that it contains both the motor starter and a power-source disconnect (switch). Usually in industrial control, combination motor starters are used.

The motor control circuit in the elementary diagram is divided into two distinct circuits, the load circuit and the control circuit. The load circuit, which provides the utilization power to the motor, is drawn horizontally from left to right. The control circuit, which contains the operating coil of the contactor and provides the motor control, is drawn so that devices are arranged horizontally

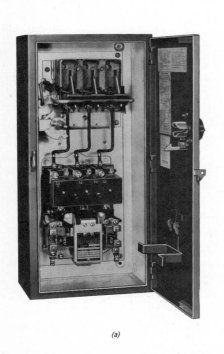

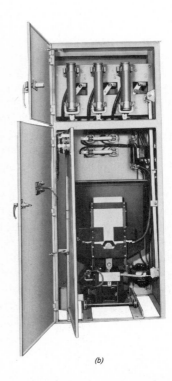

(a)   (b)

Figure 8·1 Two types of motor controllers. *(a)* Switch-type combination starter. *(b)* Synchronous motor starter. *(Allis-Chalmers Mfg. Co.)*

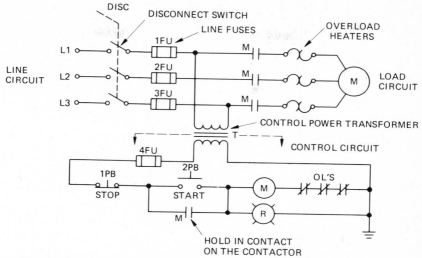

Figure 8·2 Elementary diagram of a pushbutton motor control circuit.

in vertical rows like a ladder. This drawing shows only the first "rung" of the ladder. This motor control circuit operates in the following manner. The disconnect switch (DISC) is closed and usually remains closed all the time unless it is necessary to disconnect the motor from the line circuit. With the disconnect switch closed, voltage is applied via the control power transformer to the control circuit. The start pushbutton (2PB) is depressed and energizes the contractor coil, and its power contacts apply lone voltage to the motor, which it starts and runs. In parallel with the start pushbutton is a contact, which is also on the motor starter, that keeps the contactor coil energized when the pushbutton is released—in other words, it "holds on" the contactor. The three overload relays (OLs, also called heaters) in the load circuit sense the amount of current going to the motor. The overload relays are temperature-sensitive devices. If the load current is too large or the circuit is overloaded, the temperature of sensitive coils becomes too hot, causing the temperature contacts in the control circuit to open up, and shutting down the motor. Because overload relays are temperature-actuated devices, care must be exercised in their selection, with attention to the temperature characteristics of both the relay and the motor.

There are many ways to control motor-starter contactors or relays. Figure 8·3 shows seven typical methods of industrial control. If these methods were used to control the motor in Figure 8·2, the devices—knife switch and pressure switch, for example—would appear in the control circuit. Circuits *(a)* through *(e)* are two-wire control. Circuits *(f)* and *(g)* are three-wire control, using a contact from some other device such as a relay for automatic control.

The wiring diagram in Fig. 8·4 shows exactly how the motor starter in Fig. 8·2 is wired. This type of diagram helps the electrician install and maintain

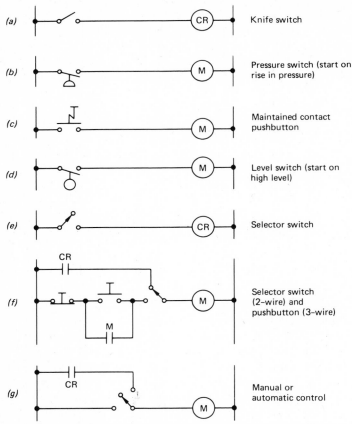

Figure 8·3 Typical methods of industrial control. *(f)* and *(g)* Three-wire control. The others are two-wire control.

the motor control circuitry. Most electrical-control manufacturers have booklets available containing many of the motor control wiring diagrams. These booklets are of invaluable aid to the designer/drafter in designing and laying out the motor control circuits.

Many times an engineering job will require that many motors be controlled from a centralized location. In order to handle this requirement, there is a convenient, economical, versatile, and designable package that allows an engineer to put a multitude of different types of motor controllers in one enclosure. The motor control center is a compact, floor-mounted assembly, comprised principally of combination starters. It consists of one or more vertical sections, with each section having a number of spaces for motor starters. The number of spaces is determined by the horsepower ratings of the individual starters. For example, a starter that will control a 10-hp motor will take up less room

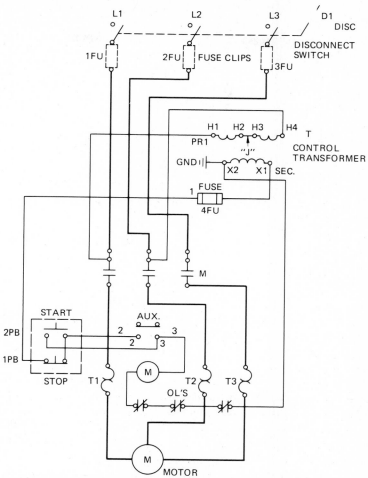

Figure 8·4 Wiring diagram of a pushbutton motor control circuit.

than a starter that will control a 100-hp motor. The motor control centers are built so that they will have starters on one side (front-of-board construction) or both sides (back-to-back construction). Basically, motor control centers have the main power busbar running horizontally and continuously through all sections and individual section busses running vertically. The motor starters plug into the section vertical busses. The drawings encountered in designing and specifying motor control centers, besides the elementary and wiring diagrams just discussed, are the motor control center layout, schedule, and single-line diagram.

Figure 8·5 shows a typical motor control center (front-of-board construction) layout. This drawing is done to scale. Each section is numbered (1, 2,

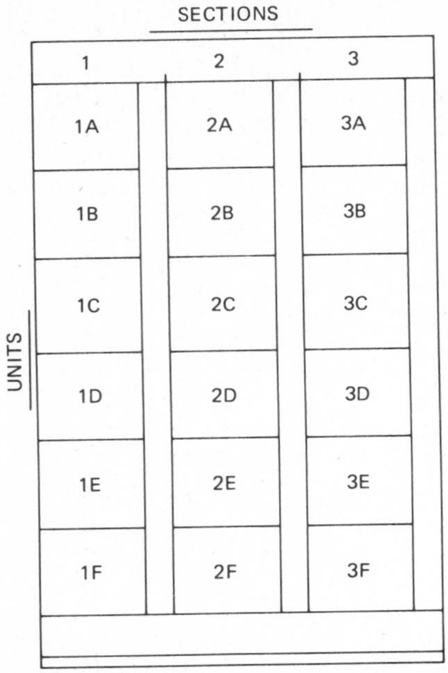

SECTIONS

1	2	3
1A	2A	3A
1B	2B	3B
1C	2C	3C
1D	2D	3D
1E	2E	3E
1F	2F	3F

UNITS

Figure 8·5 Motor control center layout having three sections.

etc.), and each space is lettered (*A, B,* etc.). If the motor control center were of back-to-back construction, two layouts would be required and an additional letter would appear in each section to designate front *(F)* or rear *(R).* The height of the space is determined by the size of the motor starter; the vertical section widths, heights, and depths are determined by the manufacturer, but are in accordance with National Electrical Manufacturers Association (NEMA) Standards. Figure 8·6 shows the motor control center depicted in the layout of Fig. 8·5. Associated with the motor control layout is the motor control center schedule shown in Fig. 8·7. This schedule is developed by the designer to show the "nuts and bolts" of what type of motor controllers go into the motor control center. The motor control center schedule usually appears on the same drawing as the layout. Along the left margin is listed the unit or space number, and adjacent to this number are all the particulars associated with the motor controller going in that space. For example, the motor controller going in space 1*B* is a combination motor starter (CMS), full-voltage (FV), nonreversing (NR), with 3 poles, a 30-A switch, 1.4-A fuses, 1 normally open contact-on contactor, a red "ON" pilot light, and a three-position selector switch for three-wire control (probably hand and thermostat). The controller is for a motorized steam unit heater.

The single-line diagram is a drawing which represents three-phase controls as a single line. Its use and format will be discussed at length in Chap. 9.

Figure 8·6 Photo of a motor control center.

BUS	UNIT NUMBER	DESCRIPTION	SWITCH RATING (AMPS)	FUSE	NEMA SIZE	MOTOR H.P.	AUX. CONTACTS N.O.	AUX. CONTACTS N.C.	START-STOP P.B.	PILOT LT. (COLOR)	SEL. SWITCH 2 POS.	SEL. SWITCH 3 POS.	NAMEPLATE	REMARKS	WIRING DIAGRAM NUMBER	DRAWING NUMBER
													MOTOR CONTROL CENTER No. 40-A			
	¹A	INCOMING LINE	–	–	–	–	–	–	–	–	–	–	INCOMING LINE		–	
	1B	CMS FV NR	3P-30A	1.4	1	3/4	1	–	–	R	*	–	UNIT HEATER NO. 1		WD-2	
	1C	CMS FV NR	3P-30A	1.4	1	3/4	1	–	–	R	*	–	UNIT HEATER NO. 2		WD-2	
	1D	CMS FV NR	3P-30A	–	1	–	2	–	–	R	*	–	SPARE		–	
	1E	FUSED SWITCH	3P-60A	60	–	–	–	–	–	–	–	–	WELDING RECEPT.		WD-4	
	1F	BLANK	–	–	–	–	–	–	–	–	–	–	–		–	
	2A	CMS FV NR	3P-30A	8	1	5	2	–	–	R	*	–	ROOF EXH. FAN NO. 1		WD-8	
	2B	CMS FV NR	3P-30A	8	1	5	2	–	–	R	*	–	ROOF EXH. FAN NO. 2		WD-8	
3φ 3W 600A MAINS	2C	CMS FV NR	3P-30A	8	1	5	2	–	–	R	*	–	ROOF EXH. FAN NO. 3		WD-8	
	2D	CMS FV NR	3P-30A	5	1	3	2	–	–	R	*	–	MACH. NO. 5 OIL PUMP		WD-6,-7	
	2E	CMS FV NR	3P-30A	5	1	3	2	–	–	R	*	–	MACH. NO. 6 OIL PUMP		WD-7	
	2F	BLANK	–	–	–	–	–	–	–	–	–	–	–		WD-7	
	3A	CMS FV NR	3P-60A	15	1	10	2	–	–	R	–	–	REFRIG. PUMP NO. 1		WD-9	
	3B	CMS FV NR	3P-60A	15	1	10	2	–	–	R	–	–	REFRIG. PUMP NO. 2		WD-9	
	3C	CMS FV NR	3P-30A	8	1	5	2	–	–	R	*	–	PUMP ON. UNIT		WD-5	
	3D	CMS FV NR	3P-60A	15	1	10	2	–	–	R	–	–	REFRIG. PUMP NO. 3		WD-9	
	3E	CMS FV NR	3P-60A	15	1	10	2	–	–	R	–	–	REFRIG. PUMP NO. 4		WD-9	
	3F	BLANK	–	–	–	–	–	–	–	–	–	–	–		–	

Figure 8·7 Motor control center schedule.

8·3 Electromechanical controls

Basically, an electromechanical control is a device that is electrically operated and has mechanical motion, such as a relay, servo, solenoid, etc. The motor starter contactors discussed in the previous section are electromechanical devices. Today, the majority of the devices used in industrial control are electromechanical, although, as mentioned, a larger percentage of the devices is becoming solid-state. Builders and users of industrial equipment have long realized that the safety of the operator and the continuance of production are of primary importance in the design and purchase of such equipment. Good drawings help promulgate such equipment.

In electromechanical design, the standards for the drawings that are most widely accepted and most frequently encountered are the Joint Industrial Council (JIC) "Electrical Standards for Mass Production Equipment" (EMP-1–67) and "General Purpose Machine Tools" (EGP-1–67). Although these standards are strictly advisory and their use is entirely voluntary, they do provide an excellent and logical basis for industrial control drawings. These standards will be the basis for doing electromechanical industrial control drawings in this text. There are some slight differences in these standards, and they should be consulted before using them.

To begin with, the JIC standards describe the drawing's size, the device designations and symbols to be used, the drawing arrangements, and the drawings required. The drawing's size should be a multiple of $8\frac{1}{2} \times 11$ or 9×12 in. with a maximum size of 24×36 in. Multiple drawings must be cross-referenced.

Device designations are shown in Table 8·1. Device designations cannot be used for other purposes. If special observations must be used, they should be listed on the diagrams. The device symbols are found in Appendix C. If no symbol is listed, then the prevailing symbol used in ANSI 32.E should be used. Logic symbols should be in accordance with NEMA Standard IC-1.

The drawings for a complete industrial control job should include an elementary (schematic) diagram, a block diagram, a logic diagram, a sequence of operations, a panel layout, an interconnection diagram, a stock list (bill of materials), an electrical layout, and a foundation drawing. Each one of these will be shown for the same piece of equipment in order to provide the reader with a sense of continuity and deeper understanding.

The elementary (schematic) diagram, shown in Fig. 8·8, is the key to a set of industrial control drawings. The elementary diagram symbolically represents the control circuit and graphically shows the observer how the control system works. The elementary diagram is similar to the elementary diagram of the motor starter, with the source of power starting on the top left, is drawn horizontally, and connects the source circuit device [a disconnect switch, the contactor contact, and the overload coils (heaters)] and the loads (motors) at the top right. The control part of the diagram is drawn horizontally between two vertical lines, which represent the control power, from the source of the

Table 8·1 Device Designations

A	Accelerating contactor or relay	MTR	Motor
AM	Ammeter	MN	Manual
AU	Automatic	OL	Overload relay
BR	Brake relay	PB	Pushbutton
CAP	Capacitor	PL	Plug
CB	Circuit breaker	PLS	Plugging switch
CR	Control relay	PS	Pressure switch
CRA	Control relay automatic	R	Reverse
CRE	Control relay electronic energized	REC	Rectifier
CRH	Control relay manual (hand)	RECP	Receptacle
CRM	Control relay master	RES or R	Resistor
CT	Current transformer	RH	Rheostat
CV	Counter voltage	S	Switch
D	Down	SOC	Socket
DB	Dynamic braking contactor or relay	SOL	Solenoid
DISC	Disconnect switch	SCR	Series control relay
DS	Drum switch	SS	Selector switch
ET	Electron tube	SSW	Safety switch
F	Forward	T	Transformer
FLS	Flow switch	TB	Terminal board
FS	Float switch	THS	Thermostat switch
FTS	Foot switch	TR	Time delay relay
FU	Fuse	TVM	Tachometer voltmeter
GRD	Ground	Q	Transistor
IOL	Instantaneous overload	U	Up
LS	Limit switch	UCL	Unclamp
LT	Lamp	UV	Undervoltage
M	Motor starter	VM	Voltmeter
MB	Magnetic brake	VS	Vacuum switch
MC	Magnetic clutch	WM	Wattmeter
MCS	Motor circuit switch	X	Reactor

Relays:	General use	Examples:	CR 1CR 2CR
	Master		CRM
	Automatic		CRA
	Electronically energized		CRE 1 CRE 2CRE
	Manual (hand)		CRH
	Latch		CRL 1CRL 2CRL
	Unlatch		CRU 1CRU 2CRU
	Timers		TR 1TR 2TR
	Overload relay		OL 1OL 2OL
	Motor starters		1M 2M etc.

Source: JIC "Electrical Standards for Industrial Equipment."

control power to the last control device required in the circuit. As can be seen, this method of drawing control circuits between two vertical control-power lines produces a diagram similar to a ladder; thus this is often called a ladder diagram. The control sequence starts on the top rung of the ladder and ends on the bottom rung, always going left to right.

Under selected control devices and on the right-hand side of the ladder diagram, there are terms describing the command and status functions, e.g., master stop. Associated with the right-hand-side control device and found under

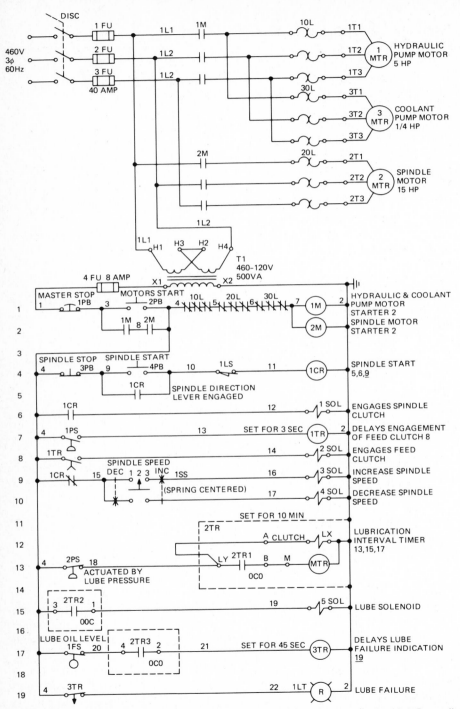

Figure 8·8 Elementary or ladder diagram. [*Courtesy of Joint Industrial Council (JIC).*]

262

the descriptive terms and on the left-hand side of the ladder diagram are numbers called line numbers. They increase consecutively from top to bottom. Line numbers are used for reference: control devices are referenced to the circuits they control by these line numbers, in order to make it easier to understand and follow the circuitry. The numbers between the devices are conductor (wire) numbers, and these numbers should *not* be repeated. Experience has shown the authors that minimum spacing should be $\frac{3}{8}$ in. between lines and $\frac{1}{2}$ in. between components and that minimum component size (circle diameter) should be $\frac{1}{4}$ in., for clarity and ease of reading.

In order to draw an elementary diagram, a number of rules must be observed. Some of these rules are as follows. Actuating coils of control devices, i.e., relays, solenoids, etc., are to be shown at the right-hand side. All contacts should be shown between these coils—not necessarily on the same line—and the left vertical line. If subassemblies are used and their internal wiring diagrams are shown on separate drawings, the subassemblies should be represented on the elementary diagram as rectangles, with actuating coils or contacts shown. The control-device symbols are shown and numbered in the order in which the control sequence takes place. Services *must* be shown in the deenergized position, with the utilities (e.g., water to a flow switch) turned off and the equipment at its normal starting position. Only device contacts that are to be used should be shown. Contacts of multiple-contact devices should be on the line of the elementary diagram, where they are connected in a circuit, and all except contactors and control relays should be connected with dotted lines. Continuity of device descriptions should be maintained through all drawings required. Device functions should be shown adjacent to the respective device symbol.

Descriptive terms for command and status terms should be written in present or past tense. The values of electronic components, i.e., resistors, capacitors, etc., should be listed. If electronic diagrams are used, pertinent information for maintenance troubleshooting should be listed. Although quite a number of rules for elementary diagrams are presented here, do not become anxious or overconcerned. These rules are really just good common sense and will become "old hat" after you have done several elementary-diagram drawings.

Sometimes block diagrams are used to supplement elementary diagrams, when the complexity of the control systems warrants. On a block diagram, each block should be identified and cross-referenced in such a manner that circuitry can be easily located on the elementary diagram.

Logic diagrams are used in conjunction with the elementary diagram when static control or logic modules are provided. These diagrams will be discussed in detail in the next section of this chapter.

Associated with, and on the same drawing as, the elementary diagram is a verbal description, explaining how the circuit works. This description, called the *sequence of operations,* is shown in Fig. 8·9. Typing of the sequence of operations is preferred. Basically, the sequence of operations indicates the progression of operation of the devices shown on the elementary diagram and, thus, tells how the control circuit works. Graphical representations, such as bar charts, may be used to supplement the written description.

SEQUENCE OF OPERATION

A. MACHINE OPERATION: PRESS "MOTORS START" PUSHBUTTON "2PB". MOTORS START.

B. SELECT SPINDLE SPEED BY TURNING SELECTOR SWITCH "1SS" TO "INC", ENERGIZING "3 SOL", TO INCREASE OR TO "DEC", ENERGIZING "4 SOL", TO DECREASE SETTING.

C. WITH CORRECT SPINDLE DIRECTION SELECTED LIMIT SWITCH "1LS" IS ACTUATED. PRESS "SPINDLE START" PUSHBUTTON "4PB" ENERGIZING RELAY "1CR" WHICH ENERGIZES "1 SOL". SPINDLE STARTS AND PRESSURE SWITCH "1PS" IS ACTUATED. "1PS" ENERGIZES "1TR" AND AFTER A TIME DELAY "2 SOL" IS ENERGIZED PERMITTING MOVEMENT OF MACHINE ELEMENTS AT SELECTED FEED RATES.

D. PRESSING "SPINDLE STOP" PUSHBUTTON "3PB" STOPS SPINDLE AND FEEDS MOVEMENTS SIMULTANEOUSLY.

E. LUBRICATION OPERATION:

F. PRESSURE SWITCH "2PS" IS CLOSED.

    1. TIMER "2TR" CLUTCH IS ENERGIZED WHEN MOTORS START.

    2. CONTACT "2TR-1" CLOSES AND ENERGIZES TIMER MOTOR "MTR" STARTING LUBE TIMING PERIOD.

    3. CONTACT "2TR-3" CLOSES AND ENERGIZES TIMER "3TR".

G. TIMER "2TR" TIMES OUT.

    1. CONTACT "2TR-1" OPENS, DE-ENERGIZING TIMER MOTOR "MTR".

    2. CONTACT "2TR-2" CLOSES, ENERGIZING "5 SOL".

    3. CONTACT "2TR-3" DE-ENERGIZING TIMER "3TR".

    4. LUBRICATION PRESSURE ACTUATES PRESSURE SWITCH "2PS", DE-ENERGIZING AND RESETTING TIMER "2TR". CONTACTS "2TR-1", "2TR-2" AND "2TR-3" OPEN.

    5. CONTACT "2TR-2" OPENING, DE-ENERGIZES "5 SOL".

H. REDUCED LUBRICATION PRESSURE DE-ACTUATES PRESSURE SWITCH "2PS" AND SEQUENCE REPEATS.

SWITCH OPERATION

1LS (4) ACTUATED BY SPINDLE DIRECTION LEVER ENGAGED
1PS (11) OPERATED WHEN SPINDLE CLUTCH ENGAGED
2PS (13) OPERATED BY NORMAL LUBE PRESSURE
1FS (16) OPERATED BY ADEQUATE LUBE SUPPLY

Figure 8·9 Sequence of operations. [*Courtesy of Joint Industrial Council (JIC).*]

The panel layout and interconnection diagram are shown in Fig. 8·10. These diagrams are of the same control system shown in the elementary diagram of Fig. 8·8. The panel layout shows the general physical arrangement of all devices inside the control-panel enclosure. The panel layout is used to tell the tradesperson how to lay out and construct the control system. Devices are represented by squares, rectangles, and, sometimes, circles, and bear the same identification they had in the elementary diagram. Spare panel space is dimensioned. The panel layout also shows the layouts of the remote console or operator stations. These stations contain the switches that the operator uses to work the control system and are often called pushbutton stations.

The interconnection diagram, also shown in Fig. 8·11, shows all terminal boards and their associated identification within the electrical control system. Terminal-board numbers must correspond exactly to the number of the wire which attaches to the board.

The authors have seen several variations of the panel layouts that have been used successfully. One is to lay out each component on the panel, showing

PANELS AND CONTROL STATION LAYOUT

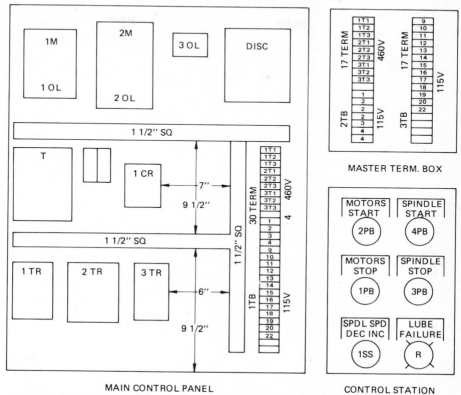

MASTER TERM. BOX

MAIN CONTROL PANEL

CONTROL STATION

Figure 8·10 Panel layout and interconnection diagram. [*Courtesy of Joint Industrial Council (JIC).*]

the coils, contacts, etc., of each device and the wire number associated with it. Another method is to show the interconnect wiring between the main control panel and the various operator stations. This method uses parts of a highway diagram and labels the wires' numbers at the individual stations.

The stock list, or bill of materials, shown in Fig. 8·11, tells what devices to use for the construction and maintenance of the control system. The stock list shows each device by designation, quantity, manufacturer's name and model (and, sometimes, serial) number, and any other information necessary for construction and maintenance replacement. The stock list should appear on the same drawing as the elementary diagram or panel layout, but it may appear on a separate drawing because of its size. If it appears on the elementary diagram or panel layout, entries should be made from bottom to top to avoid running out of room. If it appears on a separate sheet, entries should be made from top to bottom, left to right.

SYM.	QUAN.	STOCK NO.	DESCRIPTION
2TR	1		TIMER, 15 MIN, EAGLE NO. HP518
1FS	1		FLOAT SWITCH, ALLEN BRADLEY NO. 840–1A2
1TR	1		TIMER, 60 SEC, TDOE, GE NO. CR122 A04022AA
1CR	1		RELAY, 4PST, 2NO–2NC, A-B NO. 700–N400
1–5 SOL.	5		SOLENOID VALVE, 125 PSI, ASCO NO. 8210D1HW
1LS	1		LIMIT SWITCH, SPDT, MICROSWITCH NO. LSA1A–1A
1PS	1		PRESSURE SWITCH, 0–30 PSI, A-B NO. 836T–T251J
1PB	1		OPERATOR SQ–D NO. KR–5R
3PB	1		OPERATOR SQ–D NO. KR–1R
1,3PB	2		CONTACT BLOCK SQ–D NO. KA–3
4FU	1		FUSE, 5A, 130V, BUSS NO. KAA–5
2MTR	1		MOTOR, 15HP, 460V, 3600 RPM, GE NO. 5K284BN322
1MTR	1		MOTOR, 5HP, 460V, 1200 RPM, GE NO. 5K215BL305
1M,2M	2		STARTER, ALLEN-BRADLEY, SIZE 1, NO. 709–B00103
T	1		TRANSFORMER, 750VA, SORGEL NO. 750SV1B
1–3FU	3		FUSE, 40A, 600V, BUSS NO. FRS–40
DISC	1		SAFETY SWITCH, 3P, 60A, 600V, SQ–D NO. ARD–13
	1		PANEL, HOFFMAN NO. A30P24
	1		ENCLOSURE, HOFFMAN NO. A30S2508LP

Figure 8·11 Stock list or bill of materials.

The electrical layout, in Fig. 8·12, shows the outline of the equipment that the control system is to control. It shows the location of the main control panel, the operator's console, and other remote-control devices, such as limit switches, whose location cannot be readily determined from the elementary diagram. Also shown are accessory units, such as hydraulic power units, which are not attached to the equipment, and their relative locations. All devices shown on the electrical layout have the same identification as they have on the elementary diagram.

8·4 Solid-state logic control

All control design is done in logical process, whether the thought process is deliberate or subconscious. Electromechanical devices, such as magnetic relays, have been used for many years to provide decision or logic circuits for the electrical industry. The continued wide use of electromechanical devices can be expected because of their proven reliability, flexibility, low cost, and large current-carrying abilities. Solid-state control, on the other hand, became a reality after the invention of the transistor. Solid-state control was not intended to replace or make obsolete conventional relays, but rather to meet the increasing demands for reliability, speed, and life, required to control the increasingly complex industrial processes.

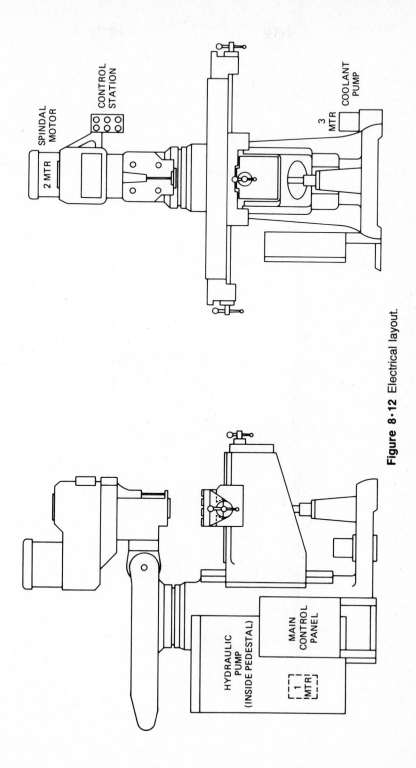

Figure 8-12 Electrical layout.

SPINDAL MOTOR

CONTROL STATION

2 MTR

COOLANT PUMP

3 MTR

HYDRAULIC PUMP (INSIDE PEDESTAL)

MAIN CONTROL PANEL

1 MTR

Solid-state logic used for industrial control is generally high-threshold logic (HTL), having the characteristics of being highly noise immune, easily compatible with discrete solid-state and electromechanical devices, able to withstand higher temperatures than many solid-state devices, and easily interfaceable with industrial voltages. Solid-state logic circuitry for industrial control generally uses discrete transistors or integrated circuits with several logic circuits mounted on one printed circuit board (called logic modules or cards).

Figure 8·13 shows a logic module. Observe the integrated circuits and printed circuitry. Logic modules are sometimes encapsulated, although this one is not. This logic module contains four OR logic functions. The NEMA symbol of an OR function is used. The logic diagram of this logic module is shown in Fig. 8·14. In some instances, logic functions may be represented by more than one symbol. In fact, some logic functions are so complicated that the

Figure 8·13 Photo of a logic module containing four OR logic functions. *(Courtesy of Allen-Bradley Co.)*

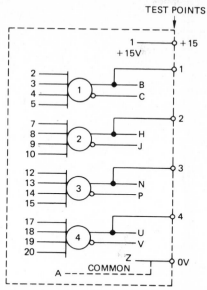

Figure 8·14 Schematic of logic module containing four OR logic functions. *(Courtesy of Allen-Bradley Co.)*

function is represented by a rectangle or square, with the function described, abbreviated, or symbolized inside. Logic modules generally mount into some type of rack that has connections in the rear. This type of mounting makes for easier wiring and troubleshooting. Figure 8·15 shows logic modules mounted

Figure 8·15 Photo of logic modules in mounting rack. *(Courtesy of Allen-Bradley Co.)*

in a rack. Most racks are mounted in some type of metal enclosure for their protection and personnel safety.

Figure 8·16 shows a block diagram of the device and circuit function in a solid-state logic control system. It must be emphasized that this functional block diagram could be made with different devices. The input devices are used for sensing or as the information source. Input devices include pushbutton switches, limit switches, pressure switches, photocells, pulse generators, etc. The input converters change input-level signals to logic-level signals. The logic-function circuits are the "brains" of the control system. They act upon the input information, make decisions, and provide the desired outputs in accordance with the designed control scheme. The logic-circuitry type (not function) used in industrial control, as mentioned previously, is usually the high-threshold-logic (HTL) type.

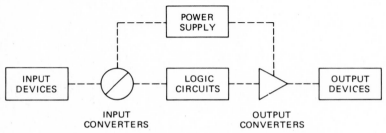

Figure 8·16 Block diagram of logic functions.

The output converters change the logic signal level to the signal level required by the output devices. The output devices carry out the commands of the logic circuits and, effectively, the control scheme. Typical output devices are starters, amplifiers, meters, solenoids, pilot lights, solid-state contactors, etc. The power supply provides the power source (usually 15 to 25 V dc) to the logic circuitry and input and output converters, and, sometimes, other voltage levels (usually dc) to input or output devices.

The drawings required for an industrial control job using solid-state logic are the logic diagram, the module layout, the panel layout, an interconnection diagram, a sequence of operations, a stock list, and an electrical layout. The drawings not discussed in the previous section will be discussed in this section. The elementary diagram used in solid-state logic design is actually a logic diagram. The logic diagram shows the logic-function arrangement of the control sequence and does not show all the wiring connections, as an elementary diagram does.

Figure 8·17 is a logic diagram of a ramp control used on a truck dock. The controls work in the following manner. The motor is a hydraulic oil pump. The up pushbutton (5PB) and the down limit switch (1LS) must be closed to actuate the up solenoid (1SOL), which, in turn, pushes up the ramp hydraulically. The down pushbutton (6PB) and the up limit switch (2LS) must be closed simultaneously to start the ramp down.

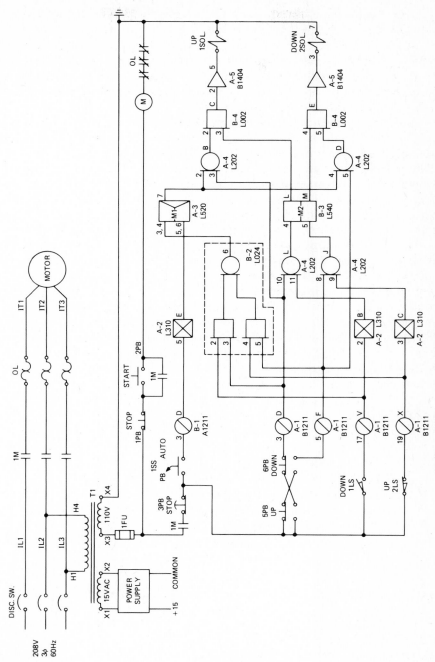

Figure 8·17 Logic diagram of an industrial ramp control.

In examining this diagram, there are several generalizations that can be made. The logic symbols used are from the NEMA IC-1 "Electrical Standard." The logic power sources are shown in the upper left corner just below the load circuit. The logic element symbols are arranged so that the inputs are on the left side and the outputs are on the right side. Single-line diagramming is used, and thus return circuits are omitted. The logic circuits are composed of one or more logic-function symbols.

The control sequence, represented by the logic circuits, goes from left to right, top to bottom; thus the starting point in the control sequence is found in the upper left corner of the logic circuitry. If the logic element is composed of two or more logic symbols, the entire element may be encased in a dashed rectangular box. Within or adjacent to the symbols or rectangular boxes, there are two letter-number combinations: one represents the manufacturer's logic-module number, the other the position the logic module occupies in the rack. Adjacent to the line entering/leaving the logic element is the logic pin number.

Figure 8·18 is a logic-module layout which shows where the logic modules are mounted in the rack. This particular logic-module layout corresponds to the logic diagram of Fig. 8·17, showing exactly where the modules are mounted in accordance with their identifications. Because of the small number of logic modules or cards used in the logic diagram, the mounting rack could have been one row high. However, this two-row-high mounting rack is shown to demonstrate the method of identification. For example, the bottom input converter on the left side of the logic diagram is identified as A_1, which means it is found in the row A, column 1 position on the module layout.

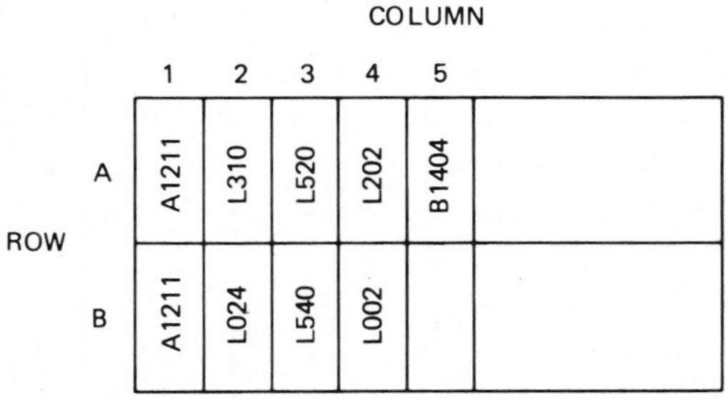

COLUMN

	1	2	3	4	5		
A	A1211	L310	L520	L202	B1404		
B	A1211	L024	L540	L002			

ROW

Figure 8·18 Logic-module layout for the logic diagram in Fig. 8·17.

Although solid-state logic has a place in the industrial control scene, its use is now diminishing because of the use of programmable controllers—the topic of the next section.

8·5 Programmable controllers

Programmable controllers (PCs) are probably one of the hottest pieces of equipment to hit the industrial control market in years. The programmable controller is a programmable solid-state replacement for "relay-type" sequencing panels

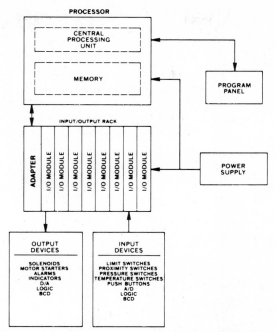

Figure 8·19 Block diagram of programmable controller. *(Courtesy of Allen-Bradley Co.)*

or solid-state logic control systems. PCs came into being in the late 1960s to meet the requirements of the automobile industry, which wanted to reduce the time and expense of building and testing new relay control panels between model years. Thus, the PC contains the relays in programmable logic; therefore, new control sequences require only reprogramming, and not modifying existing or building new relay control panels.

The heart of the programmable controller is the processor, which is either hard-wired solid-state logic, keyboard-programmable solid-state logic, or a microprocessor. The recent trend is toward greater use of microprocessors. A typical PC consists of four components—the processor, the power supply, input-output (I/O) modules, and the program console. Figure 8·19 is a block diagram that shows the arrangement of these components.

Figure 8·20 shows a programmable controller. Within the enclosure, the I/O modules are in the upper center and right, the processor is at the bottom, and the power supply is in the upper left. A card-input programmer is shown on the table. It should be noted that input and output wiring is not shown in this picture so that emphasis may be placed on the components of a programmable controller.

Because the PC replaces relay ladder-type logic, it is actually programmed in a relay ladder-type language. Many programmers (CRT, keyboard panel) have relay symbols on the entry buttons. Some PCs have the ability to be

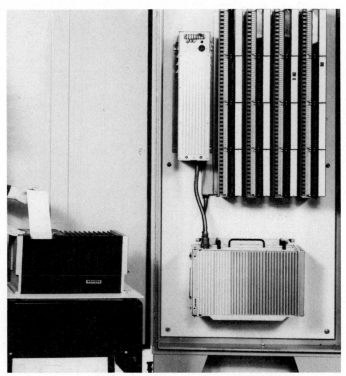

Figure 8·20 Photo of a programmable controller mounted in an industrial enclosure. *(Courtesy of Modicon Division/Gould, Inc.)*

programmed in a boolean-algebra-type language or other assembly-type languages. Many PCs can be interfaced with a computer, and some have the ability to print out their stored "logic" in a ladder diagram. Presently, there are no drawing standards governing the use of programmable controllers in control work. The drawings associated with an industrial control job involving PCs should include an "elementary," a sequence of operations, a panel layout and interconnection diagram, a bill of materials, and an electrical layout. The drawings discussed in the previous sections will not be discussed in this section. Because the relays of a PC are contained within the processor and are not discrete devices, there are varied opinions on what schematic-type or logic-type drawings are required for engineering, construction, and maintenance of the control system. Therefore, the authors will show a number of typical examples of the elementary or schematic-type drawing.

Figures 8·21, 8·22, and 8·23 show a method of drawing a PC in an industrial control system. The system shown is the controls of Load Station B of a large shipping conveyor system in an industrial plant. Essentially, the Load

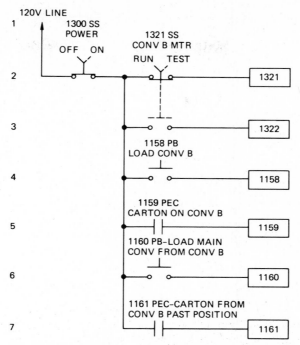

Figure 8·21 Input diagram of a conveyor control system using a programmable controller.

Station B controls allow an operator to transfer cartons from conveyor A to conveyor B, load conveyor B, and transfer cartons from conveyor B to the main conveyor. All is done by controlling the motors of conveyors A and B and the raise solenoid of conveyor B. There are many interlocks to permit logical and safe operation and lockouts to permit testing the system. There is even a timer (1TMR) to stop the system until a carton stops shimmying.

Figure 8·21 is the input diagram showing the connections from the input device to the input modules of the programmable controller. This diagram is similar to a ladder diagram, but only one side of the power source is shown. Connections to external devices, such as photocells (e.g., 1159PEC), are not shown, but their associated contacts that control a PC input are shown. The inputs to the PC are shown as rectangles and are identified with a four-digit number.

Figure 8·22 is a relay-ladder logic representation of the internal PC logic used. This is the logic used to program the PC. The PC "outputs" are shown as circles and are identified with a three-digit number. These outputs are either internal (400-series number), driving other internal relays, or external (100-series number), driving output devices, such as solenoids and motors. This logic diagram presents a couple of new twists in the standard relay-ladder diagram.

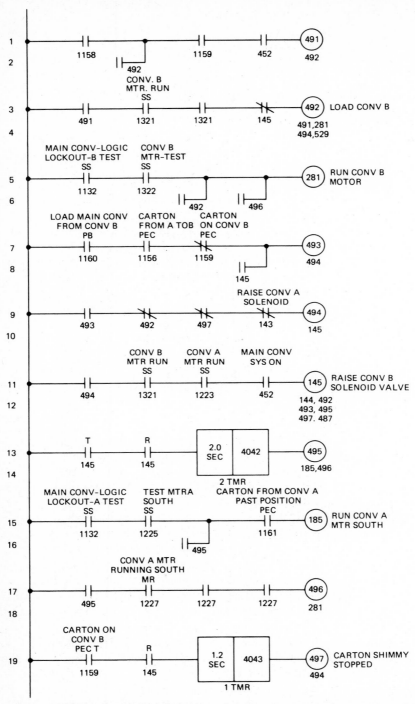

Figure 8·22 Logic diagram of a conveyor control system using a programmable controller.

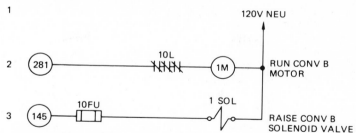

Figure 8·23 Output diagram of a conveyor control system using a programmable controller.

For example, contact 492 on line 2 is actually connected to the left side of the ladder and is thus in parallel with contact 1158 on line 1.

Figure 8·23 is the output diagram showing the connections from the output modules of the programmable controller to the output devices. Again, this diagram is similar to a ladder diagram, except that only the neutral of the power source is shown. The PC logic of this system controls two devices, namely the B conveyor motor and raise solenoid.

Figures 8·24, 8·25, and 8·26 show another method of drawing a PC control system. This system marks codes on relay cans. Essentially, it works by a revolving feeder initially loading seven relays in a magazine, which has been placed in the marker chute. The actuation of the initial start pushbutton (6PB) causes the advance solenoid (1SOL) to advance the magazine and the advance-up solenoid (4SOL) to engage the magazine and move it seven positions. Then, the escapement solenoid (3SOL) and the in-place solenoid (2SOL) allow the feeder to load and set another relay in the magazine. After 17 relays have been loaded and advanced, the marker clutch (1CLUTCH) is energized and starts marking the relays as they continue down the chute until all relays are marked.

Figure 8·24 is basically a ladder-type elementary diagram, showing all controls exterior to the programmable controller, the power connections, and the connections to the PC input module and from the PC output modules. The inputs to the PC are identified with an X and a number (e.g., X2 is Start Print Operation) and the outputs from the PC with a Y and a number (e.g., Y1 is an advance solenoid). The remainder of the circuit is self-explanatory.

Figure 8·25 is a logic *representation* of the PC logic used for the control system. Note the word *representation*—the circuit is not the actual logic circuit inside the PC, but one engineer's way of logically depicting how the control system should operate. It should be emphasized that this diagram could just as well have been done using relay symbols, as in our previous example. However, it is easier to optimize using logic-function symbols. The PC inputs and outputs are the interface between this diagram and the preceding diagram. The control relay numbers are fictitious, but invaluable in helping the engineer depict the logic and enter the logic program into the PC. The squares shown with titles in them are actual program subroutines designed to operate the control system.

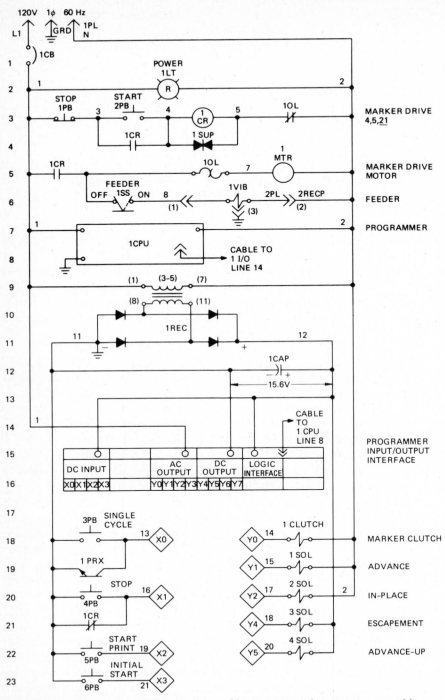

Figure 8·24 Elementary drawing of a marking system using a programmable controller.

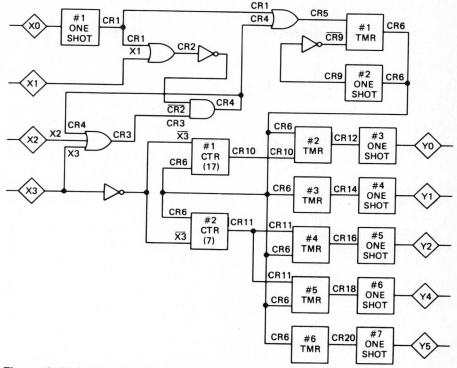

Figure 8·25 Logic diagram of a marking system using a programmable controller.

STORAGE LOCATION	DATA	I/O DEVICE
41	STR	CR6
42	STR/NOT	X3
43	CNTR	—
44	—	17
45	—	—
46	OUT	CR10

Figure 8·26 One of the programs for the programmable controller used in the marking system.

For example, the box containing the #1 counter is actually a program used for counting.

Figure 8·26 is the actual program that must be entered into the PC for the #1 counter program. The program works in this way: the outputs of X3 are inverted (set and reset), CR6 (counts) set, the timer counted up to 17 counts, at which time its output (CR10) is entered in #2TMR.

8·6 Computer control

Computer control in industry has become a reality from the computerizing of the process itself to the computerizing of shipping and warehousing operations. With the invention of the microprocessor, and thus the microcomputer, many engineers are finding it more economical to use a computer for an industrial control job. Computers are generally used for jobs whose requirements include at least one of the following: many control points, data processing, fast scanning, or fast control speeds. For the most part, computers, unlike PCs, cannot be placed in the harsh ambient conditions of many manufacturing facilities without providing some type of climatic correction (air conditioning, filtering, etc.).

As mentioned at the outset of this chapter, as industrial control circuits become more complex, the drawings required become simpler. Such is the case with computer control. Other than the individual computer internal drawings or software (programming) flowcharts, the drawings particularly required for a computer control system are a system flow diagram and interface drawings. Many other drawings may be used in computer control, such as layouts, wiring diagrams, stock lists, etc.; however, these drawings have been covered in previous sections and need not be described in detail here.

Figure 8·27 shows a flow diagram of a digital numerical control (DNC) system. The concept of using digital computers to do numerical control work

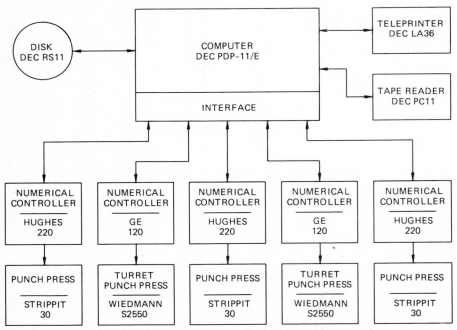

Figure 8·27 Flow diagram of a computer-controlled numerical control system.

by replacing the tape readers is relatively new. In the system shown on this diagram, the computer was added; the existing tape readers were left to replace the computer in case of computer failure. The numerical control programs are read into the computer and stored on the disk. The teleprinter provides online access to the system and thus enables the operator to monitor or change programs. The computer interface is the electronics that lets the computer communicate with the machine controls, which in turn operate the numerically controlled punch press. The arrangement of flow-type diagrams should be as simple as possible to clearly present the "big" picture of the overall system. Generally, control flow should go from left to right or top to bottom. Symbols should be simple (rectangles, circles, etc.) and should be identified.

Computer interfaces can be relatively simple or complex, custom engineered or stock manufactured items, depending on the computer, the control equipment, and the type of control desired. Thus, computer interface drawings can also be relatively simple or extremely complex. In fact, the authors have seen computer interface drawings over 10 ft long, although they could probably have been made smaller. The interface drawings usually consist of logic diagrams and logic-module layouts similar to those found in Sec. 8·4. However, in complex interfaces, the individual logic circuits often are not shown on the logic diagram— it shows only the external connections, the module outline, and certain internal circuits represented by rectangles.

Figure 8·28 is an interface diagram showing an address selector. An address is a unique number that a control device is assigned by a designer so that it may receive/transmit its data from/to the computer. Inside the rectangle representing the address selector, there are a number of squares, each of which represents two or more logic functions. This address selector has the ability to select four addresses, namely, 764200, 764202, 764204, and 764206 (octal numbers). On the left-hand side of the diagram there are a number of circles with "B" inside, signifying the data bus. From the data bus, connections are run to the module card. The numbers associated with each bus connection tell where the bus is found in the module rack (e.g., B_8-S_1, refers to the module in row B, column 8, with connection at pin S_1). Adjacent to the address selector are its terminals, which are numbered. Inside the address-selector module and next to the terminals are alphanumeric combinations, which are acronyms for the computer bus functions (e.g., A_{17} is address line 17, CO is the data transfer mode). On the right side of the diagram there are connections with arrowheads going to other interface logic modules. For example, SELO selects address 764200 and, in combination with OUT HIGH or OUT LOW, selects the high or the low byte. The number/letter combination on the bottom of the address selector represents the location in the module layout (A_7) and the manufacturer's model number (M105, Digital Equipment address selector). As can be seen, computer interfaces can become complicated; however, like all drawings, a clear and logical layout of the diagram will provide better readability and understanding.

Computer control of industrial processes is definitely the future trend, and thus its associated drawings will continue to be with us in some form.

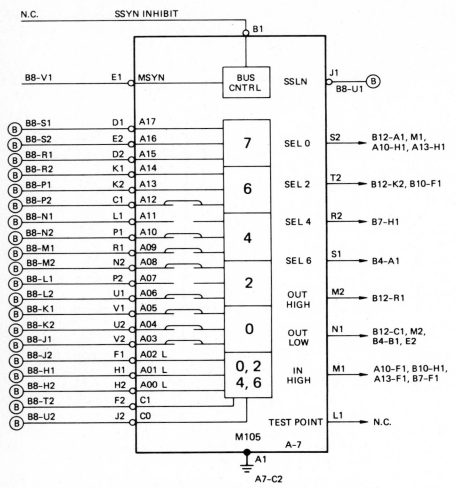

Figure 8·28 Interface diagram of an address selector.

8·7 Instrumentation

Many industrial control systems have associated instrumentation which either measures or controls, or does both to, the process variables. As the complexity of industrial processes has increased, the complexity of instruments and instrument systems has also increased; thus, the complexity of associated drawings has also increased.

Besides the usual drawings associated with most industrial control (see chapter introduction), there is one particular drawing unique to the instrumentation field, namely, the balloon drawing. The balloon drawing is associated with the Instrument Society of America Standard ISA-S5.1, titled "Instrumentation

Symbols and Identification," which has been adopted by ANSI as Y32.20-1975. The balloon drawing is essentially an instrumentation flow or logic diagram. Unlike most of the drawings encountered previously in this text, the balloon drawing can involve mechanical (hydraulic, pneumatic) signals and supplies (gas, air, steam, etc.), as well as electrical signals (current, voltage) and service (voltage). However, this text will cover primarily electrical and electromechanical devices. The balloon is approximately a $\frac{7}{16}$-in.-diameter circle, used to represent an instrument or instrument tagging. An instrument is defined as a device used directly or indirectly to measure or control, or do both to, a variable. The term includes control valves, relief valves, annunciators, pushbuttons, etc. Tagging is the process of identifying other instrument symbols with a code. Figure 8·29 shows the balloon tag of an indicating voltmeter and will be used to demonstrate how identification is done in accordance with the ISA Standards.

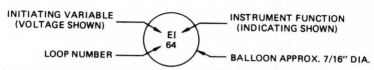

Figure 8·29 Balloon-tag diagram of an indicating voltmeter.

The first letter represents the initiating or measured variable—in this case, voltage. The second letter represents the instrument function—in this case, indication. The numbers below the letters are the loop numbers. A loop is a combination of one or more interconnected instruments arranged to measure or control a process, or both. Table 8·2 lists the instrument-function code letters and the associated measured or controlled letter. The standard should be consulted before using these codes because there are numerous qualifiers. They may have the alternative of identifying relays with function symbols shown in Table 8·3.

Figure 8·30 shows a partial balloon drawing of a combustion air control system. Figure 8·31 shows an alternative and equally acceptable method of drawing the balloon drawing of the same system. Here, most of the balloons represent instruments with two instrument tags, shown at the bottom. The logic sequence of measuring to controlling flows from top to bottom; however, this orientation is optional. Instruments are arranged in terms of logic flow and will not necessarily correspond to signal-correction sequence. Signal lines may enter the balloons at any angle, and symbols may be drawn at any orientation. However, drawing neatness and organization leads to better understanding. Electrical signals are shown as dashed lines, and pneumatic signals are shown as solid lines with double crosshatch (╫) interspaced. Directional arrowheads are added to signal lines when clarification of the flow of intelligence is needed. Explanatory notes are sometimes added adjacent to the symbols in order to clarify the function of an instrument or process device.

Individual instruments, like instrumentation systems, have become exceedingly complex in recent years. Thus, the drawings associated with instruments

Table 8·2 Meanings of Identification Letters

This table applies only to the functional identification of instruments. Numbers in table refer to notes in the standard.

	First Letter		Succeeding Letters (3)		
	Measured or Initiating Variable (4)	Modifier	Readout or Passive Function	Output Function	Modifier
A	Analysis (5)		Alarm		
B	Burner flame		User's choice (1)	User's choice (1)	User's choice (1)
C	Conductivity (electrical)			Control (13)	
D	Density (mass) or specific gravity	Differential (4)			
E	Voltage (eMF)		Primary element		
F	Flow rate	Ratio (fraction) (4)			
G	Gaging (dimensional)		Glass (9)		
H	Hand (manually initiated)				High (7, 15, 16)
I	Current (electrical)		Indicate (10)		
J	Power	Scan (7)			
K	Time or time-schedule			Control station	

	Measured or Initiating Variable	Modifier	Readout or Passive Function	Output Function	Modifier
L	Level		Light (pilot) (11)		Low (7, 15, 16)
M	Moisture or humidity				Middle or intermediate (7, 15)
N (1)	User's choice		User's choice	User's choice	User's choice
O	User's choice (1)		Orifice (restriction)		
P	Pressure or vacuum		Point (test connection)		
Q	Quantity or event	Integrate or totalize (4)			
R	Radioactivity		Record or print		
S	Speed or frequency	Safety (8)		Switch (13)	
T	Temperature			Transmit	
U	Multivariable (6)		Multifunction (12)	Multifunction (12)	Multifunction (12)
V	Viscosity		Well	Valve, damper, or louver (13)	
W	Weight or force		Well		
X (2)	Unclassified		Unclassified	Unclassified	Unclassified
Y	User's choice (1)			Relay or compute (13, 14)	
Z	Position			Drive, actuate, or unclassified final control element	

Table 8·3 Function Designations for Relays

The function designations associated with relays may be used as follows, individually or in combination. The use of a box enclosing a symbol is optional; the box is intended to avoid confusion by setting off the symbol from other markings on a diagram.

Symbol	Function
1. 1-0 or ON-OFF	Automatically connect, disconnect, or transfer one or more circuits provided that this is not the first such device in a loop 8·2
2. Σ or ADD	Add or totalize (add and subtract)†
3. Δ or DIFF.	Subtract†
4. $\pm$ $+$ $\boxed{-}$	Bias*
5. AVG.	Average
6. % or 1:3 or 2:1 (typical)	Gain or attenuate (input:output)*
7. $\boxed{x}$	Multiply†
8. $\div$	Divide†
9. $\boxed{\surd}$ or SQ. RT.	Extract square root
10. x^n or $x^{1/n}$	Raise to power
11. f (x)	Characterize
12. 1:1	Boost
13. $\boxed{>}$ or HIGHEST (MEASURED VARIABLE)	High-select. Select highest (higher) measured variable (not signal, unless so noted).
14. $\boxed{<}$ or LOWEST (MEASURED VARIABLE)	Low-select. Select lowest (lower) measured variable (not signal, unless so noted).
15. REV.	Reverse
16.	Convert
a. E/P or P/I (typical)	For input/output sequences of the following:

Designation	Signal
E	Voltage
H	Hydraulic
I	Current (electrical)
O	Electromagnetic or sonic
P	Pneumatic
R	Resistance (electrical)

Symbol	Function
b. A/D or D/A	For input/output sequences of the following:

A	Analog
D	Digital

Symbol	Function
17. $\int$	Integrate (time integral)
18. D or d/dt	Derivative or rate
19. 1/D	Inverse derivative
20. As required	Unclassified

* Used for single-input relay.
† Used for relay with two or more inputs.

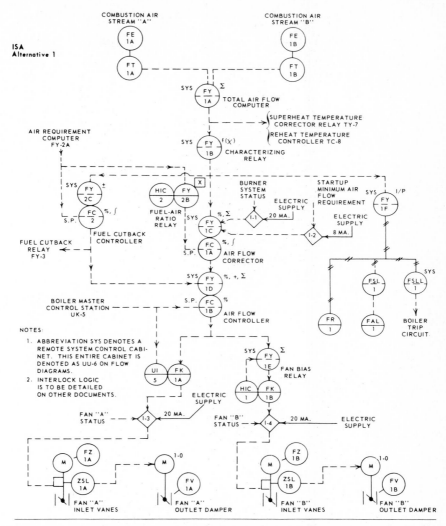

Figure 8·30 Balloon drawing of a combustion air control system—Alternative 1. *(Courtesy of Instrument Society of America.)*

have also become complex. Many instruments are a mixture of sophisticated electronics, electromagnetic devices, and motors. Figure 8·32 shows a three-pen recorder, and Fig. 8·33 shows a block diagram of this recorder. This instrument may be used in measuring or measuring and controlling a variable, such as voltage or temperature. The basic parts of this recorder, as shown in the block diagram, are: (1) the measuring circuit (shown on the left side), (2) the

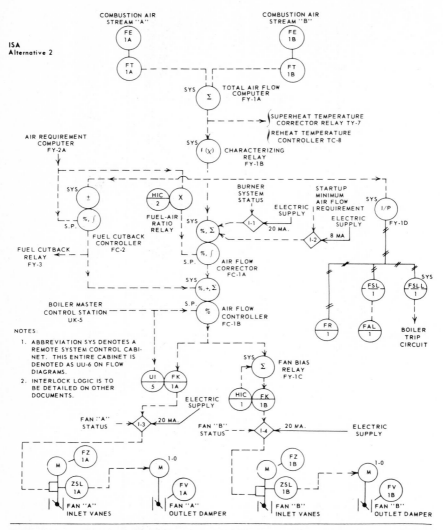

Figure 8·31 Balloon drawing of a combustion air control system—Alternative 2. *(Courtesy of Instrument Society of America.)*

detector/amplifier, (3) the pen or balancing motor, and (4) the display or recording and indicating devices. The power input to the recorder is shown on the right side of the diagram.

The recorder works in the following manner. The measuring circuit for each input compares the incoming voltage signal to a known voltage, and the difference or error signal is detected and amplified by the amplifier. After amplification, the error signal is applied to the balancing motor, which, in turn, adjusts its position to nullify the error signal. Then, the position of the balancing or

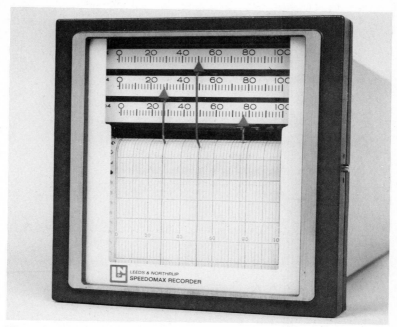

Figure 8·32 Photo of a three-pen recorder. *(Courtesy of Leeds & Northrup Co.)*

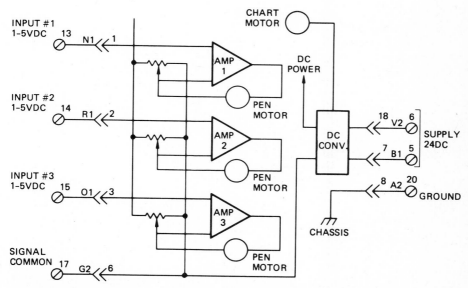

Figure 8·33 Block diagram of three-pen recorder. *(Courtesy of Leeds & Northrup Co.)*

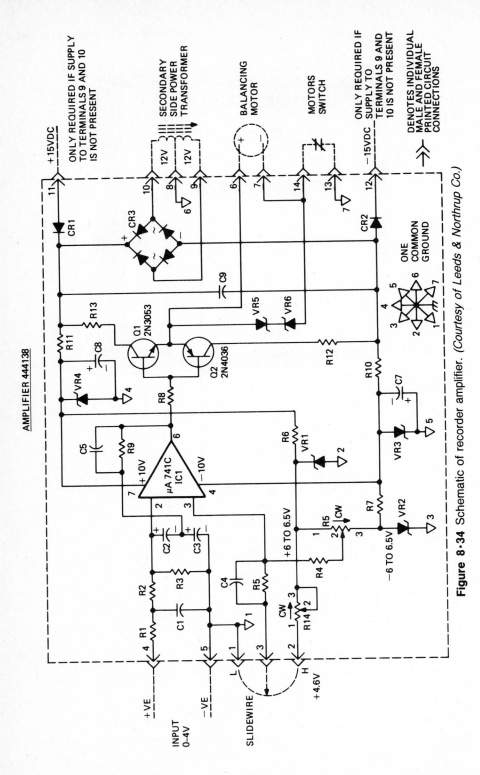

Figure 8·34 Schematic of recorder amplifier. *(Courtesy of Leeds & Northrup Co.)*

pen motor is indicated by an associated marker and recorded on the chart by an associated pen. This position is calibrated to the value of the variable measured.

Figure 8·34 shows a schematic of one of the recorder's three detector amplifiers. This amplifier is on a separate printed-circuit card, which is represented by the dashed lines. The detection and preliminary amplification is done in the integrated circuit (IC_1), and final amplification is done in the transistorized (Q_1 and Q_2) push-pull amplifier. The amplified error signal drives the balancing motor. Note the symbol for a ground in the lower right-hand side of the drawing.

Figure 8·35 is a block diagram showing how this recorder might be used in a closed-loop control system in which completely automatic control is required.

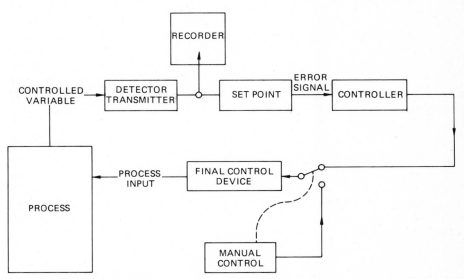

Figure 8·35 Block diagram of automatic control system. *(Courtesy of Leeds & Northrup Co.)*

The system consists of: (1) a detector to produce a dc voltage in proportion to the controlled variable, (2) a set-point unit to provide a control set point to which the process is to be controlled (and to compare the voltage from the detector with the set-point voltage to provide an error voltage), (3) a controller to provide an output related to the detector and error voltage, (4) a final control device, such as a valve drive motor or SCR, etc., and (5) a recorder to provide the process operator with information about how well the process is being controlled and to show trends with time of the process variable. The two- and three-pen recorders may be used to provide operator information from two or three control loops. When used with an associated input selector unit, each pen may be switched into any one of several different control loops selected by the process operator.

8·8 Power semiconductors

In the last two decades, the thyristor family of power semiconductors has taken and is taking over some of the control functions traditionally assigned to electromechanical devices. A thyristor is a semiconductor switch whose on/off action depends upon positive feedback. Basic functions being taken over include conversion (ac to dc), rectification (ac to dc), inversion (dc to ac), and cycloconversion (higher-frequency ac to lower-frequency ac). Thyristors are found in a wide range of controls, including solid-state relays, motor starters, motor-speed controllers, and furnace controllers. Thyristors offer a designer longer life, more precision, and faster control, and have no moving parts compared with electromechanical devices; however, thyristors are less efficient and consequently radiate more heat. But the future for the thyristor is extremely rosy: projections show them being used in applications ranging from common household appliances to large power-handling applications.

The best known and most representative of the thyristor family is the SCR or silicon controlled rectifier (sometimes called the semiconductor controlled rectifier). The SCR is a reverse blocking triode; it operates in only one direction. For operation during the entire cycle, two SCRs (back to back) are required. SCRs can be bought either separately for independent design or in prepackaged control systems.

SCR controls used in industry are generally complex, with many components. The drawings encountered include: elementary layouts, wiring diagrams, stock lists, etc.; only the elementary diagram will be discussed in this section because of its differences. The only standards associated with SCR control drawings are the ANSI Standards on symbology and diagrams. The authors have found that most companies have unique ways of drawing SCR elementary control diagrams, but generally the elementary diagrams involving SCRs are hybrids of electronics schematics and electromechanical elementary diagrams.

Generally, dc motors are more adaptable to speed control than ac motors; however, most industrial plants do not have a dc source. Therefore, ac must be converted to dc. Figure 8·36 is a partial schematic of a dc, SCR variable-speed drive. This type of unit has virtually replaced the shunt-field-rheostat and armature-circuit-resistance methods of speed control. This speed controller operates in the following manner: Three-phase ac power enters the fuse, is fed through the line reactor (FC) and the line contactor (M), and enters the SCR power-conversion module, where six SCRs, working independently but in a predetermined sequence, rectify the ac to an adjustable dc voltage. The dc power is fed through a shunt resistor (RT) to the motor armature. The speed of the motor is proportional to the dc voltage applied. The dc voltage is adjusted by changing the positive trigger voltages applied to the gates of the SCRs, which changes their conduction period and, thus, the amount of dc voltage. The trigger circuit, power supply, and miscellaneous control and interface circuits are not shown here owing to their complexity.

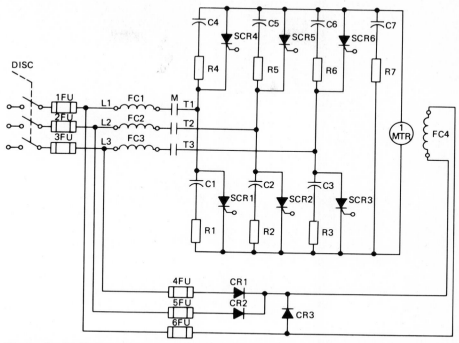

Figure 8 · 36 Schematic of the load (power) circuit of an SCR variable-speed drive for dc motors. The control schematic is not shown. *(Courtesy of General Electric Co.)*

This schematic of a speed controller has the power circuit going from left to right. Generally, in SCR control schematics, the power circuit goes in this direction or from top to bottom.

Figure 8·37 shows an SCR power module for this speed controller. Note the large heat sinks to radiate and convect the heat generated by the SCRs. The capacitor and resistor that shunt each SCR may also be observed.

Figure 8·38 shows a dc speed controller. The SCRs are in the lower right-hand corner of the enclosure. Note the printed circuitry through the entire enclosure and the motor contactors in the upper half of the enclosure.

SUMMARY

The drawings in this chapter have been selected from among thousands of control systems that are employed in United States industry today. These drawings represent many different types of industrial control systems, from basic motor control to more sophisticated computer control. Different types of controls may stand alone, or they may be integrated with other types of controls. Formats have been suggested for doing each type of control drawing. Several drawings that are common to almost all types of control work include an elementary

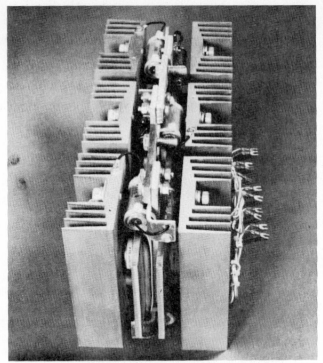

Figure 8·37 Photo of an SCR module. *(Courtesy of General Electric Co.)*

Figure 8·38 Photo of a dc SCR variable-speed drive. *(Courtesy of Allen-Bradley Co.)*

or schematic diagram, a panel layout, a parts list, and assembly drawings. Symbology varies from control type to control type, and it even varies within control type. (There are six different sets of logic symbology in use today.) Some symbology used in this chapter in certain types of diagrams is somewhat different in appearance from that in electrical diagrams earlier in the book. In some cases, the types of devices used, such as relays and contacts, have their own special symbols. Sometimes new and special symbols, such as instrument balloon tags, have been introduced. Standard identification of components and the numbering of wires and terminal boards is extremely important in industrial control systems due to their complexity-vast number of components. And, finally, the trends of industrial controls have been presented, with their indications of more emphasis on programmable controllers and computers and less on electromechanical devices in the future.

QUESTIONS

8·1 Name five different drawings generally encountered in all types of control work.

8·2 Name four functions of motor control.

8·3 Describe the load circuit in a motor controller; in the control circuit.

8·4 What is the main difference between Figs. 8·2 and 8·4?

8·5 What is three-wire control? Two-wire control?

8·6 What is the difference between a contactor and a motor starter? Between a circuit breaker and a line switch?

8·7 Name two drawings strictly associated with motor control work.

8·8 What standards are generally used when doing electromechanical control drawings? Is the use of these standards mandatory?

8·9 For what words do the following abbreviations stand: PB, PS, CR, LS, OL, TR?

8·10 Why is a ladder diagram called a ladder diagram?

8·11 What is the purpose of a sequence of operations?

8·12 Describe a panel layout and its function; an interconnection diagram.

8·13 What information appears on the bill of materials (stock list)?

8·14 What type of solid-state logic component configuration is used in industrial control work?

8·15 What are the advantages of solid-state logic?

8·16 Name the six functional parts of any solid-state logic control system.

8·17 What are the two main differences between an elementary diagram (schematic) and a logic diagram?

8·18 If a logic symbol is identified with B_3, where would its module be found in the module layout?

8·19 Name the four basic components of a programmable controller (PC).

8·20 Name the two types of symbology generally used in PC elementary diagrams.

8·21 Name the control system requirements that practically dictate the use of a computer.

8·22 How should control flow be drawn on flow diagrams?

8·23 Describe the information that is found in a balloon tag.

8·24 What is the different aspect of balloon diagrams (not balloons) compared with any other type of drawing discussed in the text?

8·25 Which of the alternative methods of drawing balloon diagrams would you prefer and why?

8·26 What is a thyristor? An SCR?

8·27 What basic functions do thyristors perform?

PROBLEMS

8·1 Redraw Fig. 8·2 using three-wire control with a Hand-Off-Auto (HOA) selector switch; A pressure (drop-in-pressure-to-operate) switch to be used in the automatic mode. Use $8\frac{1}{2} \times 11$ paper.

8·2 Redraw Fig. 8·2 using three-wire control with a HOA selector switch. A temperature (rise-in-temperature-to-operate) switch to be used in the automatic mode. Replace line switch and fuses with a circuit breaker. Use $8\frac{1}{2} \times 11$ paper.

8·3 Make a drawing of the ladder diagram for the sequence timer shown in Fig. 8·39. The following changes might be in order: Add a stop pushbutton switch and selector switch to the appropriate part of the diagram. Add overload protection for the delay-period circuits and in other places you deem advisable. Use $8\frac{1}{2} \times 11$ or 11×17 paper.

8·4 Adjustable-speed drives for dc motors are of four types: (a) motor generator, (2) electron tube, (3) magnetic amplifier, and (4) silicon controlled rectifier. The tube and magnetic amplifier are obsolete. The motor generator is becoming obsolete. Figure 8·40 shows in diagrammatic form the two types that are still popular. Draw these two diagrams and label all parts. (3-ϕ means 3-phase.) Use $8\frac{1}{2} \times 11$ paper.

8·5 Draw the elementary diagram of an automatic control system that has three 10-hp pumps, which come on with pressure drops to 85, 72, and 60 psi, respectively. Use a HOA selector switch to provide a hand operation for each pump. Use $8\frac{1}{2} \times 11$ or 11×17 paper.

8·6 In Prob. 8·5, we don't want to run one pump more than another; therefore, we want to modify the circuit by installing a manually operated pump-sequence control switch. When it is in position 1, 2, or 3, the pump sequence would be 1–2–3, 2–3–1, or 3–1–2, respectively. Use a one-pole selector switch and three relays. Use 11×17 paper.

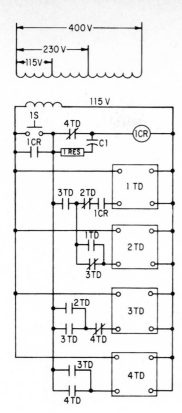

Figure 8·39 (Prob. 8·3) A ladder diagram for a sequence timer.

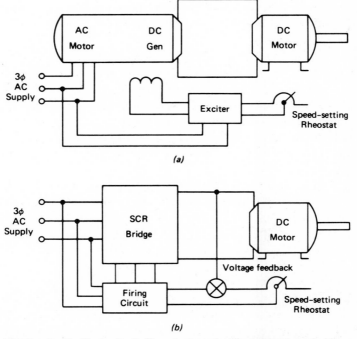

(a)

(b)

Figure 8·40 (Prob. 8·4) Two types of adjustable-speed drives.
(a) Motor-generator. *(b)* Controlled rectifier drive.

8·7 Figure 8·41 is the electromechanical and solid-state equivalents of an on/off pushbutton control circuit. Redraw them using $8\frac{1}{2} \times 11$ paper.

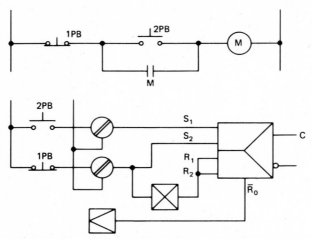

Figure 8·41 (Prob. 8·7) Electromechanical and solid-state equivalents of an on/off pushbutton control circuit.

8·8 Figure 8·42 is an example of a ladder diagram. Convert it to solid-state logic and redraw as a logic diagram. Use $8\frac{1}{2} \times 11$ or 11×17 paper.

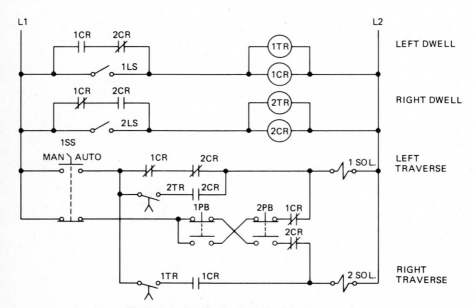

Figure 8·42 (Prob. 8·8) Ladder diagram.

8·9 Redraw Fig. 8·8 as a programmable-controller-based control system, using relay-ladder logic as in Fig. 8·21, 8·22, and 8·23. Put all three drawings on one 11 × 17 sheet.

8·10 Redraw Fig. 8·8 as a programmable-controller-based control system, using logic symbols, but not symbols for programs. Use Fig. 8·24 and 8·25 as a guide. Use 11 × 17 paper.

8·11 A computer-based control system numerically controls a rotary drill press and a multiple-turret boring machine. In addition, the peripherals are a keyboard CRT (cathode-ray tube) and a line printer. Draw the flow diagram, using Fig. 8·27 as a guide. Use $8\frac{1}{2}$ × 11 paper.

8·12 Draw a balloon tag of an indication ammeter on loop 101. A level alarm on loop 37.

8·13 Figure 8·43 is a balloon drawing of a pressure transmitter feeding a pressure indicator, a pressure recorder, and a low-pressure switch, which in turn feeds a low-pressure alarm. Redraw this circuit. Use $8\frac{1}{2}$ × 11 paper.

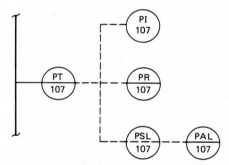

Figure 8·43 (Prob. 8·13) Balloon drawing.

Chapter 9
Drawings for the electric power field

Drawings that are made for electric power installations, whether they are for electric utilities, industry, or large commercial buildings, are as complex as those made in various electronics areas. In fact, a complete set of drawings for a typical electric generating plant just about covers the entire field of electrical drawing. Included in such a set of drawings would be one-line circuit diagrams, three-line diagrams, control and elementary schematics, logic diagrams, general arrangements, and connection diagrams, as well as plans, elevations, sections, and details of structures and equipment. In general, the ANSI Standard Y14.15, "Electrical and Electronic Diagrams," presents a good format and reference for single-line, schematic-type, and connection diagrams, and one should consult it, if beginning to draw in this field. Electrical design and associated drawings should take into account the electrical safety requirements required by the "National Electrical Code," which is discussed at length in the next chapter.

Figure 9·1 Photo of a substation. *(Kansas Power & Light Co.)*

9·1 The one (single-) line diagram

One of the first drawings made in a set of drawings for the design of a large project is a one-line diagram of the entire plant or system. Such a drawing shows by means of single lines and symbols the major equipment, switching devices, and connecting circuits of a plant or system.

Figure 9·2 is a good example of a one-line diagram. The single line running down through the figure actually represents three lines in this three-phase system. The power path can easily be traced from the shielded ACSR (aluminum cable, steel reinforced) downward past a grounded lightning arrester, through a GOAB (trade name) disconnect switch, fused disconnect switches, and a step-down transformer. It continues on down through an oil circuit breaker and auxiliary equipment. However, part of the electricity is sent through potential transformers, where it is stepped down to a small voltage for metering. In addition, current transformers near the OCB are used for current-measuring meters, none of which is shown in the drawing, although each is referred to in a note. From here, the power goes through a disconnect switch, past another lightning arrester, and out through a 12.47-kV FCWD (type F Copperweld) line.

Salient features which should be observed in the making of one-line drawings, as shown in Fig. 9·2, are:

1. Use of standard symbols, abbreviations, and designations
2. Highest voltage lines placed at the top or left of the drawing, if possible, with successively lower voltages placed downward or to the right
3. Main circuits drawn in the most direct and logical sequence
4. Lines between symbols drawn either vertically or horizontally, with minimum crossing of lines
5. Generous spacing to avoid crowding symbols and *notes*
6. When pertinent, information included on rating of equipment, feeder lines and transformers, vector relationship of equipment, ground or neutral connection of power circuits, protective measures, types of switches and circuit breakers, and instrumentation and metering

If the above points are observed in the making of a one-line diagram, the drawing will impart a clear picture of the system so represented. The one-line drawing, which is used for many purposes, is a distinctive drawing peculiar to the electrical field and as such has made a great contribution to the industry.

Approximately one-half of a one-line diagram for a part of a large electronics manufacturing plant is shown in Fig. 9·3. This diagram shows the electric energy coming from the power company after having been reduced to 12.47 kV. A small substation on plant property reduces this supply to 4.16 kV by means of an Askarel-filled (A-F) transformer. The 600-MCM conductor has an area equal to 600,000 circular mils. A circular mil is a unit used in specifying the cross-sectional area of round conductors. It is equal to the area of a circle whose diameter is 1 mil (0.001 in.).

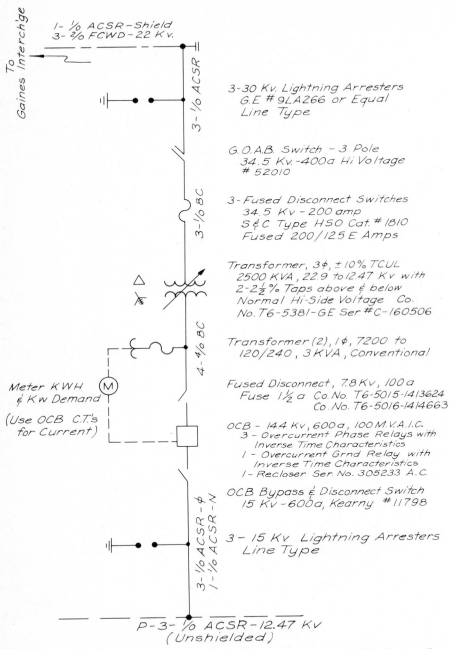

To Gaines Interch'ge

1- 1/0 ACSR – Shield
3- 2/0 FCWD – 22 Kv.

3- 1/0 ACSR

3- 1/0 BC

4- 4/0 BC

Meter KWH
& Kw Demand

(Use OCB C.T.'s
for Current)

3- 1/0 ACSR – φ
1- 1/0 ACSR – N

P-3- 1/0 ACSR – 12.47 Kv
(Unshielded)

3- 30 Kv. Lightning Arresters
G.E # 9LA266 or Equal
Line Type

G.O.A.B. Switch – 3 Pole
34.5 Kv. – 400a Hi Voltage
52010

3- Fused Disconnect Switches
34.5 Kv – 200 amp
S & C Type HSO Cat. # 1810
Fused 200/125 E Amps

Transformer, 3φ, ±10% TCUL
2500 KVA, 22.9 to 12.47 Kv with
2 - 2½% Taps above & below
Normal Hi-Side Voltage   Co.
No. T6-5381-GE Ser #C-160506

Transformer (2), 1φ, 7200 to
120/240, 3 KVA, Conventional

Fused Disconnect, 7.8 Kv, 100a
Fuse 1½ a Co. No. T6-5015-1413624
Co. No. T6-5016-1414663

OCB – 14.4 Kv, 600a, 100 M.V.A.I.C.
3 – Overcurrent Phase Relays with
Inverse Time Characteristics
1 – Overcurrent Grnd Relay with
Inverse Time Characteristics
1 – Recloser Ser. No. 305233 A.C.

OCB Bypass & Disconnect Switch
15 Kv – 600a, Kearny # 11798

3 – 15 Kv Lightning Arresters
Line Type

Figure 9·2 A one-line diagram of a substation. (*Southwestern Public Service Company.*)

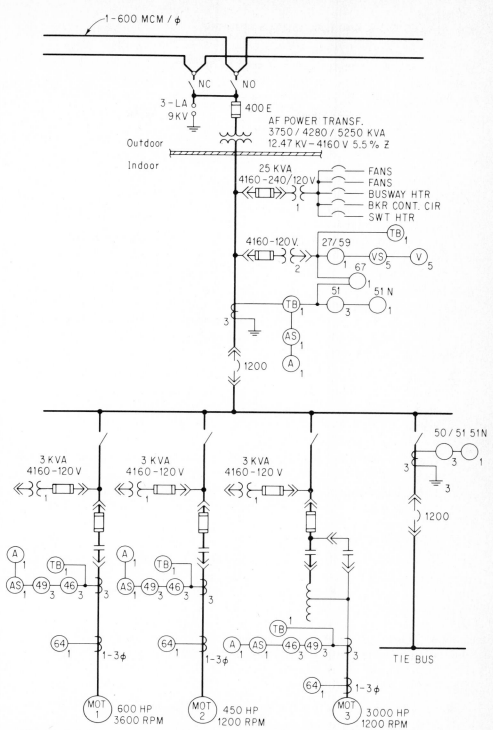

Figure 9·3 Part of a one-line diagram for an industrial plant. *(Western Electric Co.)*

The path of the energy can be traced to the motors at the bottom. These motors turn compressors and chillers for the plant air conditioning. To facilitate reading the diagram, a legend showing the abbreviations was printed on the diagram. It is reproduced below:

A	Ammeter	50/51	Overcurrent (instantaneous and time) relay
AS	Ammeter switch		
V	Voltmeter	5IN	Overcurrent (ground) relay
VS	Voltmeter switch		
TB	Test block	64	Ground-fault relay
27/59	Under/over voltage relay	67	Directional overcurrent relay
46	Phase-failure relay	87	
49	Overcurrent-overload relay		Differential-protective relay

To the right of and below most of the symbols is a number. This number indicates how many of the devices are represented by the symbol. A_3, for instance, means that there will be three devices (one on each line) at that location.

In accordance with ANS Y14.15, "Electrical and Electronics Diagrams," the primary conductors have been shown with heavy lines. Some other details recommended by this standard are:

1. Winding connection symbols should be shown for all power equipment.
2. Neutral and ground connections should be shown for all grounded equipment.
3. Rating and type of load, when available, should be shown for each feeder circuit.
4. Current transformers should show the ratio of transformers and the ampere ratio. (Example: $3 - 600/5$.)
5. Industrial control single-line diagrams may omit equipment ratings when they are used as standard drawings applying to more than one rating.

9·2 Three-line diagrams

A schematic-type drawing used in the electric power field is the three-line diagram, which is an expansion of the one-line diagram showing all three phases. Its primary purpose is to graphically illustrate the electrical connections and functions of the metering, relaying, and control power circuitry. It does not show the control of power equipment (e.g., circuit breakers) by this or other circuitry. Three-line diagrams are usually done for generating stations, substations, or any electrical installation that has metering or relaying circuitry. Three-line diagrams are an invaluable aid for construction and maintenance of electrical facilities in that they provide:

1. A complete three-wire schematic, showing all devices, wiring, and connections and their conceptual operation
2. A clean, concise picture of the point-to-point connections

3. The correct way to connect polarity-sensitive devices, such as current and potential transformers

Figure 9·4 shows a three-line diagram of a main breaker and its associated relaying and metering in a 15-kV distribution center in a large industrial complex. The breaker is one of several breakers in an enclosed, indoor, metal-clad switchgear installation. Power conductors in a three-line diagram are generally

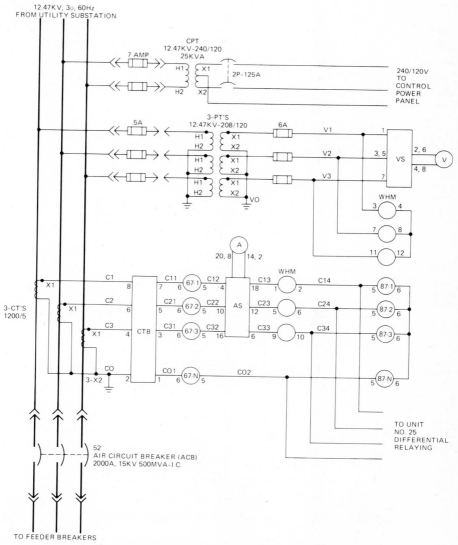

Figure 9·4 Three-line diagram of 15-kV main breaker.

drawn at the top from left to right or on the left side (as in this case) from top to bottom. The power conductors, like those in the single-line diagram, are shown in heavier lines than the relaying and metering conductors. Device ratings are shown next to the device, and, if it is multirated, the maximum rating and the connected rating should both be shown. Device-function abbreviations and/or numerical codes are shown (e.g., 52 is a circuit breaker). Where two or more devices with the same number function designation are present, they are distinguished with lettered or numbered suffixes (e.g., 87–1 means device 87 in A-phase). Device terminal numbers are generally shown to the left or below the conductor, while wire numbers are shown to the right or above the conductor. Wire numbers are generally a series of numbers or characters, such as $C1$, $C11$, $C12$, which provides easier troubleshooting and intelligent conceptual continuity. If two arrangements of the same devices are used, for example, of a current transformer (CT), one CT's wire numbers would be $1C1$, $1C12$, . . . , and the other CT's wire numbers would be $2C1$, $2C12$, . . . , in order to differentiate between the wirings.

Figure 9·5 shows a similar, but 5-kV class, air breaker and cubicle (cell). Notice how the breaker is on roller wheels and withdraws from load and line stabs (not shown) in the cubicle.

Figure 9·5 Photo of 5-kV switchgear with breaker withdrawn. *(Courtesy of Westinghouse Electric Corp.)*

Three-line diagrams should be done with good organization and attention to detail and spacing in order to present a clear, accurate picture of the metering and relaying circuits. The three-line diagram plays an important part in presenting the total picture of a power system.

9·3 Logic diagrams

Logic diagrams provide an effective means of communication among persons involved in design, construction, operation, and maintenance of electric generating stations. The five basic symbols are shown and explained in Fig. 9·6. Also shown are typical equivalent electric circuits. Some supplementary logic symbols are shown in Fig. 9·7. These symbols convey necessary information used by the electrical control designer, and with their associated descriptive information (panel item, numbers, etc.) aid the overall project design coordination.

The symbols shown do not agree entirely with any one manufacturer's standard, but they are similar to the NEMA (National Electrical Manufacturers' Association) Standard. No particular device is represented by one of these sym-

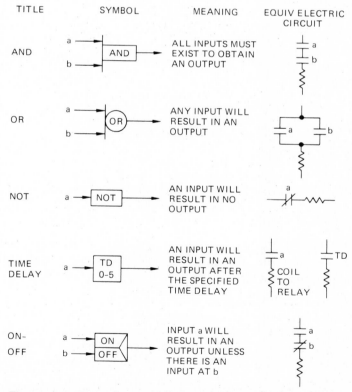

Figure 9·6 Basic logic symbols. The time-delay symbol numbers give the range of adjustment desired in seconds.

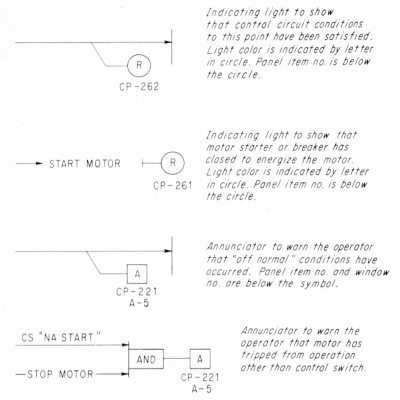

Indicating light to show that control circuit conditions to this point have been satisfied. Light color is indicated by letter in circle. Panel item no. is below the circle.

CP–262

Indicating light to show that motor starter or breaker has closed to energize the motor. Light color is indicated by letter in circle. Panel item no. is below the circle.

START MOTOR   R   CP–261

Annunciator to warn the operator that "off normal" conditions have occurred. Panel item no. and window no. are below the symbol.

A   CP–221   A–5

Annunciator to warn the operator that motor has tripped from operation other than control switch.

CS "NA START"   AND   A   CP–221   A–5   –STOP MOTOR

Figure 9·7 Supplementary logic symbols.

bols. This type of diagram, therefore, leaves the design of the circuit (including arrangement and selection of devices) up to the electrical control designer.

9·4 Relation of logic and schematic diagrams

Figure 9·8 shows the logic diagram(s) and schematic diagram of a system that provides correct operation and supervision of an auxiliary oil pump. The logic diagram is interpreted as follows:

> If the control switch is in the start position or in the automatic position *and* the turbine hydraulic-oil pressure is low, the pump starts. When the starting device is actuated or closed, a red light will go "on" on the turbine start-up panel (TSP) and on the control panel (CP). (The control switch is on the TSP, incidentally.)
>
> If the control switch is in the stop position *or* if there is a motor overload, the pump stops. In this case a green light will be energized on the CP and the TSP.

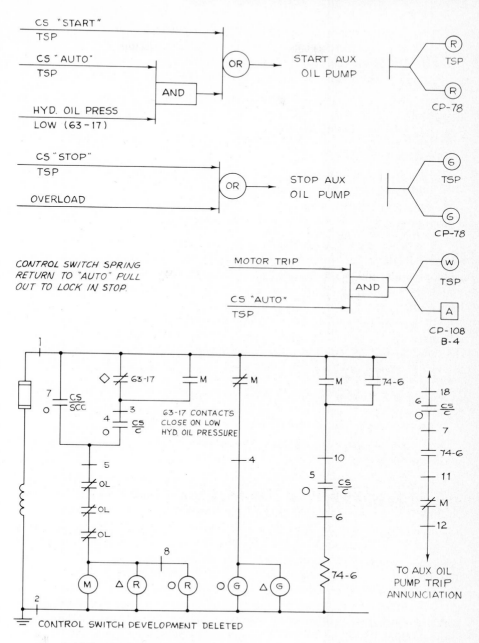

Figure 9·8 Example of a logic and a schematic diagram for an auxiliary oil-pump control system. *(Black & Veatch, Consulting Engineers.)*

If the motor had started and then tripped for some reason other than a control-switch action, an alarm would have sounded on the visual annunciator (item CP_{108}, window B_4) and a white light would have been energized on the TSP.

Certain conventional functions or actions are understood in the reading of these diagrams.

1. Unless otherwise indicated, control switches, pushbuttons, and selector switches are understood to be spring-return to normal.
2. Electrical interlock devices such as cell interlocks are not shown but are understood to be present in accordance with the clients' or office standards.
3. "Motor trip" (as contrasted to "stop") is understood to mean that the motor has stopped for some reason other than a control-switch action.

The schematic diagram which applies to the logic diagram in Fig. 9·8 explains in detail all the items necessary for accurate electrical construction and operation, and also becomes a valuable maintenance tool. The control-switch development normally shown on this type of drawing has been deleted for simplicity.

It is possible, but not practical, for the system designer to dispense with schematic diagrams and to employ only logic diagrams in communicating with the fabricator of the control systems. However, according to one designer, the following reasons justify making of schematic diagrams for most systems:

1. To show a complete system with all the desired electrical features
2. To control the selection of alternatives in the circuitry and components to be used by the manufacturer
3. To permit early determination and retain control of interconnecting cable requirements

Figure 9·9 is the control schematic of a substation feeder breaker which is used in a power generating station. The schematic shows the trip and close circuits that control the operation of the breaker. The trip and close devices which are on the breaker are represented on the diagram by rectangles bearing the labels "Tripping Devices" and "Closing Devices." These devices are usually solenoids that actuate spring-charged, mechanical operators on the breaker. The breaker and its associated contacts are represented by function number 52. The actual breaker, which is a power circuit device, is also not shown on the control schematic; however, the authors have seen control schematics on which it has been represented. The drawing also contains a logic diagram and control-switch development, which shows the switch position and the associated contact arrangement. Some of the abbreviations which may not be familiar to the reader are: CS, control switch; BKR, circuit breaker; BTGB, boiler turbine generator board; SCC, station control cabinet; PSS, permissive selector switch; RCS, remote-control switch; MOC, mechanically operated contact; and ϕA, phase A.

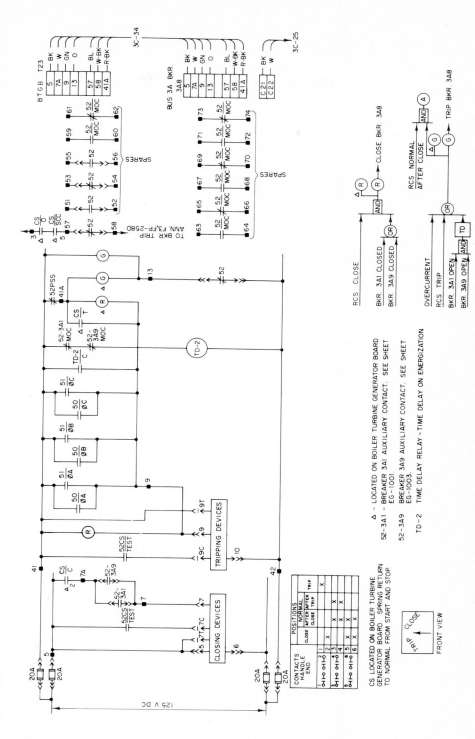

Figure 9·9 Schematic drawing of a substation feeder circuit. Also included are control-switch development (lower right) and logic diagram (upper right).

311

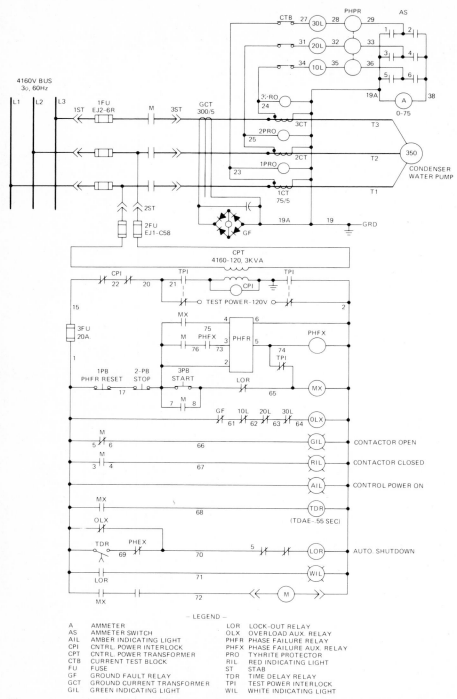

Figure 9·10 Elementary diagram of a motor controller for a condenser water pump. *(General Electric Co.)*

Many power generating stations and industrial plants use large (up to 20,000-hp) motors for pumps, such as condenser, tower, or boiler water; fans, such as boiler and tower; and auxiliary operations, such as coal pulverizers. The motors are characterized by their higher voltages (14 and 4 kV) and their larger and more complex controls. The schematic-type drawing used for the control circuitry is the elementary or ladder diagram, discussed in depth in Chap. 8. Figure 9·10 shows an elementary diagram of the motor controls for a small (350-hp) condenser water-pump motor. The contactor shown is of the removable 4-kV class. Many times, breakers are used in place of contactors for large-motor control. In addition to the normal fuse and overload protection, this motor also has phase-failure (PHFR) and ground-fault (GF) protection. Notice that current transformers are used in the overload and metering circuits. The legend explains the devices not familiar to the reader.

Schematic-type drawings, such as the control schematic and the elementary diagram, used in electric power work provide insight into the necessary functions of the controls along with the necessary electrical connections. As described earlier, the complexity of the controls required in electrical power systems sometimes dictates that logic diagrams be done to aid, but not replace, schematic-type drawings.

9·5 General arrangement

The *general arrangement* drawing shows how the equipment will be physically arranged when it is manufactured and installed. It is an extremely important drawing in a complete set of electrical drawings because it is the only one that shows the actual equipment. General arrangements show equipment exteriors and dimensions; the location of devices such as relays, meters, etc.; and pertinent details necessary for field installation, such as service entrances, mounting channels, etc. Such a drawing always contains an elevation and may also contain a plan or top view, section views, details, and, sometimes, small, simple one-line or three-line diagrams. Because the general arrangement shows the equipment as it will physically exist, it is used by engineers to tell manufacturers how they would like the equipment fabricated and installed.

Figure 9·11 shows a general arrangement of a primary-selective, 15-kV-480Y/277-V, 100-kVA indoor secondary unit substation. This substation consists of three basic assemblies, namely the two primary (15-kV) switches, the transformer, and the low-voltage switchgear. The primary switches are primary-selective because they are interlocked so that only one switch may be closed at any time. The transformer is Askarel-filled (presently not used due to environmental regulations). The switchgear has main, tie, and several feeder breakers. This general arrangement contains an elevation, a plan view, and a simple single-line. This drawing is extensively dimensioned and shows some details. The single-line shows the key interlocking system that is used for personnel safety and proper operating procedures. Figure 9·12 shows an indoor substation similar to the one shown in the previous general arrangement.

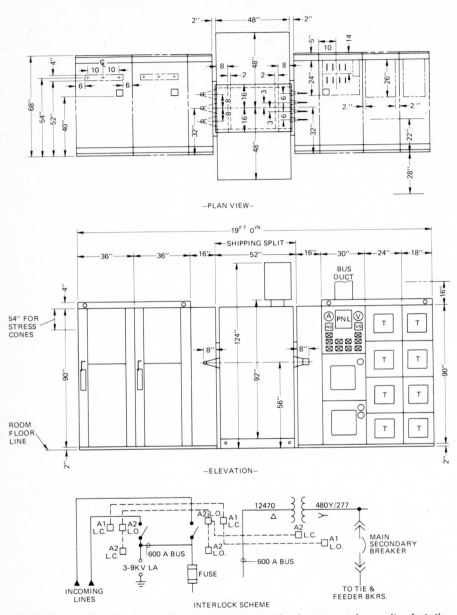

Figure 9·11 General arrangement of a primary-selective, secondary unit substation. The primary switches are on the left, the transformer is in the middle, and the secondary switchgear is on the right. *(Gould-Brown Boveri.)*

Figure 9·12 Photo of a secondary unit substation. *(Gould-Brown Boveri.)*

Figure 9·13 shows a detailed section view of a large outdoor substation. This is one of three section views of the plan on the original drawing. This drawing is partially dimensioned.

There are no accepted standards in drawing general arrangements; however, they should be done in such a manner as to give an accurate representation of the physical structure and how it will be installed. These drawings should be as simple as possible, without superfluous detail, neat, and good examples of orthographic projection as it applies to the electrical industry. Occasionally an isometric projection or some other type of pictorial view is used as a supplement to, or a substitute for, the views shown here.

9·6 Connection diagrams

Connection diagrams are usually necessary and desirable, for electrical installations and their complexity directly depend upon the complexity of the electrical installation itself. Connection or wiring diagrams show the physical wiring connections of control, metering, relaying, and auxiliary devices and do not generally involve the power conductors or devices. There are two basic types of wiring diagrams: the diagrams associated with the internal wiring of pieces of equipment, commonly called *wiring diagrams,* and those diagrams associated with the external wiring connecting the individual pieces of equipment to one another, called *interconnection diagrams.*

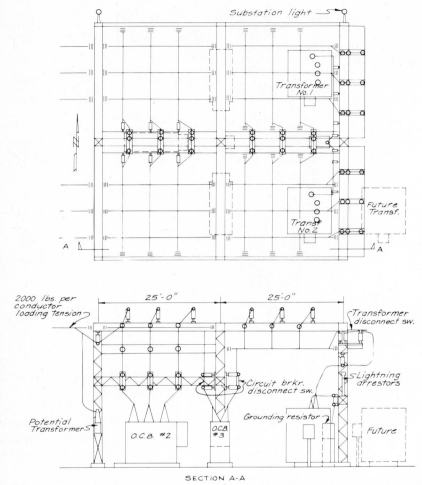

Figure 9·13 Part of general layout of a substation. *(Black & Veatch, Consulting Engineers.)*

Figure 9·14 shows the wiring diagram of a 15-kV air-circuit-breaker (ACB) cubicle that is used as a feeder breaker for a large industrial plant. It is one of several feeder breakers in a line-up (row) that make up a 15-kV distribution center. This wiring diagram shows the whole cubicle opened up from left to right, showing the rear view of the front door, the inside left side of the cubicle, the inside front of the cubicle, and the inside right side of the cubicle; it also shows a simplified three-line diagram, showing CT connections. All devices (relays, meters, etc.) are shown in their proportional sizes. This diagram uses highway-type connections, described previously in the text. The use of highway-type connections is predominant in the industry; however, the authors have

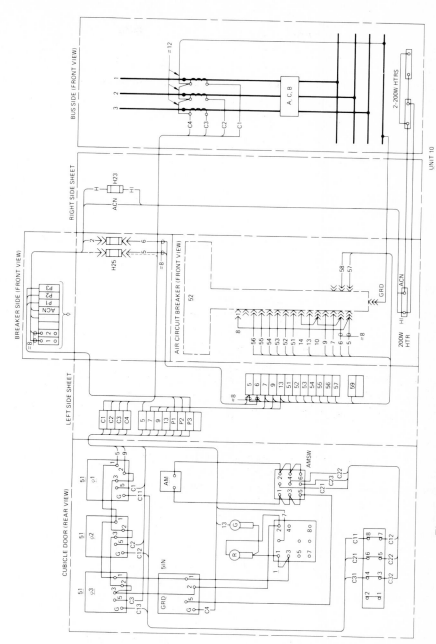

Figure 9·14 Wiring diagram of a 15-kV feeder breaker. *(Gould-Brown Boveri.)*

317

seen wide variations. Note that the wires are numbered as described in the three-line-diagram section.

Figure 9·15 shows a partial interconnection diagram for a 15-kV switchgear line-up, which includes the breaker in the previous drawing. Again, highway-type connections are used. The interconnecting wires carry the same number all through each piece of equipment. The common or interconnecting wires associated with this figure involve ac control power (ACN, H_1), dc control power (1, 2), and the 120-V metering voltage (P_1, P_2, and P_3).

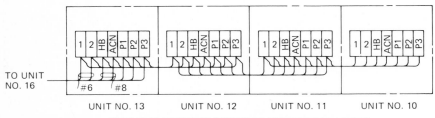

PARTIAL INTERCONNECTION DIAGRAM OF SWITCHGEAR LINEUP
(ONLY INTERCONNECTION TERMINAL BOARDS SHOWN)

Figure 9·15 Partial interconnection diagram of a line-up of 15-kV metal-clad switchgear, showing only interconnections. *(Gould-Brown Boveri.)*

Connection drawings, as these examples have illustrated, should show the best perspective and identification of the actual wiring. Routing of wiring must be shown exactly as it is to be done. Connection drawings are an important part of the electrical drawing package in that they show the actual wiring.

9·7 Power distribution plans

Power distribution plans or layouts are plan views showing the electrical distribution equipment (e.g., substation, bus ducts, etc.), the utilization equipment (e.g., pumps, furnaces, etc.), and the general routing of the electrical services between them. Because these plan views show equipment and are generally to scale, they are also used in the physical planning and installation of the equipment and services. Equipment is usually shown as an outline with little detailing. Distribution plans may also contain equipment ratings, exact service routings, service specifications (conductor information, raceway information), mechanical services, or other miscellaneous information. What should be included will depend upon the firm for which you work. Experience has shown that if power layouts are done in areas that are congested or are rearranged quite often, they should be done in $\frac{1}{4}$- or $\frac{3}{8}$-in. scale; smaller scales have proven slightly more difficult to work with.

Figure 9·16 shows a power distribution plan of a mass-production machine shop. Here the electrical distribution equipment is 480- and 208/120-V, 3-ϕ enclosed bus ducts, which are mounted over the equipment near the ceiling. Bus ducts contain solid metal bars (3 or 4), conductors which are insulated from each other, and the enclosure housing. The bus duct receives its electric

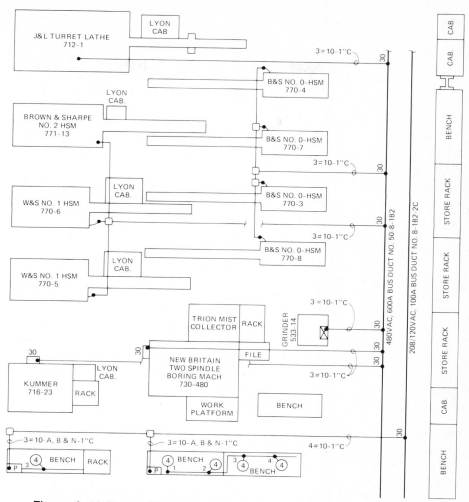

Figure 9·16 Power distribution plan of a mass-production machine shop.

service from a substation (480 V) or a transformer (480 V–208Y/120 V) connected to a 480-V bus duct. The bus ducts are plug-in. This allows bus-plug disconnects to be plugged into them much as plugs are plugged into common house receptacles. From the bus ducts, the electric service is run via the bus plug and conduit to the disconnect on the machine. This machine shop only requires 30-A bus plugs. Each service to the machines shows only the number of conductors, conductor size, conduit size, and phasing, if less than three-phase; voltages are obvious because the service source is shown. Conductor material (copper), conductor insulation (THWN), and conduit material (rigid metal-steel) are not shown because they are standard within this factory. This plan shows equipment names and numbers to locate it more accurately and/

or differentiate it from similar equipment of the same model. Some of the symbols are standard—for example, a disconnect or motor controller—while other symbols are not—for example, a circled number 4, which is a 115-V, 15-A two-outlet receptacle, and a P, which is a 1-φ, 208/120-V, 4-pole power panel. The number next to the circled number is the panel circuit number.

Power distribution plans are an integral part of the electrical drawing package in that they are the only drawings that show service routings and equipment space occupancies.

9·8 Details and specifications

No set of drawings covering a power installation of any size is complete without detail drawings. There may be several large sheets filled with small detail drawings. These may be roughly divided into two categories: (1) special, and (2) standard details.

Special details, which cover the particular project being delineated, may include many different items, installations, or interconnections.

Examples of these drawings appear in Fig. 9·17. Two views showing the arrangement of trays are given (see Fig. 9·17a). These are rectangular metal ducts in which conduits are run throughout the structure. The conduits themselves are usually not shown, but their tray placement can be determined elsewhere, such as in the circuit routing designation in the specifications. Figure 9·17b shows a section of a *duct bank*, which is an underground encasement with six banks of 3-in. Korduct in which cables are run. Korduct is a thin-walled pipe used for encasement in concrete. Transite pipe is used for direct burial in the ground, as are ducts of fiber and plastic.

These are just three of the many types of details that are encountered in drawings for the electric power industry. Standard details, which may be used for more than one project, are usually drawn just once, but reproduced many times. Sometimes they are reduced to $8\frac{1}{2} \times 11$-in. size, then bound in the printed specifications.

The preparation of specifications is an important part of the engineer's job. The person who makes the electrical drawings can expect to do much work on the specifications. Specifications are usually done for installations, equipment, or material. Specifications generally consist of a legal portion, which engineers do not write; specific requirements; general provisions; and general requirements. The specific requirements include such things as scope of work, work by others, contract drawings, information to be submitted with bid, etc. The general provisions, which are usually written by cutting and pasting from previous specifications, contain specifications on such things as codes, interferences, quality of work, materials, tests, contractor's required facilities (office, telephone, etc.), trucking, toilet facilities, etc., or items of a general nature that need to be spelled out and clarified. The general requirements would contain the specifications for equipment and installation materials and methods required for the electrical installation itself. Beside requirements for an electrician's work, this portion of the specifications would spell out the work for other trades.

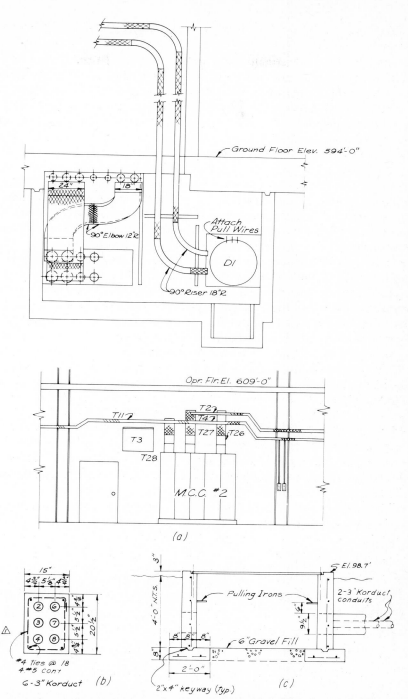

Figure 9·17 Special details that are part of a set of drawings for a power-generating station. *(a)* Tray details. *(b)* Duct bank. *(c)* Pull box. *(Black & Veatch, Consulting Engineers.)*

9·9 General trends

The number of control circuits in generating-station construction is so large that drawing them on one-line diagrams has been largely abandoned in favor of logic diagrams. Similarly, the practice of drawing circuits on electrical plans has given way to showing raceway numbers (trays, conduits, etc.) only. All details are organized in the raceway and conduit schedules and by interconnection diagrams or schematics. Figure 9·10a shows tray details in elevation; similar drawings are used to show tray and raceway details in plan (top) views.

Another trend is that some manufacturers are furnishing internal wiring diagrams of their equipment that show only device terminals and wire numbers, and not the routing of the wiring. These diagrams, therefore, are less cluttered than the typical wiring diagrams; however, the value of this trend is debatable. Another trend is the use of computer-generated drawings by manufacturers and engineering firms. This trend is most apparent in bills of materials, standard details, control schematics, and wiring diagrams. These drawings are generally clear, very well organized, and of good quality.

As in electronics, there have been great advances in technology in the electrical field. Some of these advances are better insulating materials in components (transformers, breakers, wire, etc.); larger voltages in generation (26 kV), transmission (765 kV), and distribution (161 kV); larger and more efficient generating plants; and, of course, nuclear generating stations. But the largest and yet most subtle technical advance in the electric power field is the tremendous expansion in the use of semiconductor-controlled equipment. A partial list is as follows:

1. Static exciters with capabilities up to 3000 kW for electric generators
2. Solid-state protective devices, such as relays, which give more and better protective characteristics for the power-handling devices, such as breakers
3. SCRs for rectification and inversion, one of the prime reasons that dc transmission over long distances has become a practical, economic reality
4. Solid-state speed controls for motors, which have all but replaced their mechanical predecessors

However, the complete online computer control of power generating facilities represents the ultimate in this direction, and it has been accomplished by some utilities and industries. All these technical advances will have an effect on the number and complexity of drawings needed to show the complete picture of an electric power installation.

SUMMARY

The explanations and drawings in this chapter give a representative, but not complete, picture of the drawings used in the electric power field. It is felt that the drawings are typical enough and the coverage broad enough that the reader will have a reasonable knowledge of the drawings done in this area.

The one-line diagram is almost unique to the electric power field, showing in abbreviated form the course or energy flow of the power circuit and associated devices. Logic diagrams are an invaluable aid in developing an overall control scheme. Schematic-type drawings, such as the three-line, control, and elementary diagrams, have the functions of illustrating metering and relaying, as in power circuits, breaker control, and motor control, respectively. General arrangements show how the electric power equipment will be physically arranged when manufactured and/or installed. Connection-type diagrams show the physical wiring of individual pieces of equipment or the interconnection between pieces of equipment. Power distribution plans show the physical layouts of distribution and utilization equipment and the routing of electric services between them. Special and standard detail drawings are necessary and round out the set of drawings required for any large installation. Printed specifications, sometimes volumes of them, will accompany sets of electric power drawings. Engineers and drafters engaged in the preparation of such drawings may expect to participate in the preparation of these specifications.

QUESTIONS

9·1 Describe a one-line diagram and its function.

9·2 How can one indicate that more than one relay of the same type is used at one location on a one-line diagram?

9·3 At which end of a one-line diagram are the higher-voltage lines and equipment placed?

9·4 Name five different kinds of relays that might be shown on a one-line diagram for an electric substation.

9·5 Describe briefly what the following abbreviations stand for: CT, PT, MH, OCB, FCWD, R, MCM, VS, AWG, ACB.

9·6 What is the purpose of a three-line diagram? Describe its arrangement.

9·7 How is more than one relay of the same type, connected to different lines via PTs or CTs, identified on a three-line diagram?

9·8 Describe how a logic diagram is used in electric design, and its interrelation to control schematics.

9·9 What are the five main functions used in logic diagrams for the electric power field?

9·10 What type(s) of circuits can be appropriately described by means of logic diagrams?

9·11 What is understood about pushbuttons and control switches in reading the logic diagrams shown in this chapter?

9·12 Why is it impractical to use only logic diagrams and not have control schematics?

9·13 Describe a control diagram. What is its function? What does the number 52 refer to?

9·14 What is the function of a general-arrangement drawing? What are the typical views found on the general arrangement?

9·15 Name two types of connection diagrams used in the electrical field. What is the purpose of each?

9·16 What method of drawing connection or wiring diagrams is generally used in the electric power field?

9·17 Describe an electric power distribution plan. What is its primary function? What information is found on it?

9·18 What is meant by raceways in electric power construction? By duct banks, trays, and conduits?

9·19 Describe the two types of detail drawings presented in this chapter.

9·20 What are specifications? Do drawings accompany them?

9·21 What are some of the trends in this field? How do you feel they will affect the electrical drawings?

PROBLEMS

9·1 Complete the one-line diagram of the substation shown in Fig. 9·18 by adding the missing symbols. Place suitable identification adjacent to each symbol.

9·2 Starting at the top and going down, draw the symbols in the following list (a.–k.) on a vertical line (except where noted). Place identification of each symbol at the right of symbol on the one-line drawing.

a. 69-kV feeder line

b. 1200-A–69-kV disconnect switch

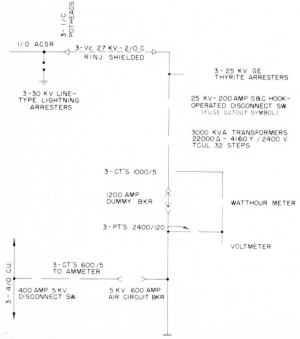

Figure 9·18 (Prob. 9·1) Incomplete one-line diagram of substation.

c. 69-kV arresters (Show leading away on a horizontal line.)

d. 69-kV oil circuit breaker

e. 69-/14.4-kV 21/28-MVA transformer, wye-delta

f. 15-kV 600-A air circuit breaker with male and female connectors

g. 5000-A 15-kV disconnect switch

h. 1000/5 current transformers (Show leading horizontally to ammeter and watthour meter.)

i. 22,000-kW turbogenerator

j. 15-kV/120-V potential transformers (Show leading horizontally to voltmeter.)

k. Ground

9·3 Complete the one-line diagram of the substation shown in Fig. 9·19 by adding the missing symbols. Place suitable identification to the right of each symbol, except for meters and fuse.

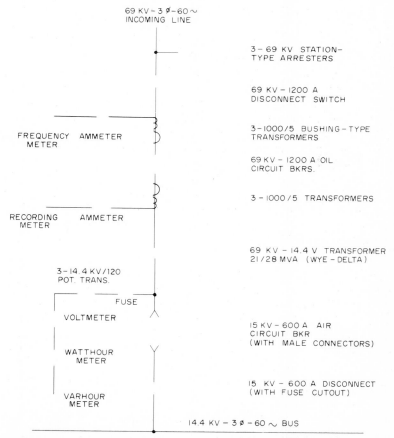

Figure 9·19 (Prob. 9·3) Incomplete one-line diagram of substation.

9·4 Figure 9·20 shows two incomplete sketches of sections through two different substations. The right-hand part in section *A-A,* which contains transformer No. 2, is identical to the left-hand part shown. In section *B-B,* the left-hand span is complete, the center span is complete—except for the circuit-breaker arrangement, which is identical to that shown at the left—and the right-hand span is identical to the center span. Make a mechanical drawing of each of the sections, completing where necessary.

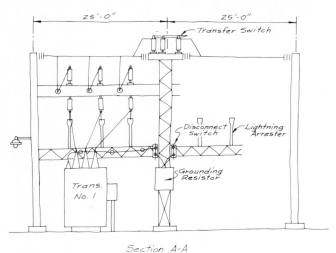

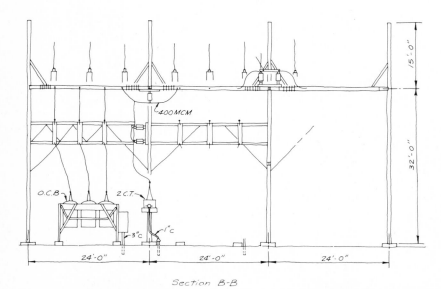

Figure 9·20 (Prob. 9·4) Sketches of substation sections. *Section A-A* (incomplete) through substation yard. *Section B-B* (incomplete) through substation yard.

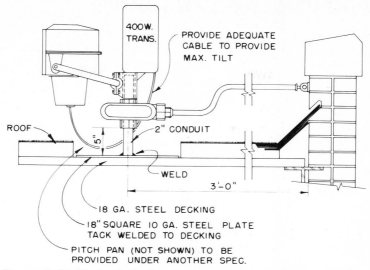

Figure 9·21 (Prob. 9·5) Detail of a 400-W floodlight. *(Black & Veatch, Consulting Engineers.)*

9·5 Redraw the floodlight detail shown in Fig. 9·21 to a size approximately two or three times that shown in the figure.

9·6 An installation has several motors and a control panel. Draw the following logic diagrams for part of this, as follows:

  a. *Control Switch Automatic* and *Safety* must both be on. *Motor No. 1* will start if the preceding is in the on-state *or* if the manual control switch is on. When motor runs, a yellow light goes on at location 2, control panel.

  b. *Motor No. 1* will stop if the *Stop Control Switch or* the *Overload, or* the *Motor Trip* is on. When the above happens, a green light goes on at location 2, control panel, and a buzzer sounds at the motor (MP_1).

  c. For the *Auxiliary Motor* to start, both the *Manual Control Switch* and the *Safety* must be on. Also the *Visual* indicator must be on. Use $8\frac{1}{2} \times 11$ paper for each drawing.

9·7 Figure 9·22 shows a gas-burner valve functional diagram, which is a type of diagram used by some consulting engineers. Make a logic diagram for this valve and add it to the functional diagram shown.

  a. For *fast close* of valve, *Fuel Gas SS valve* must be open.

  b. For *slow close* of valve, *CS* on, *Min Air Flow,* and *Pilot Flame* must *all* be in the on-state.

  c. For *opening* of *Gas Burner Valve,* the events of *both* a and b must all be in the on-state. *Note:* Use NOT symbols for closing of the valve. When valve is closed, a green light shows at 33–13. When open, a red light shows at 33–12. Use $8\frac{1}{2} \times 11$ or 11×17 paper for each drawing.

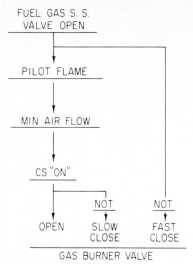

FUEL GAS S. S.
VALVE OPEN

PILOT FLAME

MIN AIR FLOW

CS "ON"

OPEN NOT SLOW CLOSE NOT FAST CLOSE

GAS BURNER VALVE

Figure 9·22 (Prob. 9·7) Functional diagram
of a gas-burner valve.

9·8 Draw a complete one-line diagram of the electric power distribution in
an industrial plant, of which Fig. 9·23 is a diagrammatic sketch. Circuit
breakers (oil) are at locations 1 through 14, and 500-hp motors are at A
and B. Possible location of fault currents are at locations a through h.
Use standard symbols throughout. Use $8\frac{1}{2} \times 11$ paper.

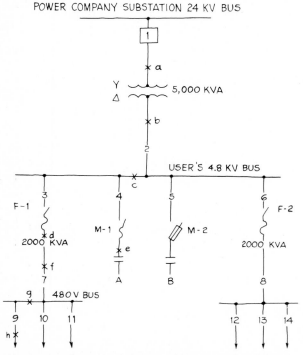

Figure 9·23 (Prob. 9·8) A diagram of electric power
service to a large plant.

9·9 About 60 percent of the one-line diagram for the power distribution of BART is shown in Fig. 9·24. Make a complete one-line drawing of the system with a layout looking like that shown at the upper left. Stations on the Berkeley-Richmond line are (1) RRY, (2) RRI, (3) RCN, (4) RCP, (5) RNB, (6) RBE, and (7) RAS. Stations on the Mission-Market line are (8) MDC, (9) MBP, (10) MGP, (11) MTF, (12) MSS, (13) MPS, and (14) BTW. Configurations of these two lines are similar to the lines shown. Include all lettering. Use 11 × 17 or 12 × 18 paper.

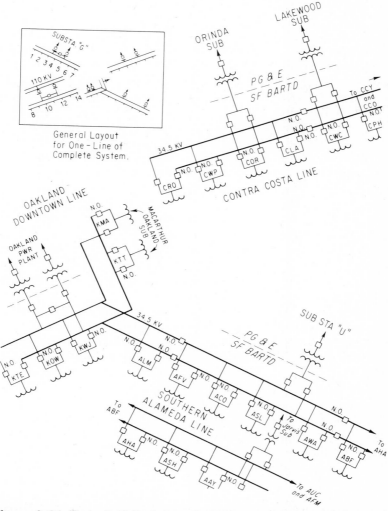

Figure 9·24 (Prob. 9·9) Partial one-line diagram of electrification of S.F. Bay Area Rapid Transit System. *(Parsons-Brinckerhoff-Tudor-Bechtel, Engineering Consultants.)*

Chapter 10
Electrical drawing for architectural plans

An integral part of any set of drawings for the construction of a building is the wiring plan or layout. Several standards apply to this type of design and graphical presentation. Symbols for the drawings (other than those used previously in this text) are shown and explained in ANS Y32.9, "Graphical Electrical Symbols for Architectural Plans," Mil Std 15–3, "Electrical Wiring Symbols for Architectural and Electrical Layout Drawings," and in the *Residential Wiring Handbook* published by the Industry Committee on Interior Wiring Design.

The "National Electrical Code" (NEC), published by the National Fire Protection Association (NFPA) and the American National Standards Institute (ANSI), provides the minimum design criteria necessary to safeguard persons and property practically from the hazards arising from the use of electricity. The Code is voluntarily written by knowledgeable persons in all diverse groups associated with the electrical industry, including unions, manufacturers, inspection agencies, users, technical societies, contractors, utilities, insurance underwriters, and governmental agencies. Many of these organizations are represented by associations or societies. The Code is not intended as a design specification or instruction manual for *untrained* personnel.

The NEC covers electrical conductors and equipment installed within or on public and private buildings, structures, mobile homes, recreational vehicles, industrial substations, and other premises (yards, carnivals, parking lots, etc.). It also covers the conductors that connect the installations to a supply of electricity and other outside conductors. In general, the NEC does not cover installations in ship, water craft, railroads, aircraft, automobiles, or mines, nor does it cover communication equipment used by communication utilities or installations under the direct control of electric utilities. The NEC is purely advisory as far as NFPA and ANSI are concerned, but it is offered for use in law and for regulatory purposes. Thus, the NEC is mandatory because it has been adopted almost in toto by the Occupational Safety and Health Administration (OSHA).

OSHA was created in 1970 by the Occupational Safety and Health Act, which is intended to assure working people safe and healthful working conditions by authorization and enforcement of standards developed under this act. OSHA is under the control of the Department of Labor. Because OSHA is primarily interested in personnel safety and not system or equipment design and installation requirements, they are proposing to adopt NFPA Standard No. 70E, "Electrical Safety Requirements for the Employee Workplace," and delete the nonessential (to OSHA) portions of the NEC.

Certain political entities (cities and counties, for example) often have their own building codes which are more restrictive than the NEC, and these must be met when doing electrical design within their jurisdiction.

10·1 "National Electrical Code" (NEC) definitions and contents

Because the NEC is such an important document, persons engaged in producing electrical drawings for architectural structures must be familiar with it and with other local codes. These persons should also be conversant with standard terminology and equipment. For the benefit of the reader, we give some of the definitions used in the code and a brief explanation of its contents, so that the rest of this chapter can be followed more easily. However, it should be remembered, the NEC is the standard for the minimum provisions associated with electrical installations necessary for personnel and property safety; it is *not* a drawing standard.

Some of the definitions used in the NEC are a little strange compared with their everyday use; however, they should be learned because they are peculiar and essential to the proper use of the Code. Some of the NEC definitions that are more applicable to the information contained in this chapter are as follows:

Accessible:
As applied to wiring methods: Capable of being removed or exposed without damaging the building structure or finish, or not permanently closed in by the structure or finish of the building. (See "Concealed" and "Exposed.")
As applied to equipment: Admitting close approach because not guarded by locked doors, elevation, or other effective means. (See "Readily Available.")
Ampacity: Current-carrying capacity of electric conductors, in amperes.
Appliance: Utilization equipment, generally other than industrial, normally built in standardized sizes or types, which is installed or connected as a unit to perform one or more functions, such as clothes washing, air conditioning, food mixing, deep frying, etc.
Attachment Plug (Plug Cap): A device which, when inserted into a receptacle, establishes connection between the conductors of the attached flexible cord and the conductors connected permanently to the receptacle.
Branch Circuit: The circuit conductors between the final overcurrent device protecting the circuit and the outlet(s).
Appliance: A branch circuit supplying energy to one or more outlets to which appliances are to be connected. Such circuits have no permanently connected lighting fixture not a part of an appliance.
General Purpose: A branch circuit that supplies a number of outlets for lighting and appliances.
Individual: A branch circuit that supplies only one utilization equipment.

Multiwire: A branch circuit consisting of two or more ungrounded conductors which have a potential difference between them, and an identified grounded conductor which has equal potential difference between it and each ungrounded conductor of the circuit and which is connected to the neutral conductor of the system.

Building: A structure which stands alone or which is cut off from adjoining structures by fire wall with all openings therein protected by approved fire doors.

Cabinet: An enclosure designed for either surface or flush mounting and provided with a frame, mat, or trim in which a swinging door or doors are or may be hung.

Circuit Breaker: A device designed to open and close a circuit by nonautomatic means and to open the circuit automatically on a predetermined overcurrent without injury to itself when properly applied within its rating.

Concealed: Rendered inaccessible by the structure or finish of the building. Wires in concealed raceways are considered concealed, even though they may become accessible by withdrawing them. [See "Accessible, As applied to wiring methods."]

Conductor

Bare: A conductor having no covering or electrical insulation whatsoever. (See "Conductor, Covered.")

Covered: A conductor encased within material of composition or thickness that is not recognized by this Code as electrical insulation. (See "Conductor, Bare.")

Insulated: A conductor encased within material of composition and thickness that is recognized by this Code as electrical insulation.

Dead Front: Without live parts exposed to a person on the operating side of the equipment.

Device: A unit of an electric system which is intended to carry, but not utilize, electric energy.

Disconnecting Means: A device, or group of devices, or other means by which the conductors of a circuit can be disconnected from their source of supply.

Enclosure: The case or housing of apparatus, or the fence or walls surrounding an installation to prevent personnel from accidentally contacting energized parts or to protect the equipment from physical damage.

Exposed

As applied to live parts: Capable of being inadvertently touched or approached nearer than a safe distance by a person. It is applied to parts not suitably guarded, isolated, or insulated. (See "Accessible" and "Concealed.")

Feeder: All circuit conductors between the service equipment or the generator switchboard of an isolated plant and the final branch-circuit overcurrent device.

Grounding Conductor, Equipment: The conductor used to connect the non-current-carrying metal parts of equipment, raceways, and other enclosures to the system grounded conductor and/or the grounding electrode conductor at the service equipment or at the source of a separately derived system.

Ground-Fault Circuit Interrupter: A device whose function is to interrupt the electric circuit to the load when a fault current to ground exceeds some predetermined value that is less than that required to operate the overcurrent protective device of the supply circuit.

Lighting Outlet: An outlet intended for the direct connection of a lampholder, a lighting fixture, or a pendant cord terminating in a lampholder.

Location

Damp: Partially protected locations under canopies, marquees, roofed open porches, and similar locations, and interior locations subject to moderate degrees of moisture, such as some basements, some barns, and some cold-storage warehouses.

Dry: A location not normally subject to dampness or wetness. A location classified as dry may be temporarily subject to dampness or wetness, as in the case of a building under construction.

Wet: Installations underground or in concrete slabs or masonry in direct contact with the earth; locations subject to saturation with water or other liquids, such as vehicle washing areas; and locations exposed to weather and unprotected.

Outlet: A point on the wiring system at which current is taken to supply utilization equipment.

Panelboard: A single panel or group of panel units designed for assembly in the form of a single panel, including busses, automatic overcurrent devices, and with-or-without switches for the control of light, heat, or power circuits; designed to be placed in a cabinet or cutout box, placed in or against a wall or partition and accessible only from the front. (See "Switchboard.")

Raceway: A channel designed expressly for holding wires, cables, or busbars, with additional functions as permitted in this Code. Raceways may be of metal or insulating material, and the term includes rigid metal conduit, rigid nonmetallic conduit, intermediate metal conduit, liquid-tight flexible metal conduit, flexible metallic tubing, flexible metal conduit, electrical metallic tubing, under-floor raceways, cellular concrete-floor raceways, cellular metal-floor raceways, surface raceways, wireways, and busways.

Receptacle: A contact device installed at the outlet for the connection of a single attachment plug.

Receptacle Outlet: An outlet where one or more receptacles are installed.

Remote-Control Circuit: Any electric circuit that controls any other circuit through a relay or an equivalent device.

Service: The conductors and equipment for delivering energy from the electricity supply to the wiring system of the premises served.

Service-Entrance Conductors:

Overhead System: The service conductors between the terminals of the service equipment and a point usually outside the building, clear of building walls, where joined by tap or splice to the service and to the device drop.

Underground System: The service conductors between the terminals of the service equipment and the point of connection to the service lateral.

Service Equipment: The necessary equipment, usually consisting of a circuit breaker or switch and fuses, and their accessories, located near the point of entrance of supply conductors to a building, other structure, or otherwise defined area, and intended to constitute the main control and means of cutoff of the supply.

Switchboard: A large single panel, frame, or assembly of panels on which are mounted, on the face or back or both, switches, overcurrent and other protective devices, busses, and usually instruments. Switchboards are generally accessible from the rear as well as from the front and are not intended to be installed in cabinets. (See "Panelboard.")

Utilization Equipment: Equipment which utilizes electric energy for mechanical, chemical, heating, lighting, or similar purposes.

Voltage (of a Circuit): The greatest root-mean-square (effective) difference of potential between any two conductors of the circuit concerned.

Briefly, the NEC contains standards on installation, application, construction, materials, and equipment associated with the electrical industry. Standards are found in the following areas:

1. Wiring design and protection, which includes all types of circuits (branch, feeder, etc.), protective devices (fuses, circuit breakers, lightning arrestors, etc.), and grounding.

2. Wiring methods and materials, which include all types of cable, raceways, busways, wireways, boxes, fittings, panelboards, switchboards, etc.

3. Equipment for general use, such as flexible cords, lighting fixtures, appliances, heating/ventilating/air-conditioning equipment, motors, motor controllers, generators, transformers, capacitors, resistors, reactors, and batteries.

4. Equipment and methods associated with special occupancies, such as places where fire or explosion hazards may exist (garages, bulk-storage plants, aircraft hangars), health facilities, theaters, studios, manufactured buildings, mobile homes and parks, recreational vehicles, and marinas or boatyards.

5. Special equipment such as electric signs, cranes, hoists, elevators, escalators, electric welders, sound-recording equipment, data-processing equipment, x-rays, induction/dielectric heating equipment, metal-working tools, irrigation equipment, and swimming pools.

6. Special electrical conditions, such as emergency systems; systems over 600 V; installations under 50 V; remote-control, signaling, and limited-power circuits; standby power-generation equipment; and fire-protective signaling systems.

7. Communications systems such as telephone, telegraph, central alarm stations, radio and TV receiving and transmitting equipment, and CATV systems.

The preceding has been a relatively brief description of the terminology and contents of the NEC. If questions should arise, the Code should be consulted. It is available from NFPA or ANSI.

10·2 Simplified and true wiring diagrams

A true wiring diagram shows every wire and its connection in a system, or circuit. Such a diagram is shown in Fig. 10·1a, in which four ceiling light-fixture outlets are depicted, two of which are connected to, and controlled by, individual single-pole single-throw switches. A simplified arrangement of this branch is shown at the right in the same figure. Here, approved symbols have been used for the light outlets, the switches, and the wire run, which may be of nonmetallic sheathed cables, armored cables, or any approved method of running conductors between outlets. The two parallel dashes across the wire runs indicate that a two-wire conductor is to be used. Actually, according to the standards, when a two-wire run is to be installed, the dashes may be omitted. If the conductor is to be composed of more than two wires, dashes indicating the number of wires must be provided on the drawing.

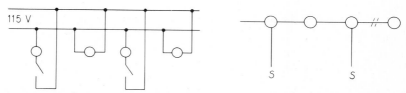

Figure 10·1 Wiring diagrams of light-fixture outlets on a circuit. *(a)* True wiring diagram. *(b)* Simplified, or installation, diagram.

10·3 Wiring symbols on a simple floor plan

The architect usually shows the location of lights, convenience and special-purpose outlets, and the desired switching arrangements on a floor plan. For small, simple structures, the required symbols and wiring arrangements may be drawn on the same floor plan (Fig. 10·2) that shows all information necessary for the erection of the building. For larger or more complicated structures, complete wiring details will probably be drawn on separate floor plans, called *electrical layouts* or *electrical plans*. In either case, the simplified type of diagram, such as that shown in Fig. 10·1b, will be used. This wiring layout will be drawn by an architect, engineer, or drafter who is familiar with the engineering and building-code requirements.

The living-room plan (Fig. 10·2) shows two circuits: (1) the three-way switching arrangement for the ceiling outlet, and (2) a similar arrangement for the two convenience outlets on the north wall. The outlet symbols, including the special-purpose outlet (indicated for TV antenna), are taken from ANS Y32.9. Also in accordance with the standard, the wire symbols for the switch-to-ceiling-light runs are drawn with a medium-weight solid line, indicating that the wires are to be concealed in the walls or ceiling above. Where no perpendicular dashes are shown across the wires, the conductors must have two wires.

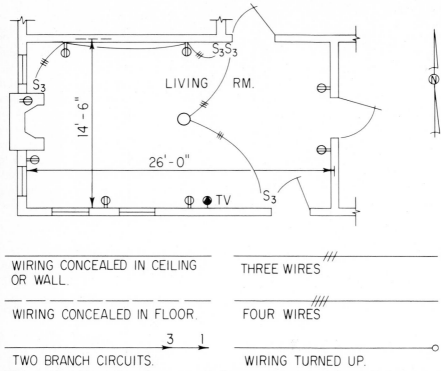

WIRING CONCEALED IN CEILING OR WALL.

WIRING CONCEALED IN FLOOR.

TWO BRANCH CIRCUITS.

THREE WIRES

FOUR WIRES

WIRING TURNED UP.

Figure 10·2 Floor plan of a room with fixtures, outlets, and switches. Standard wiring symbols are below.

With the addition of a little more information about fixtures, the floor plan, of which the living-room plan of Fig. 10·2 is a part, will yield enough information for the satisfactory installation of the complete electrical system.

10·4 Separate electrical plans

A plan for the electrical system of a small business building appears in Fig. 10·3. This drawing was one of several, including plans and details for heating, air conditioning, and plumbing, which appeared on a separate sheet of what might be appropriately called mechanical drawings.

This electrical plan shows the location of three separate distribution panelboards and the proposed location of a future one. Panel C provides power service—mainly for the motors which run the mechanical equipment; panels A and B supply electricity for lighting and the other electrical needs of offices 102, 103, 104 and the vestibule. Each branch circuit is documented with an arrow pointing in the general direction of the panel and a designation such as A-2. This means, for example, that the nine fluorescent light fixtures in office

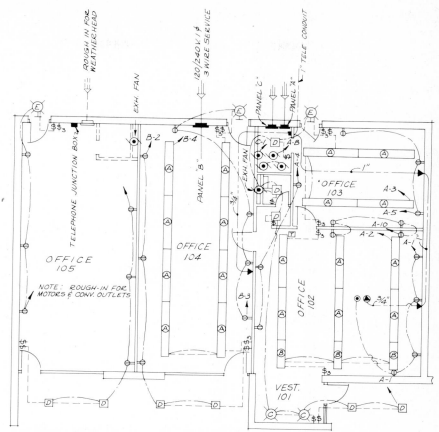

Figure 10·3 Electrical plan for a small office building. *(Brasher, Spencer & Goyette, Architects-Engineers.)*

102 are on the same circuit, No. 2, which is fed at panel A. A separate telephone circuit, enclosed in conduit, is also shown.

A number of symbols that either do not appear in or differ from those shown in ANS Y32.9 are drawn in this electrical plan. The wall bracket outlets have the four prongs which are still widely used,[1] but which are not shown in ANS Y32.9. A long-and-short-dashed line is used for circuit legs, regardless of whether the wires are run in the ceiling above, or the floor below, and the short-dash symbol is used for the switch leg. Confusion in the interpretation of these symbols is avoided by preparation of a legend.

[1] One explanation for the continued popularity of the four prongs is that many persons feel that the plain circular symbol listed in ANS Y32.9 may be easily confused with other circular symbols which may appear on drawings.

10·5 Fixture schedule and legend

Figure 10·4 shows a legend and fixture schedule that accompanies the electrical plan of Fig. 10·3. Inclusion of such schedules and legends is the customary practice of architects and consulting engineers who prepare electrical layouts and details for the construction of buildings. The installation of the electrical system is facilitated by the inclusion of a letter designation at each fixture symbol and cross-referenced designations in an accompanying schedule. The exact form of the schedules has not been standardized. A "remarks" column has been omitted from the original schedule from which Fig. 10·4 was taken in order to conserve space.

A drawing should show intent, and in the most concise manner. Time, money, and argument will be saved if the proper information is placed in the legend. Too often, the person in the field does not see the written specifications, but does see the drawings and the legend. If the symbol for a floor convenience outlet appears, for example, the worker knows what the symbol means, but not which of the 50 available combinations is required, unless it is specifically stated. Now, if the legend reads "FLOOR CONVENIENCE OUTLET—Frank Adam FB-3," the worker will know exactly what device to install.

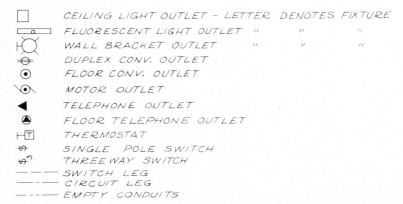

	ELECTRICAL FIXTURE SCHEDULE					
MARK	MFG.	CAT. NO.	MOUNTING	WATTS	LAMP	FINISH
A	FELCO	4039	RECESS	3-95	430 MA	STANDARD
B	FELCO	4033	RECESS	3-38	430 MA	STANDARD
C	LITECRAFT	2305	WALL BRACKET	100	I. F.	SATIN ALUM.
D	PRESCOLITE	488-6600	RECESS	100	I. F.	STANDARD
E	PRESCOLITE	WE-2	WALL BRACKET	100	I. F.	SATIN CHROME

LEGEND :

CEILING LIGHT OUTLET - LETTER DENOTES FIXTURE
FLUORESCENT LIGHT OUTLET   "   "   "
WALL BRACKET OUTLET   "   "   "
DUPLEX CONV. OUTLET
FLOOR CONV. OUTLET
MOTOR OUTLET
TELEPHONE OUTLET
FLOOR TELEPHONE OUTLET
THERMOSTAT
SINGLE POLE SWITCH
THREE WAY SWITCH
SWITCH LEG
CIRCUIT LEG
EMPTY CONDUITS

Figure 10·4 Fixture schedule and legends for electrical plan of small office building. *(Brasher, Spencer & Goyette, Architects-Engineers.)*

10·6 Example of electrical layout

Another electrical layout is shown in Fig. 10·5; this depicts the second floor of a three-story office building. This floor arrangement might be suitable for a drafting or design room. The wiring symbols conform reasonably to the American Standard code. Arrows indicate the circuit and number; half arrows indicate partial circuits.

Included with this electrical plan were a legend, lighting-fixture schedule, panel schedule, disconnect-switch schedule, telephone-circuit riser diagram, and electrical riser diagram. Of these, only the electrical riser diagram is shown in this chapter. Before it is discussed, a word about how energy is distributed throughout a structure seems to be advisable.

As electric energy is brought into a building, it is usually first passed through a meter. From here it is brought into a main load center. In a small building or residence this load center consists of a fuse box or circuit breaker to which each branch circuit is connected. Through these branch circuits energy is fed to each outlet, lamp, or appliance. In a larger structure the main load center may be a large panelboard, or switchboard, with circuit breakers, disconnect switches, and other controlling devices. From this main load center, electric energy is fed through large conductors called *feeders* to branch load centers, often panelboards, or *panels* as they are sometimes called. From these panels energy is delivered through branch circuits to each outlet, fixture, appliance, or motor. Such factors as location of panelboards, voltage and copper losses, etc., determine the method used in connecting the branch and main load centers. Many different interconnecting layouts are used. More than one panelboard may be necessary on the same floor of a large building in order that excessively long runs of branch circuits be avoided. Sometimes separate panels are used for lighting circuits only, and others for motors and machines only. Often, panelboards are used for combinations of types of circuits.

10·7 Electrical riser diagram

A riser diagram shows how electric energy is distributed through a building from the time it enters the building until it arrives at the branch load centers. It is drawn as an elevation and usually is not to scale. An electrical riser diagram for the building whose second-floor electrical plan is shown in Fig. 10·5 appears in Fig. 10·6. This diagram shows the electrical service passing through four meters, then through four disconnect switches. From switches S_1, S_2, and S_3 the current goes to lighting panels L_2, L_3, and L_B. From the latter two panelboards, energy goes to the lighting and convenience outlets in the ground floor and to mechanical equipment for the ground and first floors. Current travels from panelboard L_3 through a subfeeder to L_1, which supplies all the electrical service on the first floor, except for air-conditioning equipment. Panel L_2 supplies electric energy to most of the circuits on the second floor. However, the electrical layout for this floor (see Fig. 10·5) indicates that outside lighting and mechanical equipment on the roof are connected to panelboards on lower floors.

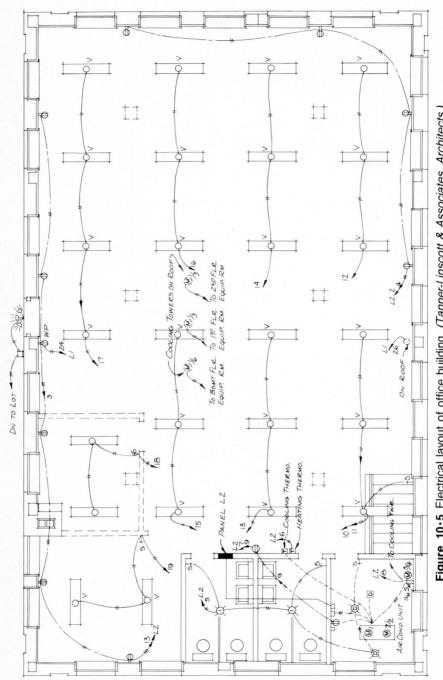

Figure 10·5 Electrical layout of office building. *(Tanner-Linscott & Associates, Architects.)*

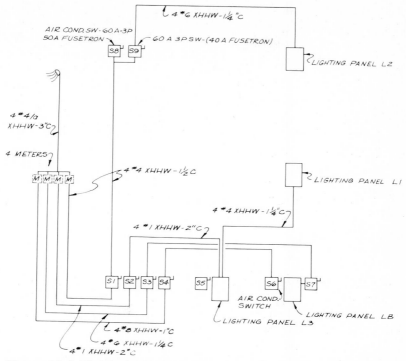

Figure 10·6 Electrical riser diagram for three-floor office building. *(J. R. DeRigne & Associates, Consulting Engineers.)*

The panelboard schedule which accompanies the riser diagram includes the following information about each panel:

1. The number of branch circuits to be served
2. Current-handling capability (amperage)
3. Name of manufacturer and catalogue number of panel

Similar information is given in the disconnect-switch schedule.

10·8 Electrical drawings for large buildings

The next series of drawings illustrates the type of drawing required for large buildings. The standards are not as much help in showing such a complicated system; hence liberal use is made of orthographic projection, pictorial drawing, schematic diagrams, notes, and specifications.[1] A very brief description of the general electrical layout of a large office building will be given by referring to the drawings, which in some instances have been simplified from the original for clarity.

[1] "The National Electrical Code" (ANSI-C1) contains information that is most helpful in the planning of electrical systems for large buildings.

The underground cables of the power and light company enter the building at the upper right, as shown in Fig. 10·7. They immediately are brought into a main service entrance, which is shown in some detail in Fig. 10·8. In the main service entrance are current and potential transformers (used for measuring current and potential), fuses, and switches, as shown in the one-line diagram. From the main service entrance the electrical service is fed to two 1000-kVA transformers (see Fig. 10·7), which step the voltage down to a four-phase state that provides 208-V three-phase service and 120-V single-phase service. From these transformers current is brought through 3000-A 4-φ (4-phase) busses to

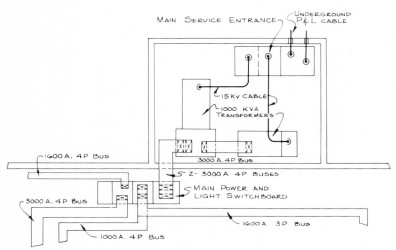

Figure 10·7 Part of basement plan of large office building, showing details of electrical service. *(W. L. Cassell, Mechanical Engineer, and Tanner-Linscott & Associates, Architects.)*

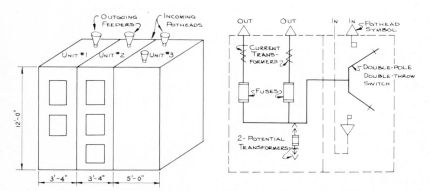

Figure 10·8 Pictorial and one-line diagrams of main service entrance of large office building.

the main switchboard, shown in pictorial form in Fig. 10·9. From this switchgear, electric power is distributed by busses to various parts of the basement, as shown in Fig. 10·7. Actually the large copper or aluminum busbars themselves are not shown. Just the bus ducts are drawn—much the same as heating and air-conditioning ducts are drawn.

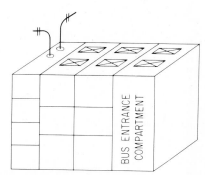

Figure 10·9 Pictorial drawing of main power and light switchboard for large office building.

Service to upper floors is fed through the large 3000-A 4-φ bus—which is shown in both Figs. 10·7 and 10·10 (the riser diagram)—to branch load centers on each floor. At each branch load center are a cable tap box, disconnect switches, and panelboards. Here, use is made of metal rectangular raceways (also called "trays"), in which connecting cables are placed. From the panelboards, service is distributed throughout the floors, as explained in the following two paragraphs.

Figure 10·11 shows a part of the electrical plan for the first floor. Shown in somewhat slight detail is the load center with its cable tap box and panels. Of special note is the underfloor duct system, shown by means of dotted lines. In these ducts, sometimes called raceways, are placed the wires for telephone service, 120-V single-phase electrical service, and 208-V three-phase service if desired. Outlets may be placed at any point along each of these ducts, and electrical service provided at each of these points. The ducts are covered with $2\frac{1}{2}$ in. of concrete, on which additional flooring material is often laid. The outlets are usually installed after the latter has been laid, but before the conductor cable has been pushed or drawn through the raceways. However, additional outlets may be emplaced along these ducts after the building has been completed and occupied, although new wiring may have to be poked through the ducts.

Figure 10·12 is a photograph of a raceway system being installed. Note that the raceways are often divided into more than one compartment. It is also possible to have a complete undergrid or cellular floor system. In this type of construction wires can be laid 12 in. apart all the way across the room

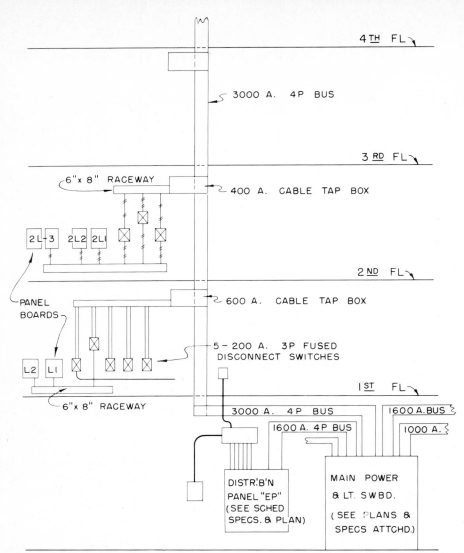

Figure 10·10 Part of riser diagram for a large office building. *(W. L. Cassell, Mechanical Engineer.)*

in both directions. This provides a high degree of flexibility because a telephone, minicomputer, or intercom connection can be placed anywhere in the room.

Details of various electrical hookups and appurtenances are often desirable or necessary. They may take several forms and have different amounts of information. Two detail drawings are shown in Fig. 10·13.

10·9 Coordination and organization of drawings

There are usually areas on every drawing which, from the electrical contractor's viewpoint, should be made more comprehensive to help in preparing bids. Also the person doing the installing would benefit by having more information about

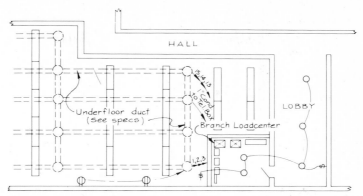

Figure 10·11 Simplified electrical plan of first floor of a large office building, showing under-floor ducts.

Figure 10·12 Under-floor duct system being installed. *(Walker-Parkersburg.)*

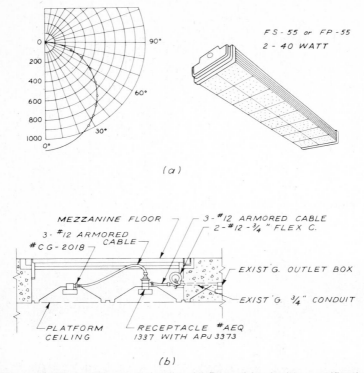

(a)

(b)

Figure 10·13 Architectural details. *(a)* Part of luminaire specification. *(b)* Cabling detail. *(S.F. Bay Area Rapid Transit District.)*

correct or desired procedure, in many situations. However, it is possible to put too much information on the drawing—in trying to answer every question that may arise—and, as a result, confuse everybody from the contractor to the final inspector. Yet when special treatment is indicated, pertinent details should be supplied in order to present a logical and definite manner for the installation.

Many offices which prepare electrical and mechanical drawings for structures have improved their coordination with all crafts doing the work on a job. Such coordination (both before work starts and during the work) avoids conflicts between various mechanical items on the project. Last-minute changes in equipment by manufacturers cause some conflicts and cannot be fully predicted in advance. However, obvious conflicts can be avoided by proper coordination while the drawings are being made.

If a consulting engineer is doing the wiring and mechanical design for an architect, their offices should also coordinate. Such things as space for raceways, ducts, and air-conditioning equipment must be provided.

10·10 Load computation

Using information supplied in the "National Electrical Code," it is possible to compute the number of branch circuits and feeders required in many types of structures. Let us assume that a residential dwelling has a floor area of 1490 sq ft, exclusive of basement, attic, and garage, and a 12-kW kitchen range. Using Tables 10·1 to 10·3, we compute the general lighting load and other requirements as follows:

Table 10·1 General Lighting Loads by Occupancies

Type of Occupancy	Unit Load per Square Foot (Watts)
Auditoriums	1
Banks	5
Dwellings (other than hotels)	3
Industrial buildings	2
Office buildings	5

Table 10·2 Calculation of Feeder Loads by Occupancies

Type of Occupancy	Portion of Lighting Load to Which *Demand Factor* Applies (Watts)	Feeder Demand Factor (Percent)
Dwellings (other than hotels)	First 3,000 or less at	100
	Next 3,001 to 120,000 at	35
	Remainder over 120,000 at	25
Warehouses (storage)	First 12,500 or less at	100
	Remainder over 12,500 at	50
Others (except hospitals, hotels, and apartment buildings)	Total wattage	100

Table 10·3 Demand loads for Household Electric Ranges, Ovens, and Counter-Mounted Cooking Units over 1.75-kW Rating

Number of Appliances	Maximum Demand (Not over 12-kW Rating)	Demand Factors	
		Less than 3.5-kW Rating	3.5 to 8.75-kW Rating
1	8 kW (8000 W)	80%	80%
2	11 kW	75%	65%
3	14 kW	70%	55%

1. General lighting load
   1490 sq ft @ 3 W/sq ft = 4470 W (see Table 10·1)
2. Minimum number of branch circuits required
   a. General lighting load (based on 115 V): 4470 ÷ 115 = 39.0 A. This can be handled by three 15-A two-wire circuits or two 20-A two-wire circuits. (NEC recommends one circuit for each 500 sq ft.)
   b. Small-appliance load: (Sec. 220–3 states that two or more 20-A branch circuits shall be installed to take care of small appliances in the kitchen, laundry, dining room, and breakfast room.)
3. Minimum-size feeders required:

Computed Load	Watts
General lighting	4470
Small appliances (computed at 1500 W for each circuit in accordance with Sec. 220–3)	3000
Total (without range)	7470
3000 W @ 100% (Table 10·2)	3000
7470 − 3000 = 4470 W @ 35%	1565
Net computed (without range)	4565
Range load (see Table 10·3)	8000
Total net computed	12565

If 115/230-V three-wire system feeders are used, the current load will be 12,565 ÷ 230, or 55 A. But, because the computed load exceeds 10 kW, service conductors shall be 100 A [Sec. 230–41(b)(2)]. Feeder size can be selected from one of several tables in the NEC, which gives ratings for types of copper, aluminum, and copper-clad aluminum conductors. The feeders might be selected as No. 3 AWG copper conductor with rubber-type insulation, such as RHW or RUH, or with a thermoplastic-type insulation, such as THW or THWN; or the feeder might be selected as No. 1 AWG aluminum or copper-clad aluminum with the same types of insulation. As mentioned earlier, city building codes may have other requirements not specified in the National Code. Kansas City requires that all wiring be run in conduit for buildings (including houses) that happen to be in the Class A fire district. In other areas, conduits must be used in duplexes and multiple-residency buildings of two stories or more.

SUMMARY

Depicting electrical requirements for residential buildings, offices, and other structures requires a combination of graphical treatment, knowledge of building codes, and written specifications. For small structures, such as houses, it is usually sufficient to show electrical outlets, fixtures, and switches right on the house plans and to indicate what switching arrangements are required. In general, the rest of the circuiting arrangement is left to the constructor. But in larger buildings, it is necessary to depict all the circuit arrangements, including panelboard locations, in order to comply with safety requirements, to allow plenty of flexibility for future equipment and expansion, and to minimize expense of installing electrical work. National drawing standards ANS Y32.9 and Mil Std

15–3 (which are identical) are followed by most engineers who prepare electrical drawings. In depicting electrical requirements for buildings, it is necessary or desirable to make wiring plans, riser diagrams, and schedules for fixtures, raceways, conductors, etc. Symbol legends are sometimes used, especially if nonstandard symbols are used in drawings. Pictorial views and orthographic views of certain features and equipment are often quite helpful. Coordination by the person making the drawings with the architect, craftspeople, and equipment manufacturers is highly desirable because it will minimize conflicts and unnecessary installation expenses. Local building codes must be followed by the designer and constructor. When the local code does not cover a situation, the "National Electrical Code" must be consulted. If there is no local electrical code, the NEC should be used.

QUESTIONS

10·1 What authority, or authorities, shows the symbols to be used in making a residential floor plan that shows the electrical arrangement?

10·2 Why are riser diagrams usually not drawn to scale?

10·3 Why do we not use bus ducts in residential structures?

10·4 Sketch two different symbols which might be used to show a TV outlet.

10·5 What column headings would you use for a fixture schedule for a small building?

10·6 Why is it not customary to show two dashes across a circuit line for a two-wire conductor, when three dashes are customarily used to show a three-wire circuit?

10·7 Can you use the same graphical symbol for an electric-range outlet and for a dishwasher outlet?

10·8 Name three different schedules pertaining to the electrical system of a large building that might be found among the drawings for that building.

10·9 Name or describe two situations in which lines for the conductors themselves are not shown on drawings describing the electrical layout of a large building.

10·10 Does ANS Y32.2 cover the graphical portrayal of the electrical system of a residence?

10·11 Briefly describe three methods of providing underfloor electrical service to an office building whereby electric current can be supplied at evenly spaced intervals throughout or across a room.

10·12 Where does a branch circuit begin? Name two kinds of branch circuits.

10·13 Where are overcurrent devices located?

10·14 What are two reasons for limiting the number of wires in a raceway?

10·15 What are the terminal points of a feeder (or feeder circuit)?

10·16 What is one difference between a switchboard and a panelboard?

10·17 What is the smallest rating a panelboard may have? Or to put it differently, what determines the minimum rating of a panelboard?

10·18 Under what conditions does the NEC say that raceways should not be installed?

10·19 What devices or systems can be used for grounding electrodes? (Name four.)

10·20 What is the minimum rating of a receptacle?

10·21 What can be done to avoid conflicts?

10·22 Why do some architects (or consulting engineers doing electrical design for architects) find it necessary or advisable to add legends to their drawings?

10·23 What sort of information would be contained in a legend for an electrical drawing?

PROBLEMS

Many of the problems are divided into two parts. Part *a* will be mainly a drawing requirement, and part *b* will involve computational requirements. In some cases, the instructor may require both parts; in other cases, only *a* or *b*. Computations do not need to be put on drawing paper. However, they should be neat, orderly, and correct.

10·1 The sketch of the floor plan of Fig. 10·14 shows all electrical outlets and fixtures and the relationships between the fixtures and switches of an apartment.

    a. Make a mechanical drawing of this plan using currently approved symbols for wire runs, fixtures, and outlets. Wiring for duplex convenience outlets does not have to be shown except where they are controlled by a switch. Use 11 × 17 or 12 × 18 paper.

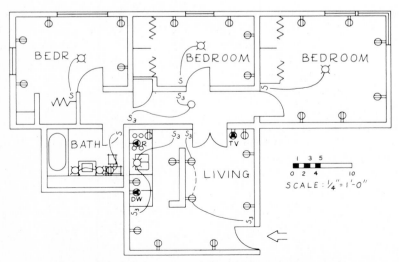

Figure 10·14 (Prob. 10·1) Plan of an apartment unit.

b. How many general 20-A and appliance 20-A branch circuits are required? If the range load is 8000 W and 110/220 three-wire feeders
are used, what will be the total current load for the first floor?

10·2 How many 15-A branch circuits will be required for the room shown
in Fig. 10·2? How many 20-A branch circuits would be required for
the office layout of Fig. 10·5? (The interior dimensions are 22 × 57 ft.
Each fluorescent luminaire is 80 W.)

10·3 Figure 10·15 shows the floor plan of an apartment with all outlets,
fixtures, and switches. Switches, for the most part, work overhead lights,
but in the living room they operate two of the four outlets.

a. Make a drawing (mechanical or freehand) of this floor plan, adding
doors and the necessary wiring. Show TV outlets in the living room
and largest bedroom. Use standard symbols. Use 11 × 17 or 12 ×
18 paper.

b. A feeder system of 115/230 V and three wires is used and the kitchen
range is rated at 11 kW. How many 15-A branch circuits will be
required for the apartment? What will be the total current load?

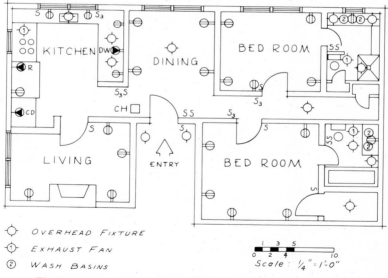

Figure 10·15 (Prob. 10·3) Floor plan of luxury apartment.

10·4 Figure 10·16 is the floor plan of a one-story residence.

a. Draw this plan to a scale of $\frac{1}{4}'' = 1'-0''$. Show all switching arrangements as you think they ought to be. At least two duplex outlets in
the living room are to be controlled by switches at the north entry.
Use standard symbols throughout the drawing.

b. The kitchen range has an 11-kW rating, and 220-V three-wire feeders are used. How many branch circuits (general and appliance) will be required? (What will the difference be if central air conditioning is installed?) What will be the total current load to be used for computing the size of the feeders?

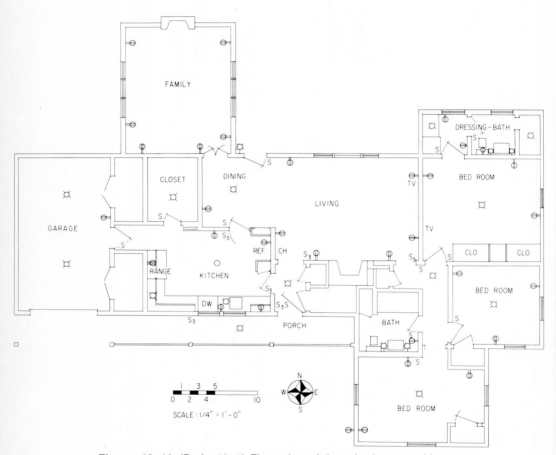

Figure 10·16 (Prob. 10·4) Floor plan of three-bedroom residence.

10·5 Figure 10·17 is the first floor plan of a two-story house with all outlets and fixtures shown.

a. Draw this plan to a scale of $\frac{1}{4}'' = 1'-0''$, and show all switching arrangements as you believe they should be made. Use correct symbols as given in the Appendix. Optional: show circuits for the convenience outlets. Use 11 × 17 paper.

b. Assume the house has three bedrooms and a hall upstairs. For an 8-kW kitchen range and the other equipment shown, how many and what kind of branch circuits would be required for the house? What is the total energy requirement?

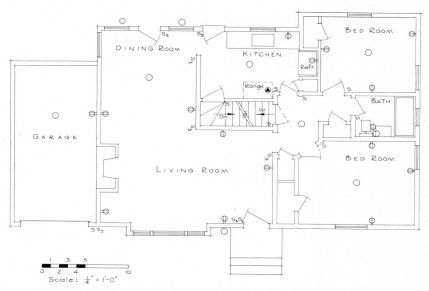

Figure 10·17 (Prob. 10·5) Floor plan of first story of residence.

10·6 The lighting fixtures for a drafting-design room are shown in Fig. 10·18. If the fluorescent lights are rated at 80 W each, estimate the number of branch circuits needed. An electrical outlet should be put near each desk. Four desks are placed under each row of luminaires, making a total of 24 drafting desks. Lights will be turned on and off at panel L_3. Other circuits go to panel L_1 on another floor. Make a scale drawing of this room, and show all circuits. Use 11 × 17 or 12 × 18 paper.

10·7 Under each luminaire shown in Fig. 10·18 will be placed a designer's desk. It is desired to have provisions for electric erasers, telephones, and electric calculators at each desk. Make a detailed drawing that includes a system of under-floor raceways which will provide this electrical service to all desks. Assume that telephone conductors and 115-V conductors can be placed together in a raceway. Use 11 × 17 or 12 × 18 paper.

10·8 Because the telephone company objected to having telephone lines in the same duct with other conductors, it was decided to install a cellular-type, under-floor wiring system for the drafting room of Fig. 10·18.

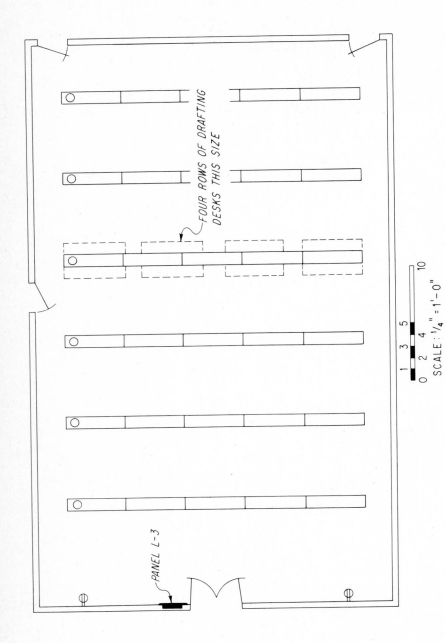

FOUR ROWS OF DRAFTING
DESKS THIS SIZE

PANEL L-3

0 2 4

1 3 5

10

SCALE: $\frac{1}{4}$" = 1'-0"

Figure 10·18 (Probs. 10·6 to 10·8) Floor plan of drafting-design room.

Draw the floor plan to scale, showing a cellular system that will supply 115-V ac energy to outlets at each desk, and separate cellular ducts for intercom and telephone service to each desk. Overhead lighting may be included in the drawing. Use 11 × 17 or 12 × 18 paper.

10·9 Figure 10·19 shows a section of a three-story building with a basement. Draw a riser diagram for this building as you believe it would be designed. Figure 10·6 will give you an idea of what conductor sizes to use. Use XHHW insulated wire. XHHW insulation is heat and moisture resistant and made of a synthetic, cross-linked polyethylene. The current-carrying capabilities of XHHW insulated copper conductors in damp locations are as follows:

Conductor Size (AWG)	Ampacity
14	15
12	20
10	30
8	45
6	65
4	85
3	100
2	115
1	130

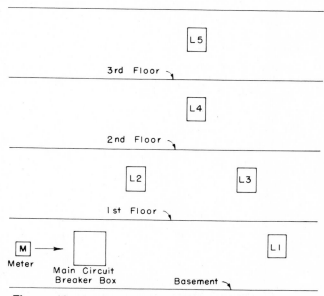

Figure 10·19 (Prob. 10·9) Section elevation of three-story office building with electrical distribution equipment.

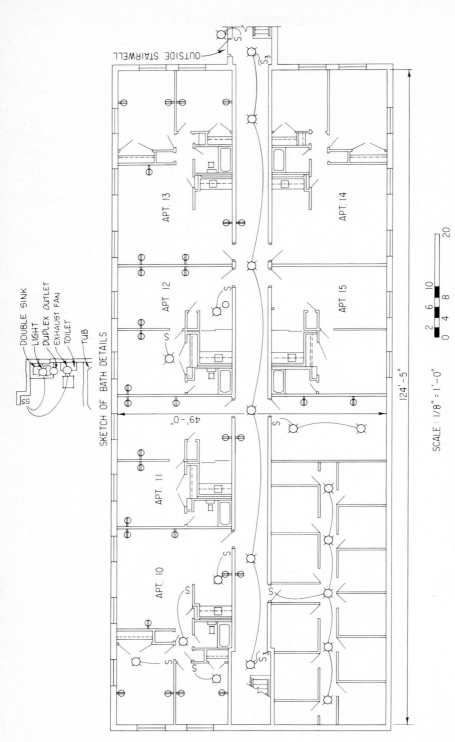

Figure 10·20 (Prob. 10·10) Electrical and heating plan of ground-floor apartments.

SCALE: 1/8" = 1'-0"

10·10 The incomplete heating and electrical plan of one floor of an apartment building is shown in Fig. 10·20. Overall dimensions are also shown. Apartments 10, 13, and 14 are identical except that they are reversed. The same is true for apartments 11, 12, and 15.

   a. Make a complete scale drawing of this plan, showing all circuiting and switching arrangements for all the rooms and the hall. Appliance loads are 8 kW for ranges, 2 kW for dishwashers, and 4.2 kW for washer-dryers. These appliances will be located in the kitchen of each apartment. Their exact locations and outlets are not shown. (Use 12 × 18 paper for a $\frac{1}{8}''$ to $1'-0''$ scale.)

   b. If a copy of the NEC is available, compute the minimum number of branch circuits and the minimum size subfeeder for each apartment. Compute, also, the main feeder for this floor and the rating of the panel, if one panel is used.

10·11 Letter a fixture schedule to accompany an electrical plan. Such a fixture schedule might include the information shown in Prob. 1·7, Chap. 1.

Chapter 11
Graphical representation of data

The graph is one of the most effective tools of communication that any technically trained person can wield. With a graph, one can bring order to a collection of data and present it in a picture form that tells a story. Also with a graph, one can compare the performance of two or more related items or processes and may be able to make certain computations not practicable by algebra, analytic geometry, or calculus. In development and research the graph is used to determine the relationships of two or more variables, to compare laboratory data with theory, and to determine if test data are accurate and reliable.

Because this information must be presented honestly, accurately, and as clearly as possible, skill and judgment are required to make a good graph. Therefore, an engineer or drafter should develop as much skill and knowledge within this area as in any other area or type of technical drawing. The fundamental principles of graphical representation are:

1. Graphs should be truthful representations of the facts.
2. Graphs should be clear and easily read and understood.
3. Graphs should be so designed and constructed as to attract and hold the attention.

Figure 11·1 is a good example of a well-drawn technical or engineering type of graph.

11·1 General concepts in preparing graphs

One of the most difficult problems in constructing a graph is choosing the horizontal and vertical scales. As an example of this problem, Fig. 11·2 is presented. In this figure the same "curve" is shown on four different graphs, each having different arrangements of scales; thus, it may be difficult to recognize that the same data are being presented. Graph *a* is the best of the four for two reasons: (1) the data are presented more honestly, and (2) better use is made of the available space than in any of the other three drawings. This problem of the selection of scales can often be solved by using commercially prepared graph paper and making wise use of the space available. However, it may not always be possible or desirable to use such paper.

Another problem in graph construction is deciding what type of graph to make. In other words, the problem might be, "What type of graph paper should be used?" There are many different types of graphs in use today, and many different kinds of graph paper. Most graph papers are printed on $8\frac{1}{2} \times 11$-in. paper, although larger sizes are available. Lines come in black, blue, green, orange, and red for many graph styles. Grids are available in rectangular, polar, and probability coordinates, to name several examples.

Most technical and engineering graphs are drawn on *rectangular coordinate* paper. (It is also called *arithmetic, rectilinear, cross-ruled,* and *square-grid* paper.) Typical spacings for this type of paper are 5, 10, and 20 lines (or spaces) to the inch and 10 to the centimeter. Other spacings such as 4, 6, and 8 lines to the inch are available, but are not in wide usage. The graph of Fig. 11·1 was made on paper that has 10 lines to the inch, as are several other examples in

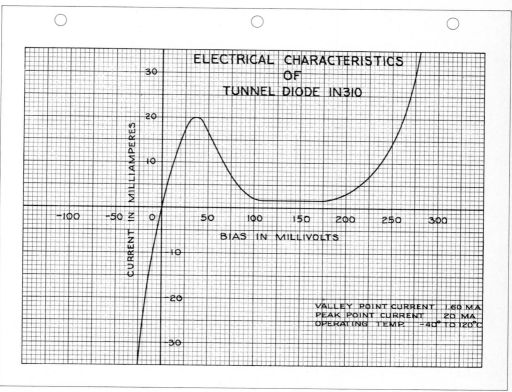

Figure 11·1 A typical graph, plotted on rectangular coordinate paper.

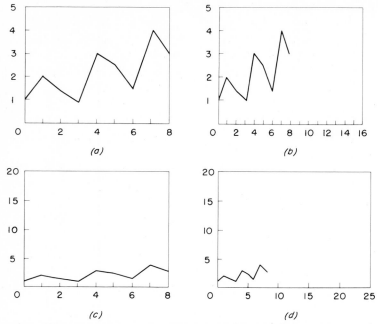

Figure 11·2 Graphs having different scales on which identical data have been plotted.

this chapter. If it is desired to make blueprints, Ozalid prints, or other similar types of reproductions, then thin, translucent graph paper can be used. If the graph paper has blue grid lines, these will not appear in the reproduction, but orange, red, or black lines will show up on a print. The graphs in this chapter which have the grid lines showing have been drawn on red-line graph paper.

In order to make good reproductions (by Ozalid, Thermofax, xerography, and other reproduction processes), lines and lettering must be made heavy and dark in order to "stand out" from the grid lines. Ink drawings give the best results, but dark, heavy pencil work will provide legible copies in some reproduction processes. In order to make lettering stand out even more, one can do the lettering on heavy, white paper, then paste the paper on the graph. This has been done with the title on Fig. 11·5 and in several other graphs in the chapter. This is effective, but not necessary, and will not reproduce on blueprints or Ozalid prints. Sometimes the thin grid lines on a graph can be used as guidelines. This has been done with the supporting data (lower right) of Fig. 11·1. This lettering should be the smallest lettering on the graph. The $\frac{1}{10}$-in. height between successive guidelines is ideal.

The tunnel-diode curve of this graph falls in two quadrants, making it necessary to arrange the vertical and horizontal scales so that negative as well as positive values can be plotted. By locating the zero point *(origin)* to the left of center, it was possible to draw the desired portion of the curve, show the scales and their captions, and provide space in the upper part for the title and in the lower right for the auxiliary or supporting data. The X and Y scale values could have been placed around the edges of the graph, but it is believed that they are more appropriately located close to their respective axes, as shown in Fig. 11·1.

11·2 Selection of variables and curve fitting

Data to be plotted graphically are generally available in tabular form, with each point having two coordinates as follows:

Point	Coordinates	
1	0	0
2	10	2.6
3	20	3.8
4	30	5.3
5	40	7.8
6	50	10.1

One set of coordinate values must be plotted on the horizontal, or X axis (or *abscissa* as it is often called), and the other set of coordinates must be plotted along the vertical, or Y axis (or *ordinate*). Standard practice is to plot the *independent* variable *horizontally* and the *dependent* variable *vertically*. The independent variable is that variable which the operator can control during a test, if one can be controlled. In some cases where a variable cannot be controlled, one variable is arbitrarily selected. *Time,* for instance, is generally considered to be the independent variable. A glance at the coordinates, appearing above, shows that the first set of coordinates progresses at even intervals of 10. It is obvious that the operator or observer was able to take readings at 10, 20, etc., either by controlling the variable or, if it were a natural phenomenon such as time, by taking readings at convenient intervals. Those coordinate values in the *left* column, then, represent the independent variable.

After the graph is laid out and the points are plotted, the problem of drawing, or "fitting," the final curve presents itself. Whether to draw a smooth curve or straight lines between successive points depends on several factors. Some of them are as follows:

1. Most physical phenomena are "continuous." This means that the curve showing the relationship between such variables should be smooth—with few inflections—and should pass through or near plotted points.

2. Data backed up by theory should be represented by a smooth curve. (However, if plotted points are not abundant, straight lines are often drawn between points, as with instrument calibration.)

3. Discrete, or discontinuous, data—representing a discontinuous variable having discrete increments—should be shown by joining successive points with straight lines.

4. Observed data not backed by theory or mathematical law should be represented by point-to-point straight lines, unless continuity can be definitely established.

Figure 11·1 includes a curve representing continuous data. Figure 11·2 has a curve which follows the discontinuous relationship of periodic observations. Figure 11·3a shows a theoretical curve and points taken from actual field data. Such points should be joined by a smooth curve. Figure 11·3b is a typical example of discontinuous data.

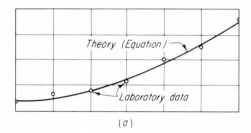

(a)

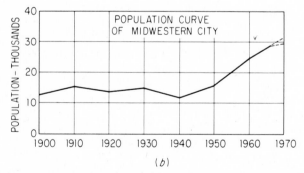

(b)

Figure 11·3 Examples of curve fitting. *(a)* A situation in which a curve should be drawn through or near points plotted from laboratory data. *(b)* Census figures, taken once every 10 years, yield discontinuous, or discrete, data. Straight lines should be drawn from point to point.

11·3 Curve identification

Not infrequently, it is necessary to put more than one curve on a graph. The curves must be drawn or labeled so that they can be easily identified and distinguished from one another. There are three methods by which this can be done:

1. Clearly label each curve
2. Use a different type of line for each curve
3. Use different plotting symbols for each curve

Often two or more of these methods are combined. For example, in Fig. 11·4 different lines and plotting symbols have been used, except for the *composite* curve, which is a sort of weighted average of the other curves. A similar identification system has been used in Fig. 11·5. The difference between the two figures is the method of labeling curves. In Fig. 11·4, a *legend* in the upper right corner identifies each line. In Fig. 11·5, each line is identified by means of a title, or caption, and a leader pointing to the curve itself. Both methods are widely used.

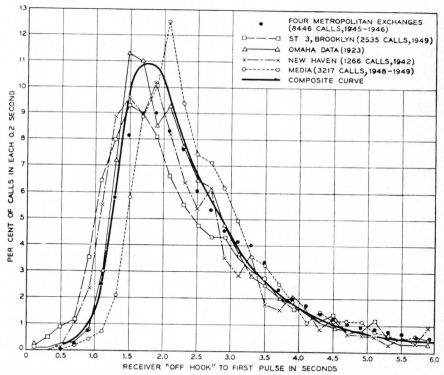

Figure 11·4 A composite curve. The heavy curve is a composite of the other four curves. *(Bell Telephone Laboratories.)*

11·4 Zero-point location

In a great majority of cases, line graphs are drawn with the *origin* at the lower left corner. Figures 11·2 and 11·4 are in this category. In other cases, the vertical scale begins at zero, but the horizontal scale does not. Figure 11·3*b*

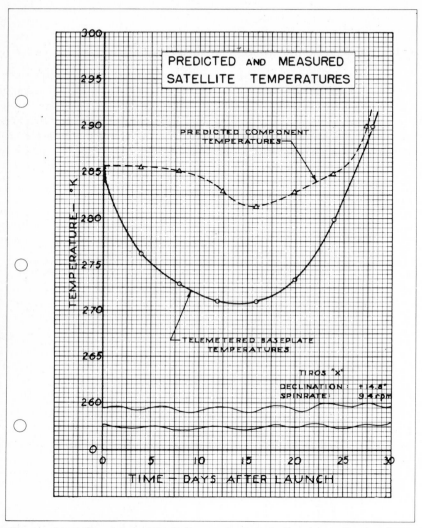

Figure 11·5 A graph in which the vertical scale has been "broken" to permit greater vertical excursion.

is a good example. Most engineers feel that the *Y* scale should begin at zero; or conversely stated, they feel that starting the *Y* scale with a figure *not* equal to zero tends to distort the picture.

Sometimes, however, the values to be plotted on the vertical scale are such that starting the scale at zero will produce a plotted curve that is too flat and inaccurate for working purposes. Figure 11·5 is an example of this problem. Note that the temperature plots of the weather satellite fall between 270 and 290 K. By "breaking" the vertical scale between 0 and 260 K, we were able to establish ordinate values which provided sufficient vertical latitude ("excursion") for accuracy and comparison of the two curve shapes. There are other techniques for accomplishing the same result, the point being to warn the reader that the vertical scale is not all there or that it starts at something other than zero.

11·5 Steps in construction of an engineering graph

The following steps illustrate how a graph, as drawn in many engineering offices, is constructed. Such a graph is shown in Fig. 11·6.

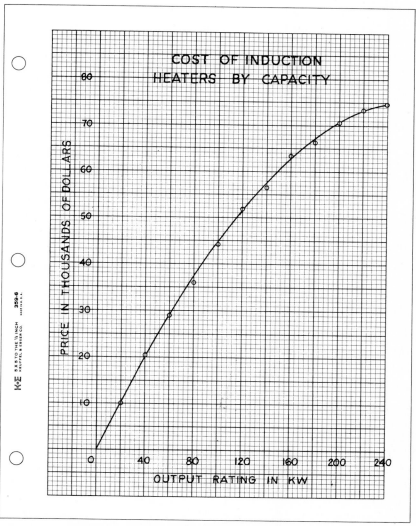

Figure 11·6 Another typical engineering graph.

1. Arrange the data in table form, preferably in a logical order, from the smallest figure to the largest. This will make plotting easier and faster, and will clearly show the upper and lower limits for each variable.

2. Determine which variable will be the independent variable and which the dependent, for the very practical reason that proper coordinate paper must be selected and a decision made whether the long dimension of the paper will be up and down or sideways on the drawing board. These will probably depend on which variable is which.

3. Select the graph paper that will best show the curve and that will accommodate the points to be plotted taking into account their extreme values. This may include a trial plotting of values along the horizontal and vertical axes. In the case of Fig. 11·6, we selected rectangular coordinate paper having fine lines spaced $\frac{1}{10}$ in. apart, and heavier lines $\frac{1}{2}$ and 1 in. apart.

4. Locate the zero point of each scale (origin of graph), allowing for margins within the grid portion of the paper at the bottom and left side. One-inch margins are common because they allow room for binding the graph at the left side and for scale values and their descriptions.

5. Letter in the values along and outside the two "baselines" provided in step 4. Standard practice is to use multiples of 1, 2, 5, or 10. In Fig. 11·6, multiples of 10 (each inch representing $10,000) were used on the vertical scale, and multiples of 2 (each half-inch representing 20 kW) were used on the vertical scale. Do not put these numbers too close together.

6. Letter in the descriptions of the scales close to the figures, as shown in Figs. 11·6 and 11·7. Vertical lettering should be used, and each description should be centered between the zero point and the figure at the other end of the baseline. Each description should tell what the scale values show, and what units the scale values represent. Typical standard abbreviations are: kW, A, Hz (cycles per second, often cps in drawings), and kHz (kilocycles per second).

7. Plot the points with a sharp pencil or pricking instrument. It is a good idea to circle these points immediately after plotting so that they will be easily seen when drawing the curve later. If the circles are to be shown permanently, as is so often done with experimental data, they should be hollow and from $\frac{1}{16}$ to $\frac{1}{10}$ in. in diameter.

8. Draw in the curve. If the curve is not of the straight-line variety, many persons prefer to sketch it lightly freehand until the desired result is obtained, then to put it in with a heavy pencil line, using an irregular curve. If plotting symbols are used, the curve should not be drawn through the symbols. If a symbol is in the path of the curve, the latter should come up to and just touch each side of the symbol.

9. Place the title on the graph. Titles—which are clear, yet as brief as possible—are commonly placed either above or below. This lettering should be vertical uppercase and larger than the numbers and letters

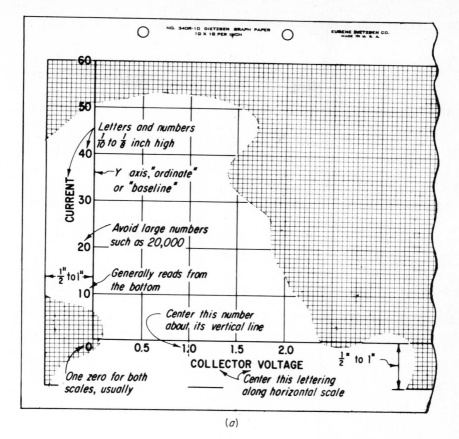

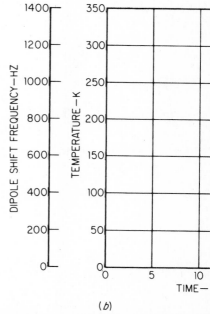

(b)

Figure 11·7 (a) Layout details for the construction of a graph. (b) When two or more scales are required.

describing the vertical and horizontal scales. Some firms require that the fine, preprinted grid lines be erased from around titles and borders drawn around the lettering.

10. Place supporting data, if desired or required, on the graph. Such data may include date of preparation, site of tests or observations, name of person or party making the tests or graph, source of data if not original, equations, and simple circuit diagrams. Such data are often placed in the lower right-hand area of the graph, but are sometimes placed elsewhere if circumstances require. (See Figs. 11·1 and 11·5.)

11. Complete the graph. This may include inking curve, borders, and lettering if inking is required. Sometimes letters and figures can be typed on graph paper with standard typewriters or varitypers if special ribbon is used.

A graph should be made interesting and clear through the use of different line weights. The curve should be the heaviest line on the graph, the border(s) or baselines the next heaviest, and grid lines (if not preprinted) the lightest.

11·6 Drawing a smooth curve

Drawing a curve generally involves two steps:

1. Sketching, freehand, a trial curve through (or near, as the case may be) the plotted points

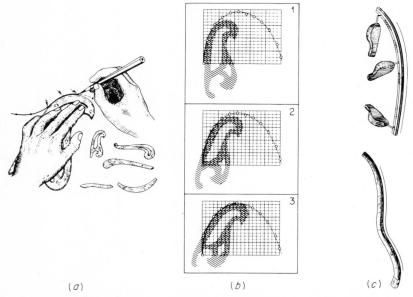

(a)   (b)   (c)

Figure 11·8 Irregular curves, splines, and use of the curve. *(a) and (c) from Frank Zozzora, "Engineering Drawing," 2d ed., McGraw-Hill Book Company, New York, 1958. (b) from Thomas French and Carl Svenson, "Mechanical Drawing," 6th ed., McGraw-Hill Book Company, New York, 1957.)*

2. Drawing the finished curve along the trial curve, using pencil or pen and a plastic curve or spline

Some of the plastic (often called *irregular* or *French*) curves and their usage are shown in Fig. 11·8. In 99 cases out of 100, it will not be possible to select a plastic curve that will match the plotted curve for its entire length. The next-best solution is to try a curve (if more than one is available) or that part of a curve (if only one is available) that will fit as much of the trial curve as is possible. (If the curve is to be inked later, it is a good idea to remember which parts of the curve were used at different locations along the final curve shape.)

Curve drawing is usually more satisfactory if the pencil is used on the edge of the curve that is away from the person who is drawing. This is somewhat analogous to using the upper (far) edge of a T square.

11·7 Scales and their titles or captions

There are good and bad ways to organize the numbers and descriptions along the baselines of a graph. Here are a few examples:

		Good Selection of Numbers			
0	10	20	30	40	50
0	0.2	0.4	0.6	0.8	1.0
0	50	100	150	200	250

		Poor Selection of Numbers			
0	10,000	20,000	30,000	40,000	50,000
0	30	60	90	120	150
0	7	14	21	28	35

Good Title or Caption Organization

RESISTANCE IN OHMS
RESISTANCE IN THOUSANDS OF OHMS
RESISTANCE—THOUSANDS OF OHMS
RESISTANCE IN MEGOHMS

Satisfactory But Sometimes Confusing

RESISTANCE—OHMS $\times 10^3$
PARTS PER FT$^3 \times 10^6$
TRIGGER CHARGE—10^{-10} COULOMBS
(The numbers shown are all positive.)

11·8 Families of curves

Figure 11·9 shows a *collector-characteristic* curve of a transistor which was obtained by varying the voltage and measuring collector current for several values of base current. This family of curves can be used for the determination of transistor performance in a common-emitter circuit and also for the calcula-

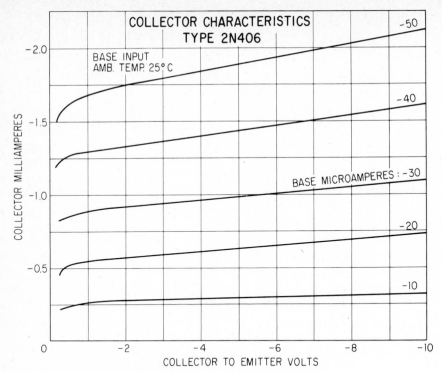

Figure 11·9 A family of curves. *(RCA.)*

tion of other useful parameters. For example, it is often desirable to know the ratio of the direct collector current, I_c, to the direct base input current, I_B. This *current gain* is typically around 49 or 50, calculated as follows:

$$\beta = \frac{I_C}{I_B} = \frac{0.981}{0.021} = 49$$

Other useful characteristics which can be obtained from such a table and other information supplied by a manufacturer are frequency cutoff, breakdown voltage, reach-through voltage, and storage time. When five or more curves are placed on a single graph, it is not very practical to draw a different type of line for each curve. Proper identification of each line is usually accomplished in the manner shown in Fig. 11·9.

11·9 Graphs for publication

When graphs are to appear in printed matter, such as books or technical journals, they are usually not drawn on commercial graph paper. ANS Y15.1, "Illustrations for Publication and Projection," is an appropriate guide for such cases. Covered are such points as minimum letter size and line weights before reduction to legible and clear size. A standard outline proportion of $6\frac{3}{4} \times 9$ in. ($\frac{3}{4}$:1) is recommended. No more coordinate lines should be drawn than are necessary to guide the eye. Because additional discussion is presupposed, supporting data and equations should be omitted. In the majority of published graphs, plotting

370

symbols are not shown, although some do include these symbols. Figures 11·9, 11·10, and 11·14 are good examples of this type of graphical presentation.

In short, the published graph is a rather simple construction, uncluttered by minute data, yet very clear and legible. In order to have clarity and legibility, it might not include all data necessary to tell the whole story.

11·10 Line graphs on other types of graph paper

As mentioned previously in this chapter, other kinds of commercially printed graph paper are available and used by many engineers and scientists for various reasons. Figure 11·10 exhibits a response curve for an amplifier over a range of frequencies going from 100 to 5,000,000 Hz. The vertical scale (not logarithmic) is measured in decibels, or units of volume. This response curve, like most amplifier-response curves, is plotted on five-cycle semilogarithmic paper, in order to plot accurately along the great ranges of frequencies. It would be impossible to plot numbers over such a large range on linear (rectangular coordinate) paper. This is one reason for using paper with one or more logarithmic scales.

Another reason for using different kinds of graph paper is to achieve a pattern that yields a "straight-line curve." Sometimes, points that yield a curved

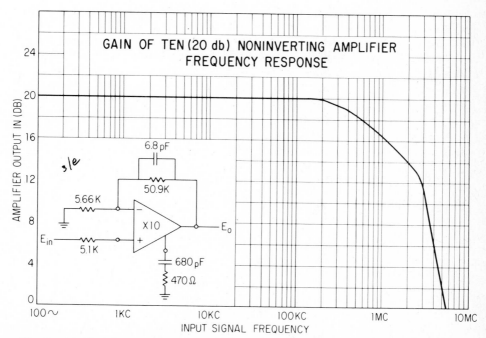

Figure 11·10 A frequency-response curve plotted on five-cycle semilogarithmic paper. The range on the horizontal scale is from 100 cycles to 10 megacycles (10 million cycles) per second. The new term for cycles per second is hertz, abbreviated Hz.

pattern on one kind of paper yield a straight-line pattern on another kind of paper. The following will result if a straight-line curve can be drawn:

1. Future prediction (extrapolation) is easier if a straight line is used.
2. Two curves can be better compared if they are straight-line curves than if they are otherwise shaped.
3. The slope and equation of the line (and therefore of the data which produced the line) can be obtained graphically.

If the logarithmic scale is oriented vertically and the linear scale horizontally, a curve representing a constant *rate* of change will plot as a straight line. Such a curve would result if we were to plot the 2.8 percent annual increase in the rate of change of our output per man-hour since World War II.

11·11 The logarithmic scale

The logarithmic scale is a functional scale in which the distances are laid out to equal the function (logarithm), but the numbers at these distances are those of the variable. A logarithmic scale from 1 to 10 is called a *cycle*. The left edge of the scale is marked 1 (the logarithm of 1 is zero), and the right edge 10 (the logarithm of 10 is 1), and we have a *unit* scale from the log of 1 (zero) to the log of 10 (1). This unit scale is called a cycle, but we can scale intermediate distances like 1.5, 2, 3, 5, etc., and show those numbers at those points. A logarithmic scale from 10 to 100 or 100 to 1000, etc., is also called a cycle, or *modulus*.

It is impossible to have a zero showing on a logarithmic scale. If the scale includes the number 1 at either end, the zero is there *graphically* because the logarithm of 1 is zero. Also, the logarithm of zero is minus infinity. This would be impossible to plot graphically.

It is also worth noting that *interpolation* between marks on a logarithmic scale must be done *logarithmically*. When using logarithmic graph paper one does not have to be concerned about this.

11·12 Equations of straight-line plots

The following equations will apply in the situations indicated:

Equation	Type of Coordinates
$y = mx + b$	Rectangular
$y = \dfrac{x}{mx + b}$	Rectangular $\left(\dfrac{x}{y} \text{ values plotted against } x \text{ values}\right)$
$y = bx^m$	Logarithmic
$y = b\,(10)^{mx}$ or	Semilogarithmic (logarithmic scale vertical)
$y = b\,(e)^{mx}$	
$y = m \log x + b$	Semilogarithmic (logarithmic scale horizontal)

In the equations on page 372, m represents the *slope,* and b the *intercept.* We will show how to get these quantities in the next two examples.

11·13 Use of logarithmic paper

Logarithmic paper (both scales are arranged logarithmically) is used for reasons similar to semilogarithmic paper. If the range of plotted values is large in both the x and y directions, logarithmic paper, with the proper number of cycles, can be used very much as semilogarithmic paper was used to accommodate the large range of frequency values used in Fig. 11·10. As in the case of semilogarithmic graph paper, logarithmic paper is available with anywhere from one to five logarithmic cycles, and anything in between. The most-used papers have the same number of cycles in the horizontal and vertical directions, but papers are available with different combinations. Usually, however, the length of one horizontal cycle is the same as the length of one vertical cycle.

Another reason for plotting data on logarithmic paper is to get a straight-line pattern. It might be found by trial and error that a certain group of data provides a straight-line plot on such paper. Or theory or previous tests with similar data may indicate that a set of plotted points will yield a straight-line pattern. Figure 11·11 contains such a situation. Plotting of points and construction of the graph follow the same techniques, previously explained, for drawing a graph on rectangular coordinate paper. One difference is that the numbers around the edge are usually already printed on the sheet. The drafter often must add zeros or decimal points to the numbers of the scale.[1] Because these numbers are outside the grid, the scale caption must also be placed outside the grid. Another difference is that the curve itself should not be very thick, if one wants to obtain accuracy in measuring the slope and locating the intercept.

In Fig. 11·11, a right triangle with a 5-in. base has been drawn. Each leg of the triangle has been measured accurately with a decimal scale to obtain Δx and Δy values. (This is *construction* work, and it is sometimes erased and does not appear on the finished graph.) The important point about measuring the legs of the triangle on logarithmic paper (where the horizontal and vertical cycles are equal in length—which is usually the case) is that a linear scale be used and that the *same* scale be used to measure each leg. Instead of using a decimal scale, we could have used a 50 scale, 40 scale, quarter scale—or any *linear* scale. Using the logarithmic scale of the graph paper to measure the legs will *not* provide the correct slope.[2] The slope is obtained as follows:

$$m = \frac{\Delta y}{\Delta x} = \frac{1.68}{5.00} = 0.336 \text{ or } 0.34$$

[1] Sometimes the printer does not leave enough room for this, or there is not much room for scale titles, presenting a rather awkward situation.

[2] If the horizontal and vertical cycles of the paper are not equal in length, one will have to select points on the line and work with their coordinates to get the slope: $m = (\log Ay - \log By)/(\log Ax - \log Bx)$.

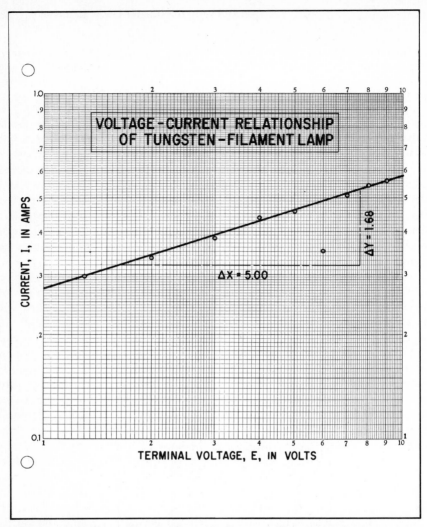

Figure 11·11 Data plotted on logarithmic paper. These data yield a straight-line pattern. $\Delta y/\Delta x$ represents the slope.

The *intercept* is found by observing where the line intersects the Y scale where $x = 1$ (remember, the log of 1 is zero) and reading the value along the Y scale. The intercept in Fig. 11·11 is

$$b = 0.272$$

The equation for this line is, therefore,

$$y = 0.272x^{0.34}$$

or, using the abbreviations for the actual units plotted,

$$I = 0.27E^{0.34}$$

The third-decimal-place values have been dropped because there probably is not enough justification for this type of accuracy. Two-decimal accuracy is appropriate, however.

11·14 Use of semilogarithmic paper

Figure 11·12 illustrates a situation in which a series of plotted points provides a straight-line pattern, as shown by the line. As in the case of the logarithmic

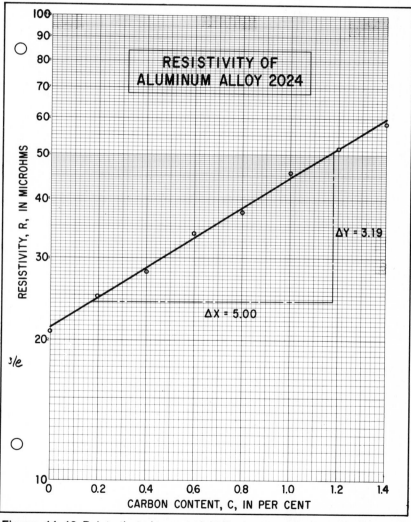

Figure 11·12 Points that give a straight-line pattern when plotted on semi-logarithmic paper.

graph, previously cited, the scales had to be marked off around the outside edges of the grid because the log scale numbers had already been printed.

The Δy and Δx values shown were measured with the same *linear* scale. But because the two scales of a semilogarithmic graph are different, we have to do a little more work to get the correct slope. The additional work (beyond what was done in the case of the logarithmic graph) is to obtain values of *unity* for the X and Y scales. Unity for the X scale is 5.00 in. (the actual distance from $x = 0$ to $x = 1.0$). Unity for the Y scale is 10.00 in. (the distance from log $10 = 1$ to log $100 = 2$), in this case the length of the logarithmic cycle from 10 to 100. Now, we are ready to determine the slope.

$$m = \frac{\Delta y/\Delta x}{y \text{ unity length}/x \text{ unity length}} \quad \text{measured with the same linear scale}$$

$$m = \frac{3.19/5.00}{10.0/5.0} = 0.319 \text{ or } 0.32$$

The intercept is found by reading the value along the Y scale where the line intersects at $x = 0$:

$$b = 21.3$$

The equation is

$$y = 21.3 (10)^{0.32 x}$$

or

$$R = 21.3 (10)^{0.32 C}$$

If it is preferred to use e instead of 10 in the equation, the slope must be divided by 0.434, which is $\log_{10} e$.

$$m = \frac{0.32}{0.434} = 0.74$$

and the equation becomes

$$R = 21.3 \ (e)^{0.74 C}$$

If one cannot reconcile the graphical method just explained, one can check the slope by selecting two points on the curve and using their coordinates as follows:

$$m = \frac{\log Ay - \log By}{Ax - Bx}$$

We have done this by selecting points at $A = (1.16, 50)$ and $B = (0.46, 30)$ to get

$$m = \frac{1.699 - 1.476}{1.16 - 0.46} = 0.319 \text{ or } 0.32$$

There are other methods which can be used to obtain, or check, these equations. One method which requires more work but which is more accurate—and which can utilize a digital computer—is the method of least squares.

11·15 Polar coordinates

Because of the directional characteristics of electrical devices such as lamps, antennas, and speakers, studies must be made of their performance in different directions. The results of these studies can be most appropriately shown on polar charts. These show two variables, one having a linear magnitude plotted on equally spaced concentric circles, and the other an angular quantity plotted radially with respect to a *pole* or origin. The plotted points are usually, but not always, joined to form a line or curve.

Figure 11·13 shows a polar plot of the power-radiation pattern of a half-

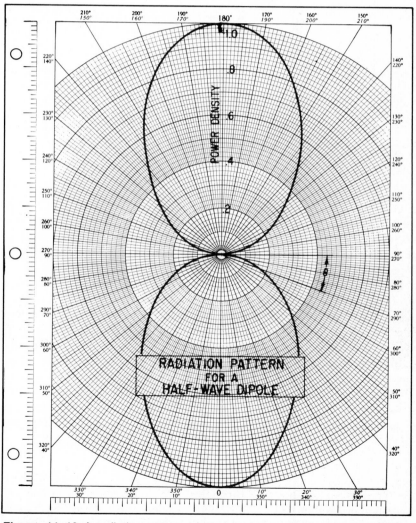

Figure 11·13 A radiation pattern plotted on polar coordinate paper. (Zero at bottom.)

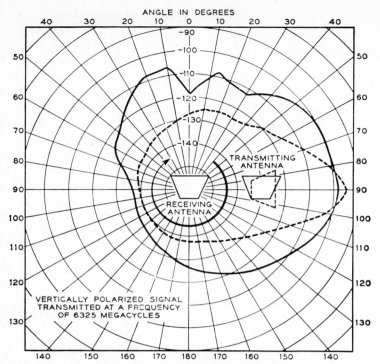

ANGLE IN DEGREES

Figure 11·14 Another graph on polar coordinates. (Zero at top.)
(Bell Telephone Laboratories.)

wave dipole, as computed by an analog computer. Notice that the linear scale is shown as units based on 1.0 being the maximum. Notice, also, that the zero is at the bottom of the graph. The curve has two large enclosures, called *lobes,* and points of zero magnitude, called *null points,* at 90 and 270°.

Figure 11·14 depicts the cross-talk coupling between two antennas at the same location. The two "envelopes" show the maximum cross-talk obtained for two positions of the transmitting antenna when the receiving antenna is rotated clockwise. Notice that zero is at the top and that degrees increase both clockwise *and* counterclockwise from zero.

In other situations, it is sometimes desirable to plot zero at the right side of the graph and go counterclockwise, much as the mathematician plots values of trigonometric functions. It is possible to buy polar coordinate paper that has the zero at the top or at the bottom. Disk recording devices use polar charting, and, instead of radial graduations being in degrees, they are in hours or days.

11·16 Bar charts

Some types of data do not lend themselves well to presentation in line graphs. They might be better suited for display in some other form, such as a bar chart. Also, data in graphical form must often be presented to clients or persons who do not have technical backgrounds. Such persons may find bar graphs, pie charts, and pictorial graphs easier to read.

Figure 11·15 is a bar chart in which the bars run up and down, for which reason it may also be called a *column chart*. Its construction is arrived at in much the same manner as a line graph that is plotted on rectangular coordinate paper. General practice is as follows:

1. Several major horizontal grid lines should appear with their scale values.
2. Bars, or columns, should be shaded.
3. Widths between bars should be no wider than the bars themselves and may be less.
4. Bars are often arranged in ascending or descending order, but this is not a requirement.

To make an attractive bar chart, one should consider using preprinted appliqués for shading. Rather than make a scale which would accommodate the United States production of 236,000,000 kW, the authors decided to use a scale that would permit the bars of the other countries to be larger and "broke" the United States bar as shown. While not generally done, this is acceptable practice as long as the values are shown at the top of the bars.

Another type of bar chart is shown in Fig. 11·16. This is only one form of a horizontal bar graph. Another form has a zero line running up the middle, with bars representing positive values extending to the right and bars with negative values to the left. Bars may also be broken down individually into

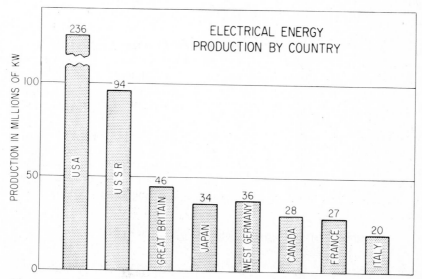

Figure 11·15 A bar chart or graph. (Titles for bars are often placed below the bars.)

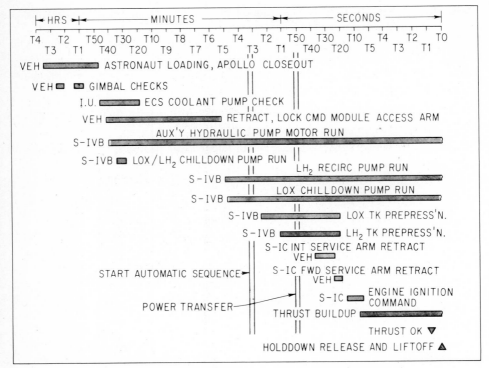

Figure 11·16 Horizontal-bar chart showing typical prelaunch events for Saturn-Apollo flight. *(NASA.)*

several parts. For example, if the information were available, we could show what percent of each country's electrical production was (1) hydro, (2) steam generating, and (3) nuclear, on the graph of Fig. 11·15.

11·17 Pie graph

A very popular type of chart, although held in low regard by statisticians, is the pie chart. It is most effective for displaying five to seven items that make up 100 percent. A well-designed pie graph has the largest item in the upper right sector, starting at 12 o'clock, followed by the next largest item, and so on in a sequence of decreasing size. The following principles of good construction apply.

1. The graph must be large enough to permit lettering within all but the very smallest sectors.
2. Arrangement of items should follow a sequence of decreasing sizes.
3. Items should have a shading sequence of ever-increasing or -decreasing darkness.

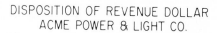

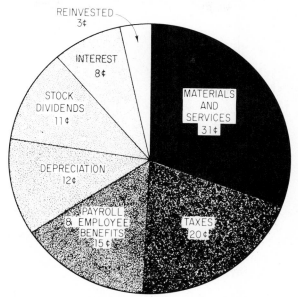

Figure 11·17 A pie graph. Notice that the material is arranged sequentially by size.

4. The largest item should start at 12 o'clock and be on the "clockwise" side.
5. A title should be placed above or below the graph.
6. The percentage or value should show in each sector.

11·18 Pictorial graphs

There is some evidence that graphs presented in pictorial form are remembered longer than those drawn in two dimensions. Many different ways of showing graphs in pictorial form have been used, including isometric, dimetric, oblique, and perspective construction. A three-variable pictorial graph has been shown in Fig. 11·18, which has been drawn in isometric projection. Making a graph like this takes quite a bit of time; for one thing, lettering is a little tricky. Therefore, one should be sure that the results will justify the extra effort. The material presented in Fig. 11·18 could have been presented in the form of the graph shown in Fig. 11·9, and vice versa.

Another pictorial drawing, shown in Fig. 11·19, is only partly a graph. This shows how ingenuity can be used to combine two different kinds of graphical representation. Other forms of pictorial representation of graphs have been used

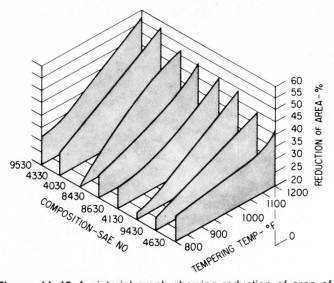

Figure 11·18 A pictorial graph showing reduction of area of various steels after liquid quenching at various temperatures. (Three variables in isometric projection.)

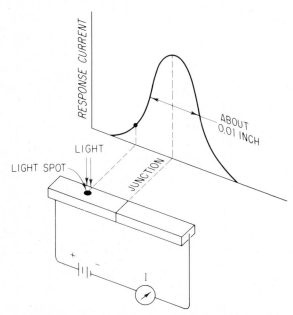

Figure 11·19 A pictorial drawing (in dimetric projection) showing the response of a PN transistor depending upon where it is illuminated. The curve is a normal frequency-distribution curve.

with success. Bar graphs have often been drawn pictorially. Oblique projection adds depth to the bars and makes possible the addition of more information in the third dimension.

11·19 Other types of graphic presentation

The graphs shown in this chapter are the type found most often (perhaps 95 percent of the time) in engineering, design, and research offices. Actually, there are a number of other ways to present material graphically, and some of these graphical forms can be used for calculation as well. Some of these graphs have been assigned names as follows:

1. Alignment charts
2. Concurrency charts
3. Conversion scales
4. Network diagrams
5. Holographs
6. Derivative curve ⎫
7. Integration   ⎭ graphical calculus

Figure 11·20 is a type of conversion scale that shows the comparative economic advantages of reducing the weights of various electrical or electronics gear. For the other types of graphs, mentioned immediately above, the reader

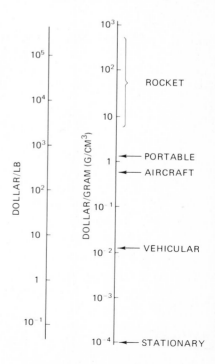

Figure 11·20 A line diagram, which is also a conversion diagram. This shows justifiable cost for removing by miniaturization one pound or gram of unnecessary weight. *(From Edward Keonjian, "Microelectronics," McGraw-Hill Book Company, New York, 1963. Used by permission.)*

should refer to a good text on engineering graphics, graphics, or graphic science. Some of the better texts have been written by Levins, Luzzader, Mochel, Paré, Rising and Almfeldt, and Vierck.

SUMMARY

Graphical construction is a very important phase of technical drawing. A careful survey of the problem is the first and most important step in successful graphic presentation. It should cover a careful consideration of items such as the following:

1. The purpose of the graph
2. The occasion for its use
3. The type of person who is to read it
4. The nature of the data
5. The medium of presentation to be used
6. The reproduction process to be used
7. The time available for preparation
8. Equipment and skill available

A graph should have a sufficiently professional appearance to inspire confidence in the facts presented. Layout and design are as important as quality of drafting. The type of data being portrayed graphically determines whether the curve should be smooth or a series of straight lines from point to point. Clarity and brevity are the two most important features of any graph. Some graphs can be used for purposes of computation. For example, equations such as $y = mx + b$ and $y = bx^m$ can be obtained if the curves appear as straight lines on the appropriate types of graph paper. Alignment charts, concurrency charts, and other graphical forms are useful in making certain computations.

QUESTIONS

11·1 What determines whether the plotted points on a graph should be connected with straight lines or be joined by a smooth curve?

11·2 What are two reasons why it is sometimes desirable to have data plotted in one straight line (by using different graph paper) instead of as a curve?

11·3 What determines whether data should be shown by means of a bar (or column) chart or by a pie chart?

11·4 Name four types of commercial graph paper that are available.

11·5 Along which axis is the dependent variable usually plotted?

11·6 In plotting the data from a laboratory experiment, how would you go about determining which variable is the independent one?

11·7 If it were necessary to draw two curves on one graph, how would you differentiate between these curves?

11·8 Would a graph that has been made into a projector slide show more or less information than a graph for the same data that will be bound in an engineering report?

11·9 Name three purposes for which graphs are or may be used.

11·10 What are three different line spacings that may be purchased in commercial rectangular coordinate paper?

11·11 Why is there not a zero point on logarithmic graph paper?

11·12 What do we mean by a "cycle," as it pertains to commercial graph paper?

11·13 List three examples of supporting data that might appear on an engineering graph.

11·14 When using commercial graph paper, where would you place the title? The supporting data?

11·15 If three different weights of lines are to be used in drawing a graph, which parts of the graph will be depicted by which weights of line?

11·16 Name four shapes of plotting symbols that are commonly used in graph work. Which symbol is the most common?

11·17 In what numbers, or multiples thereof, do we usually lay out the baselines of a graph?

11·18 List four abbreviations that are often used to indicate typical units that are placed along a baseline.

11·19 If it is desired to show each point permanently by means of a circular plotting symbol, how large would you make the symbol?

11·20 What characteristic of a heater or antenna can best be shown by means of a polar graph?

11·21 In constructing a pie chart, what two sequences should be observed and followed?

11·22 What would be a reason for using graph paper with blue lines, as opposed to using red-line paper?

11·23 List two sequences of numbers placed along the axis of a rectangular coordinate graph that are considered to be poor sequences.

11·24 Why is it possible to draw a right triangle on a straight-line curve on logarithmic paper and use the linear lengths of the two legs to obtain the correct slope?

11·25 What are six factors that should be considered in the layout of a graph?

PROBLEMS

11·1 The treasurer of Western Utility Power Company has prepared the following information about how the utility spends its "dollar" for presentation at the next stockholders' meeting. Put this information in the form of a pie chart (Table A).

Table A

Wages and salaries	$0.32
Taxes	0.20
Materials and fuel	0.08
Interest and dividends	0.26
Add. to physical plant	0.14

11·2 Prepare a bar or pie chart for the increase in capacity (in megawatts) estimated by the Public Service Company (Table B).

Table B

	Net Addition	Year-end Capacity
1974	$0.98	$ 7.58
1975	0.67	8.25
1976	1.22	9.47
1977	0.77	10.24
1978	0.99	11.23
1979	0.76	11.99

11·3 Plot the hourly demand and normal-capability curves of the interconnected system of Mid-Western Power & Light Co. (Table C).

Table C

Year	Maximum Hourly Demand	Normal Generating Capability	Year	Maximum Hourly Demand	Normal Generating Capability
	(Thousands of kW)			(Thousands of kW)	
1963	256	263	1968	411	449
1964	265	263	1969	469	437
1965	318	325	1970	501	565
1966	352	364	1971	513	584
1967	381	364	1972	580	742

11·4 The 10-year history of generation data of East States Utilities Company is shown in Table D. Plot the kWh and Btu curves on one graph.

Table D

Year	kWh Generated (Billions)	Btu Required (Billions)	Year	kWh Generated (Billions)	Btu Required (Billions)
1956	1.32	19.2	1961	1.90	23.0
1957	1.43	20.4	1962	2.14	26.3
1958	1.54	20.4	1963	2.24	26.9
1959	1.71	22.2	1964	2.43	28.6
1960	1.80	23.2	1965	2.62	30.3

11·5 Make a bar chart of the data presented in Prob. 11·3 on page 386.

11·6 Make a bar chart of the data presented in Prob. 11·4 on page 386.

11·7 The following data (Table E) show the survival probability of a nonredundant system and a redundant system with $N = 10^5$ components. Plot both curves and identify each. Failure rate is $f = 10^{-7}$ per hour.

Table E

Time (Hours)	Survival Probability, P	
	Nonredundant System	Redundant System
0	1.0	1.0
200	0.2	1.0
500	0.05	1.0
1,000	0	1.0
2,000	0	0.89
4,000	0	0.85
8,000	0	0.66
12,000	0	0.44

11·8 The following data (Table F) include the capacitance per unit area and the breakdown voltages of SiO_2—silicon structures in microcircuits for different oxide thicknesses. Plot the two curves on the same graph, identify the curves, and provide a suitable title.

Table F

Curve 1		Curve 2	
Capacitance (pFs/ Sq Mil)	Oxide Thickness (Å)	Breakdown Voltage (mV)	Oxide Thickness (Å)
0.41	500	0.055	500
0.22	1000	0.09	1000
0.14	1500	0.16	2000
0.105	2000	0.215	3000
0.07	3000		

11·9 Plot the family of curves for the 2N1490 NPN transistor having a base input on a common-emitter circuit. Curves for four base currents are shown below (Table G).

Table G Collector Characteristics

Curve 1 25 Base mA		Curve 2 10 Base mA	
Collector- to-Emitter Voltages	Collector mA	Collector- to-Emitter Voltages	Collector mA
0	0	0	0
0.5	160	0.5	160
1.0	300	1.0	250
1.5	395	1.5	315
2.0	450	2.0	375
2.5	500	2.5	420
		3.0	470

(Continued on next page)

Table G Collector Characteristics *(Continued)*

| Curve 3 3 Base mA | | Curve 4 1 Base mA | |
Collector-to-Emitter Voltages	Collector mA	Collector-to-Emitter Voltages	Collector mA
0	0	0	0
0.5	110	0.5	90
1	190	1	112
2	270	2	125
3	320		Straight
4	352		to
		4.5	141

11·10 The table below (Table H) provides data for four characteristic curves of a 2N2102 NPN transistor for an ambient temperature of 25°C. Plot the family of four curves for a common-emitter circuit having a base input. These are typical collector characteristics.

Table H

| Base Current = 2 mA | | Base Current = 8 | |
Collector-to-Emitter Volts	Collector mA	Collector-to-Emitter Volts	Collector mA
0	0	0	0
0.3	100	0.3	200
1	140	1	250
2	175	3	365
4	195	4	402
10	210	6	450

(Continued on next page)

Table H *(Continued)*

Base Current = 14		Base Current = 22	
Collector-to-Emitter Volts	Collector mA	Collector-to-Emitter Voltages	Collector mA
0	0	0	0
0.3	200	0.3	200
0.6	300	0.6	300
1	330	1	375
2	402	2	482
4	502	2.2	500

11·11 On rectangular coordinate paper, plot the depreciation-cost curve and the maintenance-cost curve for electronic heating equipment, using the data given below (Table I). A suitable title would be "Costs of 60-Hz Electric Energy." Use different plotting symbols for each curve. Your instructor may want you to add the two curves graphically to get a total cost curve.

Table I

Depreciation Cost		Maintenance Cost	
Cost/kWh (Cents)	Generator Output (kW) (Independent Variable)	Cost/kWh (Cents)	Generator Output (kW)
3	3	2	2
2	10	1.5	10
1.8	20	1.35	20
1.6	80	1.2	40
1.5	160	1.25	80
		1.6	160

11·12 Plot the family of curves for a tube having the average plate characteristics shown on page 391 (Table J). Tube type: 6AL8. $E = 6.3$ V.

Table J

$E_{c1} = 0$ Curve for No. 1 Grid		Curve for $E_{c1} = -1$		Curve for $E_{c1} = -2$		Curve for $E_{c1} = -3$	
DC Plate Voltage (Volts)	Plate Current (Milli-amperes)	DC Plate Voltage (Volts)	Plate Current (Milli-amperes)	Plate Voltage (Volts)	Plate Current (Milli-amperes)	Plate Voltage (Volts)	Plate Current (Milli-amperes)
0	5	0	5	7	13	6	5
3	16	5	16	20	21	15	10
5	25	7	20	35	25	40	16
6	35	10	30	60	26.5	80	17
10	42	15	36	80	24	28	14
25	51	24	40				
50	55	45	44				
80	56	80	45				

l·13 The annual requirements for artwork for three types of circuit manufacturing have been rising and will continue to do so. The data shown below (Table K) should be plotted on rectangular (arithmetic) graph paper to show past and estimated requirements through 1976.

Table K Number of Artwork Layouts per Year (in Thousands)

Year	Printed Wiring Boards	Thin-Film Circuits	Silicon Monolithic Circuits
1968	6.0	4.6	1.5
1970	9.0	6.5	3.0
1972	14.2	9.0	5.0
1974	19.0	13.5	7.6
1976	23.0	20.6	11.6

The curves may be extrapolated for another year or two. Estimates were made by the economics division of Atlantic Electronics Company, January 30, 1972.

11·14 On semilogarithmic paper (three cycles) make a graph for the date given below (Table L). An appropriate title might be "Input Resistance versus Load Resistance of Transistor Circuit."

Table L

Load Resistance (Ohms, Independent Variable)	Input Resistance (Ohms)
10 k	1020
50 k	820
100 k	700
300 k	600
1 M	560
10 M	540

11·15 Construct a graph, using multicycle semilogarithmic paper, for the phase response of an *RC* amplifier. Let the phase shift, below, be the dependent variable (Table M).

Table M

Response Curve without Feedback		Response Curve with Feedback	
Frequency (Units as Shown)	Phase Shift (Degrees)	Frequency (Units as Shown)	Phase Shift (Degrees)
		10 Hz	+130
10 Hz	+150	40 Hz	20
100 Hz	50	100 Hz	5
1 kHz	0	1 kHz	0
10 kHz	−20	10 kHz	0
100 kHz	−120	100 kHz	−20
1 MHz	−180	300 kHz	−100
		1 MHz	−150

11·16 Make a graph showing the two curves, one with and one without feedback, for the frequency response of the transistor amplifier (Table N).

Table N

Without Feedback		With Feedback	
Frequency (Hz)	Gain (dB)	Frequency (Hz)	Gain (dB)
10	48	10	18
30	54	35	20
100	56	100	21
10^4	55	10^3	21
10^5	43	4×10^4	21
3×10^5	35	9×10^4	22
(300,000)		2×10^5	20
		3×10^5	15

11·17 On polar coordinate paper, plot the dipole radiation pattern using the data listed below (Table O). There will be a major and two minor lobes. Do not number or circle points. This is a vertical plane pattern of a half-wave center-fed antenna. (Zero should be at bottom of chart.)

Table O

Major Lobe			Minor Lobe			Minor Lobe		
Point	Angle	Amplitude	Point	Angle	Amplitude	Point	Angle	Amplitude
0	0°	0	0	0°	0	0	0°	0
1	155°	0.2	1	120°	0.10	1	215°	0.1
2	158°	0.4	2	124°	0.20	2	216°	0.2
3	162°	0.6	3	126°	0.24	3	218°	0.24
4	169°	0.8	4	130°	0.28	4	222°	0.28
5	180°	1.0	5	134°	0.30	5	226°	0.30
6	191°	0.8	6	138°	0.28	6	230°	0.28
7	198°	0.6	7	142°	0.24	7	234°	0.24
8	202°	0.4	8	144°	0.20	8	236°	0.20
9	205°	0.2	9	145°	0.10	9	240°	0.10
10	0°	0	10	0°	0	10	0°	0

11·18 The data listed below (Table P) provide a vertical plane radiation pattern for a $\frac{3}{4}$-wave antenna. The 0-to-180° line represents the axis of the antenna. Make a polar plot, using an irregular (French) curve wherever possible. (It may be necessary to round off the tips of the lobes by careful freehand drawing.) Do not number points or put plotting symbols around them. (You may put 0° at the left side of the graph if you wish.)

Table P

Point no.	0	1	2	3	4	5	6	7	8	9	10	11	12
Angle (°)	0	15	16	19	23	27	30	33	37	41	44	45	0
Amplitude	0	0.2	0.4	0.6	0.8	0.96	1.0	0.96	0.8	0.6	0.4	0.2	0

Point no.	0	1	2	3	4			0	1	2	3	4
Angle	0	60	64	70	0			0	75	81	86	0
Amplitude	0	0.18	0.3	0.18	0			0	0.18	0.3	0.18	0

Point no.	0	1	2	3	4			0	1	2	3	4
Angle	0	92	98	104	0			0	250	256	262	0
Amplitude	0	0.18	0.3	0.18	0			0	0.18	0.3	0.18	0

Point no.	0	1	2	3	4	5	6	7	8	9	10	11	12
Angle	0	135	136	139	143	147	150	153	157	161	164	165	0
Amplitude	0	0.2	0.4	0.6	0.8	0.96	1.0	0.96	0.8	0.6	0.4	0.2	0

The following graphical plots are of such a nature that each one yields a straight-line pattern if plotted on a certain type of graph paper. If the straight-line pattern can be achieved, then the student can write the equation for the line (and the data), using one of the equations listed in Sec. 11·12. Some experimentation of a trial-and-error nature will be required, in most cases, before the student will be able to draw the graph on the correct paper. Principles of graphical presentation should be followed, and the final graph should have a title and good line work and lettering, and the equation (if required by the instructor) should be shown prominently on the graph.

11·19 The following data (Table Q) provide a single-family transfer characteristic of a transistor, which is part of an integrated circuit on a silicon wafer.

Table Q

Collector Characteristic (I_c in Milliamperes)	Gate Voltage (V_G in Volts)
−1	0.2
10	0.6
21	1.0
32	1.4
42	1.8

11·20 The following data (Table R) are from test records on 2409 integrated circuits. There are two curves, one for 90 and one for 95 percent upper-confidence limits.

Table R

90% Curve		95% Curve	
Operating Time (T in h)	Failure Rate (F in fails/h)	Operating Time (T in h)	Failure Rate (F in fails/h)
40,400	10^{-4} (1 in 10,000)	50,400	10^{-4}
10^5	4.4×10^{-4}	10^5	5.2×10^{-4}
10^6	4.2×10^{-5}	10^6	5.2×10^{-5}
4.2×10^6	10^{-6}	5.2×10^6	10^{-6}
10^7	4.0×10^{-6}	10^7	5×10^{-6}

11·21 The following data (Table S) show the optimum frequency of different skin thickness of brass and iron. Plot frequency in megahertz.

Table S

Frequency (Hz)	Skin Thickness in Centimeters	
	Brass	Iron
100 thousand	0.007	0.06
200 thousand	0.0042	0.042
400 thousand	0.0026	0.03
600 thousand	0.0019	0.024
1 million	0.0013	0.019

11·22 The following data (Table T) show the relationship of power loss to the armature voltage of a $\frac{1}{3}$-hp electric motor.

Table T

Loss (Watts)	Armature Voltage (Volts)
0.080	1.2
0.122	1.5
0.213	2
0.340	2.5
0.480	3
0.842	4

11·23 Below (Table U) are data that show the resistance in ground connections according to how deep the grounding rod is placed in the soil.

Table U

R (Ohms)	D (Feet)
90	2
68	3
51	4
33	7
24	10
17	15
11.5	25

11·24 By stressing improvement in engineering and production services and cost reduction, United States industry can maintain its constant increase in productivity. The following data (Table V) cover a 10-year period. Use 1950 for getting the intercept.

Table V

Year	Output per Worker-Hour (Dollars)
1950	81
1955	93
1960	106
1965	121
1970	139

11·25 Failure to align the length standard when making linear measurements with a laser causes a cosine error which plots as follows (Table W):

Table W

Misalignment Angle X (minutes)	Cosine Error e_c (ppm)
2	2.0
3	4.5
4	8.0
6	18.0
10	51.0

11·26 The manufacturing tolerances of thin-film resistors, according to the present state of the art, are (Table X):

Table X

Resistance Value (Ohms)	Manufacturing Tolerance (Percent)
200	±0.29%
600	0.48
1k	0.78
1.4 k	1.33

Chapter 12
Computer-aided drafting and design

It is projected that, in 10 years, most major corporations will be creating, maintaining, and communicating engineering data with the aid of computer technology. It is the object of this chapter to familiarize the designer and drafter with the operation, components, and advantages of computer technology as applied to the electrical design and drawing fields.

Interactive design and drafting, computer graphics, automatic drafting systems (ADS), and *computer-assisted design (CAD)* are different names for one thing—a design and drafting tool which, through the assistance of a computer, provides significant productivity increases to the designer and drafter. Automatic drafting systems used to have only drafting capabilities, but today, with the advent of the graphic CRT, they also incorporate design features. Computer-aided design systems have been and are being used in all types of engineering, including structural, mechanical, electrical, automotive, aerospace, civil, etc.; however, because this text is about electronics and electrical drawing, this chapter will be oriented in this direction.

In general, the term computer-aided design implies that the computer assists the designer in analyzing and modifying a previously created design with new design parameters the designer has established. Almost any designer or drafter can use graphic data processing effectively and can analyze, modify, and distribute this data with only a limited knowledge of programming. This is because the designer and computer are in two-way communication using the designer's own language of lines, symbols, pictures, and words.

In addition to taking over the drudgery of simple, repetitive drafting chores, the computer-aided design system can do numerous other tasks. It can re-create existing drawings and prepare tapes, bills of materials, wiring diagrams, and manufacturing sheets. It can make masks for integrated circuits. It can scribe and mask printed circuit boards. In many configurations, analysis and computation may be done, if desired. Some typical capabilities of a computer-aided design system are:

1. Printed-circuit artwork with dimensional accuracy beyond the drafter's ability

2. Printed-circuit-board assemblies
3. Integrated- or hybrid integrated-circuit masking
4. Schematic, logic, block, or any symbol-oriented diagram
5. Switch developments
6. Parts lists and legends
7. Physical drawings, such as plan views, general arrangements, and wiring diagrams
8. PERT charts, flowcharts, and maps
9. Plots of numerical data
10. Perspectives, isometrics, and three-dimensional views
11. Exploded views
12. Paper tape for driving numerical control (N/C) equipment for hole drilling, component insertion, and wire wrapping

A CAD has many advantages. It can reduce overall costs up to 50 percent; for certain kinds of drawings, labor costs can be cut 90 percent. Drawings are usually better. Revisions are faster and easier. Symbols, notes, line weight, lettering, and other style and format factors are consistent. Rough sketches can be turned into finished drawings quickly and automatically. The designer has the option of looking at many views, rotating the image, making design changes, and instantly seeing these changes, and has the ability to store these data or output them in many forms.

12·1 The basic system

A typical CAD system consists of a drafting table, a digitizer, an alphanumeric keyboard, a cathode-ray-tube (CRT) input-display station, a teletypewriter, a magnetic tape or disk unit, a minicomputer, and a plotter (drum or flatbed). A computer-output microfilm station, an artwork generator, a card reader, a tape reader/punch, and additional input-display stations are sometimes included. Figure 12·1 shows such a system.

A CAD system is usually housed in an area that has a controlled environment and an area of 300 sq ft or larger. Because of the intricate nature of the equipment, this area is access-controlled so that unknowledgeable personnel will not play with it or accidentally damage it. Although the system operates in normal lighting, the ability to dim the lights is desirable for operator comfort.

CAD systems use normal electric power—typically 115 V ac, 50 A, single phase. However, current requirements may exceed 80 A, depending on how much peripheral equipment is used. Large current requirements will sometimes dictate the use of larger, multiphase voltages of 240 and 208 V ac.

CAD systems are designed to minimize downtime and simplify maintenance. Modular construction allows easy replacement of electronic assemblies, drafting tables, and other components. The use of proven components in CAD systems

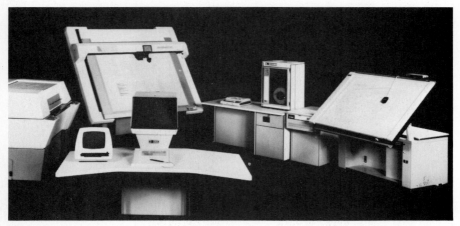

Figure 12·1 Photo of some of the components of a computer-aided design system. In the center, there is an interactive CRT design console having a graphic CRT with a tablet (menu) and a pen and a CRT with a keyboard. From left to right, there are a photoplotter, a plotter/track-type digitizer, a graphics processor with keyboard and tape drive, and a tablet with a cursor-type digitizer. *(Courtesy of Computervision Corp.)*

has resulted in reliability levels of 95 percent for 24-h/day operation. The average downtime for a malfunctioning system is approximately 4 h.

A CAD system consists basically of four functional elements, namely the input devices, the processing equipment, the output devices, and the software (or programs). Figure 12·2 shows a pictorialized block diagram of a CAD system emphasizing the four major parts. These basic parts and their operation and interaction will be discussed in the following sections.

12·2 Software

Because a CAD system uses a computer, it is controlled and run by a set of instructions called a *program* or *software* (to distinguish it from the physical equipment, such as a computer, which is categorized as *hardware*). The designer or drafter usually does not write the software or see it. However, by acting on the input and output devices, the designer executes it and knows its capabilities. Therefore, a cursory knowledge of the workings of software and software packages is necessary.

Essentially, the computer is a giant calculator, but it must be told precisely what, how, when, and where to do these calculations. It is told what to do by a set of written instructions called a program. When the program is written, it is either written in a high-level language, an assembly language, or machine code; but, no matter which language the program is written (or coded) in, it must be converted into machine code, or the operating code of the computer. Original programs are entered into the computer by card readers, tape readers,

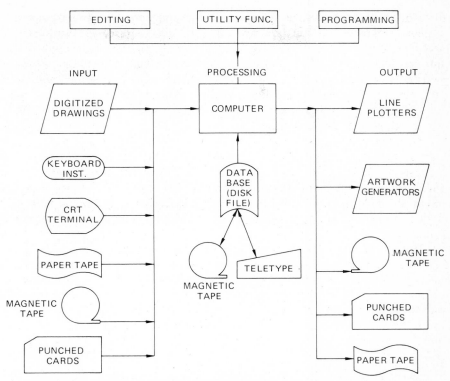

Figure 12·2 Block diagram of the major functional components of a CAD system.

and alphanumeric keyboards; if the program has been previously converted to machine code and stored on magnetic disk or tape, it can be retrieved from these devices.

Higher-level languages such as FORTRAN (FORmula TRANslation), COBOL (COmmon Business-Oriented Language), and ALGOL (ALGOrithmic Language) are easier to write programs in. They require fewer statements than programs written in assembly languages or machine codes for the same tasks. High-level language programs are converted by the computer into machine code by a program called a *compiler.* Assembly languages are resident to only one manufacturer's type or series of computers. Compared with higher-level languages, assembly languages are more difficult and time consuming to write programs in; however, when they are executed, they are generally faster and more efficient. Assembly-language programs are converted into machine code by a program called an *assembler.* It is the program that allows the designer or drafter to communicate the drawing information (data) to the computer by permitting the data to be entered, manipulated, verified, corrected, and outputted. The input and output devices will be discussed in the following sections.

Custom-made or standard ("canned") programs are available from CAD-system manufacturers, software developing houses, major manufacturers, and the government. Software packages are available for electrical design for all the items mentioned in the introduction of this chapter, including electrical schematics, logic diagrams, layouts, integrated-circuit and printed-circuit design and artwork, detailing, general arrangements, wiring diagrams, and bills of materials, plus a host of programs in other disciplines, including structural, architectural, mechanical, and civil engineering.

12·3 Input equipment

Input equipment can be as simple as an ordinary teletypewriter or as sophisticated as an automatic digitizer. Where an automatic drafting system is used as a CAD system, a CRT input-output console, complete with light pen, is almost a necessity.

Because the computer only "sees" digital information, all drawings or sketches that it sees must be in a digital format. Keyboard instruments like a teletypewriter and a keyboard communicate digitally with the computer via software programs. The same applies for drawings that are already programmed on cards, tapes, or disks. But the operator wants the computer to see on a drafting table; so the drawing lines and other graphic representations of a sketch must be converted, or *digitized,* to a series of digits which are pair-coordinate points (e.g., *X*-4) telling the computer where the graphic symbol is to be located on the drawing that is to be outputted from the computer. The device that does the digitizing is called the *digitizer* and is usually mounted on a board similar to a drafting board. There are two kinds of digitizers that mount on boards—the track type, similar to a drafting machine, and the freehand cursor type. The track type is more accurate, while the cursor type is easier to use. In Fig. 12·1 a cursor-type digitizer is shown on the table at the right of the picture and a track-type digitizer is shown on the table in the left center of the picture. After digitizing the point by pressing the digitize button on the digitizer, the operator presses one or a series of function buttons for the graphic symbol desired.

The *CRT display console* provides a visual display of the drawing just inputted or one previously stored in the computer. Because of its ability rapidly to display drawing changes entered by the operator, the CRT display console has been the major contributor to what is called *interactive graphics*—the real-time response or interaction between person and computer. This ability to interact quickly allows the operator to see the design drawing changes, additions, deletions, and corrections instantaneously without having to wait for a final printed copy.

The CRT display console usually is operated in conjunction with an alpha-numeric keyboard and a sensing pen or stylus. The keyboard allows the operator to add words, numbers, or symbols, or edit the drawings by typing in the appropriate code for these graphic functions. Sometimes the keyboard has a controllable grid used for digitizing. When using a CRT display console, the operator works from a sketch and views the work on the CRT or works directly with the picture on the CRT. In all instances, the coordinate points of the drawing must be digitized, whether by a cursor-type digitizer on the sketch, a keyboard-controlled digitizer, or a light-sensing pen (digitizer) on the CRT picture. After the point is digitized, the operator may type in a symbol code for a symbol on the keyboard or cursor or may touch a sensing pen to a symbol shown on a *menu*. A menu is a coded chart that shows graphic symbols.

In all cases, the computer receives these codes, converts them to the proper symbols, then writes the symbols in storage (tape, disk, etc.) and displays them on the CRT at the point that has been digitized. Many times, whole minischematics have been coded and may be written and displayed just like individual symbols. Graphic CRTs in conjunction with light-sensing pens provide the capability of writing directly on the face of the CRT.

Card and tape readers read into the computer design and drawing data already programmed on punched cards, magnetic tape, or paper tape. Often, the output of digitizers is sent directly to tape recorders or punches, and then all the information is read at one time by these readers for outputting on a plotter, for example. This procedure is done for efficiency, because it is unwise to tie up a very fast computer with a relatively slow process such as plotting.

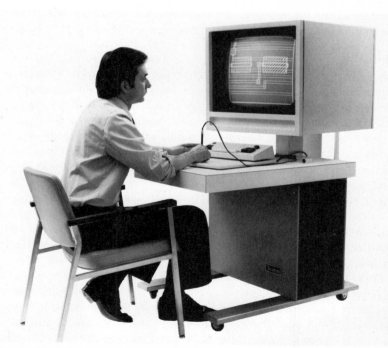

Figure 12·3 Photo of a CRT display console with operator using a sensing pen. *(Courtesy of Applicon, Inc.)*

12·4 Processing equipment

The processing equipment is the "brains" of the CAD system, consisting of a computer (central processor) and the necessary peripherals. The computer is really not a brain, as many would believe, but, rather, it is a very large (in terms of capacity) and very fast calculating machine which must be programmed. Figure 12·1 shows a processor used in a CAD system. The computer or central processor contains all the arithmetic, control, logic, and storage (called memory) circuitry. The memory is programmed to store often-used symbols, drawings, minischematics, computer programs, and specialized software for allowing the system to produce finished drawings from digitized input data. In addition to the basic control functions of tape search and read, data buffering. and velocity/acceleration control of the drafting head on the plotter(s), the computer permits operator control of data scaling, origin selection, editing, pen selection, and image rotation.

Peripheral memories include magnetic disks, tapes, and cassettes. Disks are used for storing temporary or permanent drawing data, while tape is used for storing and filing drawings. Primarily, it is done in this manner because disks are easier and faster for the computer to search and retrieve data from, while tapes are more convenient and economical for bulk storage and filing of data.

Many times, some sort of teletypewriter or teleprinter is used to communicate commands from an operator to the computer, or messages or instructions from the computer to the operator.

12·5 Output equipment

Obviously, the output is the end product of the CAD system, whether it be final vellum drawings, microfilm, photographs, or storage of information on magnetic or paper tape. Without the output, a CAD system would be worthless!

The prime output device is the *line plotter,* which converts the data from the computer or from a tape reader into a finished drawing. Essentially, there are three types of plotters, namely a flatbed, a drum, and a beltbed. Some of these plotters have stand-alone control units. The primary differences between these three pieces of equipment are that the drawing medium on a flatbed plotter is mounted on a bed and the pens or other drawing devices move in both axes; the drawing medium on a drum plotter rests on a drum and moves in the vertical axis while the pens move in the horizontal axis; and, on a beltbed plotter, either the pen and drawing medium move in opposite axes or the pens move in both axes. The flatbed plotter may be used as a photoplotter or a scribe plotter. The drawing medium may be paper for drum plotters or, depending on the plotter, paper, vellum, plastic, Mylar, film, metal, or any material with a scribe coat for flatbed plotters. The pens used may be Rapidograph; however, pressurized liquid-ink pens give better results at higher speeds.

Figure 12·4 shows a drum plotter. Notice that pens are plotting right at the top of the drum. The operator is replacing a pen. Figure 12·5 shows a flatbed plotter. Some drawings that have been done on a beltbed plotter will be shown later in the chapter.

Figure 12·4 Photo of a drum plotter. Notice that the pens are plotting on the paper at the top of the drum. The operator is replacing a pen. *(Courtesy of Calcomp.)*

Photoplotters, one type of artwork generator, are usually flatbed plotters that use a computer-controlled light source for producing graphic representations of patterns directly on photographic material, such as film. The photoexposure head on the plotter is moved to the correct locations and the light beam is turned on and off by the computer. This machine produces some very accurate artwork patterns for printed circuit boards, hybrid integrated circuits, and chemical etching. Some of these plotters have interchangeable heads, allowing film cutting, scribing, and, of course, line plotting.

Other output devices might include a hard-copy printer which could be used for quickly producing copies or checking prints without tying up the slower, more expensive plotter. A teletypewriter could be used for outputting textual information such as XY coordinates, parts lists, and messages. A computer-output microfilm (COM) is used to take the output from the computer and directly generate microfilm drawings in the form of aperture cards, microfilm, or roll film. Magnetic-tape recorders are used to record drawing information or other data on magnetic tape for mass storage. A paper-tape punch may be used to produce paper tape for driving plotters or numerical-controlled machines.

12·6 The output

This section of the chapter will show and explain a number of examples of drawings and artwork that have been produced by CAD systems. It should be reiterated that all outputs were produced in basically the same steps. The

Figure 12·5 Photo of a beltbed plotter. *(Courtesy of Calcomp.)*

initial conceptions (sketches) were digitized into the computer either by a digitizer or keyboard, or in bulk (if previous drawings) by magnetic tape. Then, the drawing or artwork was modified, changed, and edited at a CRT console. After the final revision appeared on the CRT, the drawing or artwork was sent either to the computer for outputting or to a tape reader for future outputting.

Figure 12·6 is part of a drawing showing the one-line diagram of a large ac generating station. The regulator and control circuits, which are enclosed in a boundary line, are shown as block diagrams. This drawing was done on a beltbed plotter, using pressurized-ink pens and drafting vellum as the drawing medium. First, notice the overall flowing symmetry and easy readability of the drawing, which is typical of most computer-generated drawings. The line weights are of such an even consistency—imagine yourself doing the same drawing in ink. The heavy lines are done by multiple passings or strokes of the pens, usually after everything else is completed. This drawing has been done

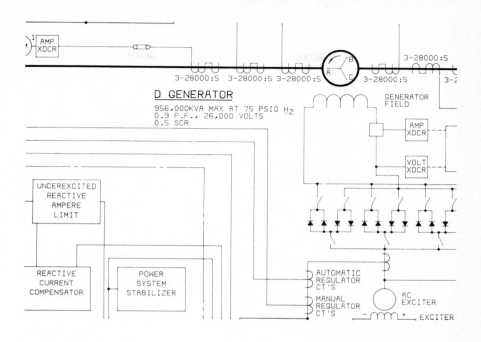

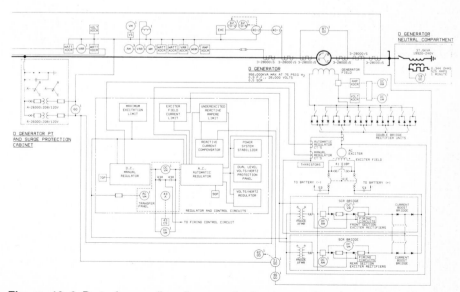

Figure 12·6 Part of a one-line diagram of a large ac generating station. (*Courtesy of Black & Veatch, Consulting Engineers.*)

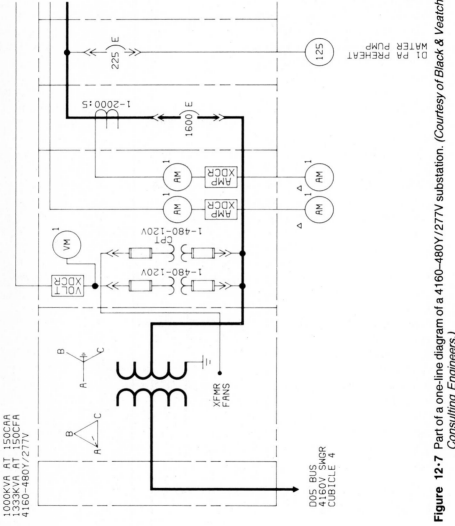

Figure 12·7 Part of a one-line diagram of a 4160–480Y/277V substation. *(Courtesy of Black & Veatch, Consulting Engineers.)*

on a CAD system that uses the dot-type wire connections preferred by the authors, but other types of connections are available. Notice how the lettering is neat and even; either vertical or 22.5°-slant letters are available. The authors watched the complete drawing (the figure shown is approximately one-fourth of the drawing), and it took approximately 10 min to complete.

Figure 12·7 shows part of a one-line diagram of a 4160 V-480Y/277 V substation. This drawing was done on the same machine as the previous figure. Again notice the symmetry and easy readability. Notice how well the mechanical grouping boundaries isolate but do not confuse the electrical components.

Figure 12·8a shows the control circuit diagram and interconnection diagram of a recycle pump and the wiring diagram of a discharge valve used on a flue gas treatment system. The original drawing had the diagrams for two pumps and valve combinations. The diagrams are easily readable; however, the control diagram is turned 90° from the conventional arrangement. Figure 12·8b shows the valve limit-switch development, which was on the same drawing as Figure 12·8a. The development explains the operation of the limit switches on the valve. The printing of the development is easy to read with virtually no possibility of error. The development is a good example of a situation in which printing must be done in a small area ($3\frac{1}{2}$ in. high $\times$ $2\frac{7}{8}$ in. wide, full size).

Figure 12·9 shows part of a logic diagram. This diagram was done on erasable Mylar with a Rapidograph pen. Because this drawing has larger amounts of open space, it is easier to read, lending itself to better and easier interpretation. The dot connections and dc-supply arrowheads are not totally inked in. Compare this drawing with the previous drawings.

Figure 12·10 shows the artwork of a two-sided PC board with feedthrough and component layouts. After the design of the PC board was done, it was sent to an artwork generator, in this case a photoplotter, where the artwork was done on 15- $\times$ 20-in. photoetchable Mylar. The artwork for the printed circuits has been optimally designed, and, therefore, it is laid out with good organization for circuit spacing and component placement. Thus it assures a functional PC board. The circuits, feedthrough, and component terminals are very accurately done; again this assures good operation of the PC board. The accuracy of artwork is a major reason for using a CAD system. To support the design of the PC board, the designer had a number of design options available, namely:

1. Board outline
2. Pads/feedthroughs
3. Component side runs
4. Solder side runs
5. Component outlines
6. Drill-hole data
7. Annotation/signature block
8. Parts list
9. Schematic drawing

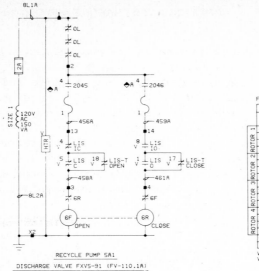
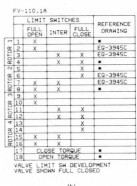
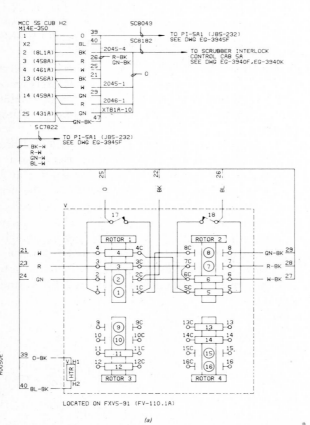

Figure 12·8 *(a)* Control circuit and interconnection wiring diagram of a recycle pump and discharge valve on a flue gas treatment system. *(b)* The limit-switch development of the discharge valve. *(Courtesy of Black & Veatch, Consulting Engineers.)*

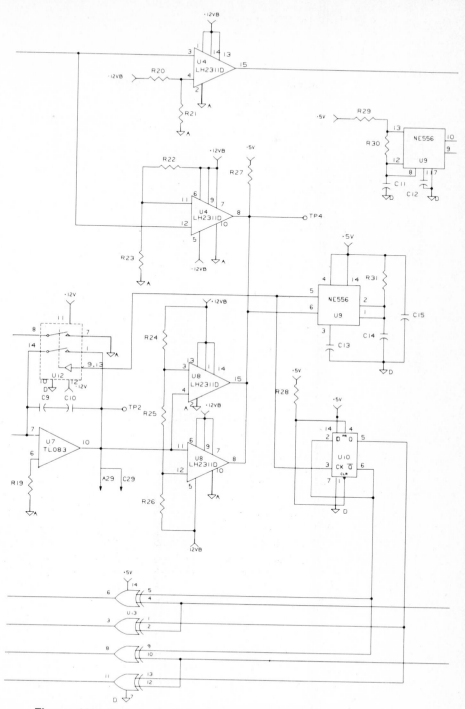

Figure 12·9 Part of a logic diagram. *(Courtesy of Computervision Corp.)*

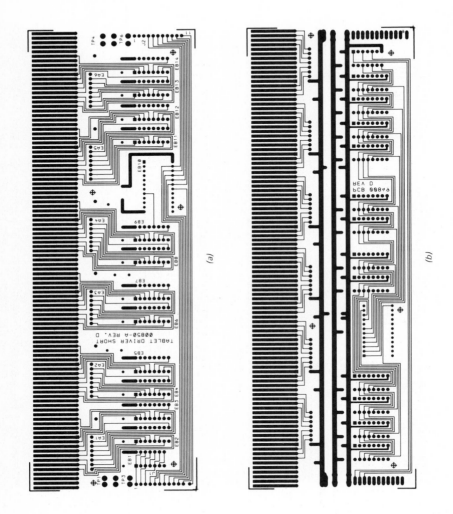

(a)

(b)

TABLET DRIVER SHORT
008S0-A REV. D

PCB 008S0 REV D
REV D

412

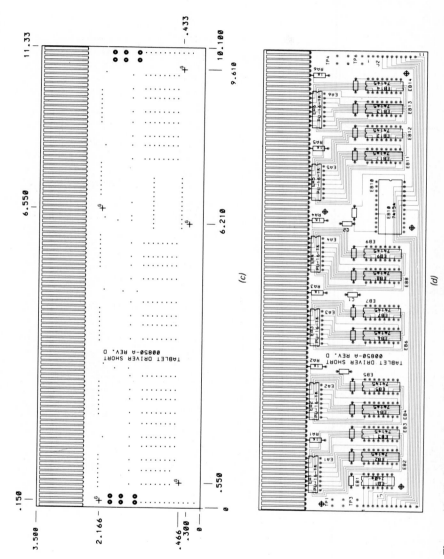

Figure 12-10 Artwork for a PC board. *(a)* Artwork for side 1. *(b)* Artwork for side 2. *(c)* View from component side. *(d)* Component layout. *(Courtesy of Applicon, Inc.)*

10. Functional flowchart

11. Ground plane (if required)

12. Signal plane (if required)

13. Assembly hardware

These options of drawings/artwork could also have been designed, and many would act like movable overlays on the PC-board artwork. If the designer had changed a component on one drawing, the others would have been automatically changed.

SUMMARY

This chapter has briefly and concisely described the operations of a computer-aided design and drafting system—from the input devices to the output devices. The CRT console has made the immediate interaction between person and computer a reality today. It should be remembered that this chapter was oriented toward computer-generated electrical drawings and artwork and that CAD systems are being used very effectively in other design disciplines, such as mechanical, civil, structural, and automotive. The future for CAD systems looks bright, for no one can deny their advantages of speed, accuracy, productivity, and quality. Time, technology (microprocessors), and competition have been rapidly removing the disadvantage of CAD systems, namely, cost.

QUESTIONS

12·1 What is CAD? Name three other terms that mean approximately the same thing.

12·2 Name at least five electrical and electronic drawings that a CAD system is used to create.

12·3 What are the four functional basic parts of a CAD system?

12·4 Name the three types of languages CAD systems are programmed in. Does a designer or drafter usually do the programming?

12·5 What is digitizing? What is a digitizer?

12·6 What is interactive graphics?

12·7 Name three input devices.

12·8 Name three processing devices.

12·9 What functions does the computer perform during CAD?

12·10 Describe the functions of peripheral storage devices.

12·11 Name three output devices.

12·12 Name three types of plotters. Describe them.

12·13 What kinds of drawing mediums are used by plotters? What kinds of pens?

12·14 What is a COM?

12·15 Discuss the advantages of CAD.

Appendix A

Glossary of Electronics and Electrical Terms

Ampere: The practical unit of current. One ampere will flow through a resistance of one ohm when a difference of potential of one volt is applied across its terminals.

Amplification: The process of increasing the strength (current, power, or voltage) of a signal.

Amplifier: A device used to increase the signal voltage, current, or power, generally composed of a transistor or vacuum tube and an associated circuit called a *stage*. It may contain several stages in order to obtain a desired gain.

Amplitude: The maximum instantaneous value of an alternating voltage or current, measured in either the positive or the negative direction.

Anode: An electrode at which negative ions are discharged or from which the forward current (diode rectifier) flows.

Appliqués: Symbols and other graphical shapes printed on a sheet, which can be cut out and affixed to a drawing by moderate pressure.

Attenuation: The reduction in the strength of a signal.

Base: One of (usually) three regions of a transistor. Also one of the terminals of a transistor. In some transistors the base acts much like the grid of an electron tube.

Bias:

Vacuum tube: the difference of potential between the control grid and the cathode.

Transistor: the difference of potential between the base and emitter and the base and collector.

Magnetic amplifier: the level of flux density in the magnetic amplifier core under no-signal condition.

Bipolar: One of two fundamental processes for the fabrication of integrated circuits. It involves making layers of silicon having different electrical characteristics.

Busbar: A primary power-distribution point connected to the main power source.

Capacitor: A device consisting of two conducting surfaces separated by an insulating material or dielectric such as air, paper, or mica. A capacitor stores electric energy, blocks the flow of alternating current, and permits the flow of alternating current to a degree depending on the capacitance and frequency.

Cathode: An electrode through which a primary stream of electrons enters the interelectrode space or to which the forward current flows (semiconductor).

Choke Coil: A coil of low ohmic resistance and high impedance to alternating current.

Circuit Breaker: An electromagnetic or thermal device that opens a circuit when the current in the circuit exceeds a predetermined amount. Circuit breakers can be reset.

Circular Mil: A unit of area equal to $\pi/4$ of a square mil, or 0.7854 sq mil. A unit of measure for wire sizes. (See Appendix B.)

Coaxial Cable: A transmission line consisting of two conductors concentric with and insulated from each other.

Cold Cathode: A cathode without a heater such as is found in fluorescent-type tubes.

Collector: That region of a transistor that collects electrons, or the terminal that corresponds to the anode (plate) of the electron tube in the normal mode of operation.

Commutator: The copper segments on the armature of a dc motor or generator. It is cylindrical in shape and is used to pass power into or from the brushes. It is a switching device.

Conductance: The ability of a material to conduct or carry an electric current. It is the reciprocal of the resistance of the material and is expressed in siemens (formerly called mhos).

Conductor: Any material suitable for carrying electric current.

Cryogen (Cryogenic): A device that becomes a superconductor (has practically no resistance) at extremely cold temperatures. A circuit having such devices.

Current: The rate of transfer of electricity. An amount of electricity. The basic unit is the ampere.

Deflecting Plate: That part of a certain type of electron tube which deflects the electron beam within the tube itself.

Detection: The process of separating the modulation component from the received signal.

Dielectric: An insulator; a term that refers to the insulating material between the plates of a capacitor.

Diffusion: A high-temperature process involving the movement of impurities into a silicon slice to change its electrical properties, used in fabrication of transistors, diodes, and ICs.

Diode:

Vacuum tube: a two-element tube that contains a cathode and plate.

Semiconductor: a material of either germanium or silicon that is manufactured to allow current to flow in only one direction. Diodes are used as rectifiers and detectors.

Drain: One end of a channel in a field-effect transistor. It compares with the collector of a bipolar transistor.

Electrode: A terminal used to emit, collect, or control electrons and ions; a terminal at which electric current passes from one medium into another.

Electron: A negatively charged particle of matter.

Electron Emission: The liberation of electrons from a body into space under the influence of heat, light, impact, chemical disintegration, or potential difference.

Emitter: That part or element of a transistor that emits electrons; it corresponds to the cathode of an electron tube, in the most common form of operation.

Epitaxial: A thin-film type of deposition for making certain devices in microcircuits. It involves a realignment of molecules and hence has a deeper significance than just thin-film manufacture.

Eyelet: Eyelets are used on printed circuit boards to make reliable connections from one side of the board to the other side.

Farad: The unit of capacitance.

Feedback: A transfer of energy from the output circuit of a device back to its input.

Filament: An electrically heated wire that emits electrons or heats a cathode, which then emits electrons.

Filter: A combination of circuit elements designed to pass a definite range of frequencies, attenuating all others.

Frequency: The number of complete cycles per second existing in any form of wave motion, such as the number of cycles per second of an alternating current.

Fuse: A protective device inserted in series with a circuit. It contains a metal that will melt or break when current is increased beyond a specific value for a definite time period.

Gain: The ratio of the output power, voltage, or current to the input power, voltage, or current, respectively.

Gate: A device or element that has one output channel and one or more input channels whose state(s) determine(s) the state of the output. The gate electrode of a FET which controls the current flow in the channel.

Grid: A wire, usually in the form of a spiral, that controls the electron flow in a vacuum tube.

Ground: A conductive connection with the earth to establish ground potential. Also, a common return to a point of zero potential. The chassis of a receiver or a transmitter is sometimes the common return and is, therefore, the "ground" of the unit.

Henry: The basic unit of inductance. The inductance of a circuit is one henry when a current variation of one ampere per second induces one volt. (The plural is henrys.)

Hole: In semiconductors, the space in an atom left vacant by a departed electron. Holes flow in a direction opposite to that of electrons, are considered to be current carriers, and bear a positive charge.

Impedance: The total opposition offered to the flow of an alternating current. It may consist of any combination of resistance, inductive reactance, and capacitive reactance.

Inductance: The property of a circuit or two neighboring circuits which determines how much electromotive force will be induced in one circuit by a change of current in either circuit.

Inductor: A circuit element designed so that its inductance is its most important electrical property; a coil.

Integrated Circuit: A circuit in which several different types of devices such as resistors, capacitors, and transistors are made from a single piece of material such as a silicon chip, and then are connected to form a circuit.

LED: A diode that emits light when current is passed through it.

Logic: The arrangement of circuitry designed to accomplish certain objectives such as the addition of two signals. Used largely in computer circuits, but also used in other equipment such as automated machine tools and electric controls.

Magnetron: A vacuum-tube oscillator containing two electrodes, in which the flow of electrons from cathode to anode is controlled by an externally applied magnetic field.

Modulation: The process of varying the amplitude (amplitude modulation), the frequency (frequency modulation), or the phase (phase modulation) of a carrier wave in accordance with other signals in order to convey intelligence. The modulating signal may be an audio-frequency signal, video signal (as in television), electric pulses or tones to operate relays, etc.

Oscillator: A circuit that is designed to generate an audio or radio frequency; a mode of amplification. Also the main device in such a circuit.

Oscilloscope: An instrument for showing, visually, graphical representations of the waveforms encountered in electric circuits.

Permalloy: An alloy of nickel and iron with an abnormally high magnetic permeability.

Photoconductive: A device or material that experiences a change in conductivity when exposed to light.

Photodiode: A two-terminal semiconductor that is sensitive to light.

Plate: The principal anode (electrode) in an electron tube to which the electron stream is attracted. Also, one of the conductive electrodes in a capacitor or battery.

Potential: The degree of electrification as referred to some standard such as the earth. The amount of work required to bring a unit quantity of electricity from infinity to the point in question.

Potentiometer: A variable voltage divider; a resistor which has a variable contact arm so that any portion of the potential applied between its ends may be selected.

Power: The rate of doing work or the rate of expending energy. The unit of electric power is the watt.

Raceway: Any channel for enclosing conductors which is designed expressly and used solely for this purpose.

Rectifiers: Devices used to change alternating current to unidirectional current. These may be vacuum tubes, semiconductor diodes, dry-disk rectifiers such as selenium and copper oxide, and also certain types of crystal.

Relay: An electromechanical switching device that can be used as a remote control, usually to open or close a current.

Resistance: The opposition that a device or material offers to the flow of current. It determines the rate at which electric energy is converted into heat or radiant energy.

Resistor: A circuit element whose chief characteristic is resistance; used to oppose the flow of current.

Resonance: The condition existing in a circuit in which the inductive and capacitive reactances cancel each other.

Saturation: The condition existing in any circuit when an increase in the driving signal produces no further change in the resultant effect.

Semiconductor: An element, such as germanium or silicon, from which transistors or diodes are made; the device itself. The resistivity of the element is in the range between those of metals and insulators.

Solenoid: An electromagnetic coil that contains a movable plunger.

Source: One end of the channel of a field-effect transistor. It compares with the emitter of a bipolar transistor.

Synchronous: Happening at the same time; having the same period and phase.

Tachometer: An instrument for indicating revolutions per minute.

Thick-film Circuit: A film-type circuit. The range of thickness of deposited patterns is from 0.01 to 0.05 mm.

Thin-film Circuit: A circuit made by depositing material on a substrate, such as glass or quartz, to form patterns that make devices, such as resistors and capacitors, and their connections. The thickness of the film forming these devices is only a few micrometers (0.001 mm).

Thyristor: A bistable semiconductor device having three or more junctions that can be switched from OFF to ON or vice versa. SCRs and triacs are in this class.

Transducer: A device that converts an input into a different type of output. Examples are microphones, speakers, lamps, vibrators, strain gages, and generators.

Transformer: A device composed of two or more coils linked by magnetic lines of force; used to transfer energy from one circuit to another (i.e., mutual coupling between circuits).

Triode: A three-electrode vacuum tube containing a cathode, control grid, and plate. Also a three-region semiconductor.

Volt: The unit of voltage, potential or emf. One volt will send a current of one ampere through a resistance of one ohm.

Voltage: Used interchangeably with "potential." (See "Potential.")

Watt: The unit of electric power. In a direct current a watt is equal to volts multiplied by amperes. In an alternating current the true power in watts is effective volts multiplied by effective amperes, then multiplied by the circuit power factor (1 hp = 746 W).

Electrical Device Reference Designations

FROM MIL STD 16B

Alarm	DS	Coupler, directional	DS
Amplifier	A	Crystal detector	CR
Amplifier, rotating	G	Crystal diode	CR
Annunciator	DS	Crystal, piezoelectric	Y
Antenna, aerial	E	Cutout, fuse	F
Arrestor, lightning	E	Cutout, thermal	S
Assembly	A	Detector, crystal	CR
Attenuator	AT	Device, indicating	DS
Audible signaling		Dipole antenna	E
device	DS	Disconnecting device	S
Autotransformer	T	Electron tube	V
Battery	BT	Exciter	G
Bell	DS	Fan	B
Blower, fan, motor	B	Filter	FL
Board, terminal	TB	Fuse	F
Breaker, circuit	CB	Generator	G
Buzzer	DS	Handset	HS
Cable	W	Head, erasing,	
Capacitor	C	recording,	
Cell, aluminum or		reproducing	PU
electrolytic	E	Heater	HR
Cell, light-sensitive,		Horn, howler	LS
photoemissive	V	Indicator	DS
Choke	L	Inductor	L
Circuit breaker	CB	Instrument	M
Coil, hybrid	HY	Insulator	E
Coil, induction, relay		Interlock, mechanical	MP
tuning, operating	L	Interlock, safety,	
Coil, repeating	T	electrical	S
Computer	A	Jack (see connector,	
Connector, receptacle,		receptacle,	
affixed to wall,		electrical)	
chassis, panel	J	Junction, coaxial or	
Connector, receptacle,		waveguide (tee or	
affixed to end of		wye)	CP
cable, wire	P	Junction, hybrid	HY
Contact, electrical	E	Key, switch	S
Contactor, electrically		Lamp, pilot or	
operated	K	illuminating	DS
Contactor,		Lamp, signal	DS
mechanically or		Line, delay	DL
thermally operated	S	Loop antenna	E

Magnet	E	or switch)	K
Meter	M	Repeater (telephone	
Microphone	MK	usage)	RP
Mode transducer	MT	Resistor	R
Modulator	A	Rheostat	R
Motor	B	Selenium cell	CR
Motor-generator	MG	Shunt	R
Mounting (not in		Solenoid	L
electric circuit and		Speaker	LS
not in a socket)	MP	Speed regulator	S
Nameplate	N	Strip, terminal	TP
Oscillator (excluding		Subassembly	A
elect. tube used in		Switch, mechanically	
oscillator)	Y	or thermally	
Oscilloscope	M	operated	S
Pad	AT	Terminal board or	
Part, miscellaneous	E	strip	TB
Path, guided,		Test point	TP
transmission	W	Thermistor	RT
Phototube	V	Thermocouple	TC
Pickup, erasing,		Thermostat	S
recording, or		Timer	M
reproducing head	PU	Transducer	MT
Plug (see connector)		Transformer	T
Potentiometer	R	Transistor	Q
Power supply	A	Transmission path	W
Receiver, telephone	HT	Tube, electron	V
Receptacle (fixed		Varistor,	
connector)	J	asymmetrical	CR
Rectifier, crystal or		Varistor, symmetrical	RV
metallic	CR	Voltage regulator	
Regulator, voltage		(except an electron	
(except electron		tube)	VR
tube)	VR	Waveguide	W
Relay, electrically		Winding	L
operated contactor		Wire	W

The above list is not the complete list of devices shown in Mil Std 16B. It contains the more commonly used devices.

Examples of the use of the above in electrical drawings would be: C_{201}, L_5, J_1, $J_1(P_{301})$, W_1P_2, T_{205}, $2A_4R_3$. The last example is explained as "the third resistor of the fourth subassembly of the second unit."

Abbreviations for Drawings and Technical Publications

FROM MIL STD 12B

Adaptor	ADPT	Dynamotor	DYNM
Air circuit breaker	ACB	Electric horsepower	EHP
Alternating current	AC	Electrolytic	ELECT.
Alternating current volts	VAC	Electronic Industries,	
Aluminum	AL	Association	EIA
American Society of		Engineer	ENGR
Mechanical		Engineering	ENGRG
Engineers	ASME	Escutcheon	ESC
Ammeter	AM.	Exciter	EXC
Ampere	AMP	Federal Communications	
Amplifier	AMPL	Commission	FCC
Antenna	ANT.	Federal Power	
Armature	ARM.	Commission	FPC
Arrestor	ARR	Field reversing	FFR
Attenuation, attenuator	ATTEN	Flat head	FH
Audio frequency	AF	Fluorescent	FLUOR
Auto frequency control	AFC	Fuze	FZ
Automatic gain control	AGC	Gage	GA
Battery	BAT	Germanium	Ge
Beat-frequency oscillator	BFO	Grommet	GROM
Bottom	BOT	Guided missile	GM
Cabinet	CAB.	Heater	HTR
Capacitor	CAP.	Heat treat	HT TR
Cathode-ray tube	CRT	High frequency	HF
Circuit	CKT	High-frequency oscillator	HFO
Coaxial	COAX.	High voltage	HV
Collector	COLL	Horizon, horizontal	HORIZ
Compress	COMP	Ignition	IGN
Condenser	COND	Indicator	IND
Conductor	COND	Induction	IND
Conduit	CND	Induction-capacitance	LC
Counterclockwise	CCW	Institute of Electrical	
Cycles per second	HZ	and Electronic	
Decibel	DB	Engineers	IEEE
Diameter	DIA	Instrument	INST
Direct current	DC	Intermediate frequency	IF
Double-pole double-		Junction box	JB
throw	DPDT	Kilocycle	KC
Double-pole single-		Kilohm	K
throw	DPST	Kilovolt	KV
Drawing	DWG	Kilovolt ampere	KVA
Dynamometer	DYNO	Kilowatt	KW

Kilowatt-hour	KWH	Reference	REF
Knockout	KO	Reference line	REF L
Lighting	LTG	Resistance	RES.
Low frequency	LF	Resistance capacitance	RC
Low voltage	LV	Resistance-capacitance	
Magnetic amplifier	MAG AMPL	coupled	RC CPLD
Magnetic modulator	MAG MOD	Resistor	RES.
Manual	MAN	Roundhead	RH
Master switch	MS	Saturable reactor	SR
Medium frequency	MF	Schedule	SCH
Mega (10^6)	MEG	Screw	SCR
Megacycle	MC	Secondary	SEC
Megohm	MEGO	Selector	SEL
Meter	M	Selenium	Se
Metering	MTRG	Series relay	SRE
Missile	MSL	Servomechanism	SERVO.
Modify	MOD	Shield	SHLD
Modulator	MOD	Signal	SIG
Motor	MOT	Single-pole, double-	
Mounting	MTG	throw	SPDT
Multiplex	MX	Single-pole, single-	
Multivibrator	MVB	throw	SPST
National Aeronautics		Slow operate (relay)	SO.
and Space		Solenoid	SOL.
Administration	NASA	Speaker	SPKR
National Electrical		Specification	SPEC
Code	NEC	Suppressor (ion)	SUPPR
Not to scale	NTS	Switch	SW
Oil circuit breaker	OCB	Switchboard	SWBD
Oscillator	OSC	Switchgear	SWGR
Oscilloscope	OSCP	Synchronous	SYN
Overload	OVLD	Tachometer	TACH
Pentode	PENT.	Technical circular	TC
Phase	PH	Technical manual	TM
Piezoelectric-crystal		Telemeter	TLM
unit	CU	Terminal	TERM.
Polarity	PO	Test switch	TSW
Potentiometer	POT.	Thermistor	TMTR
Power supply	PWR SUP	Thermocouple	TC
Quick-opening device	QOD	Three-conductor	3/C
Radar	RDR	Three-phase	3 PH
Radio	RAD.	Three-pole	3 P
Radio frequency	RF	Time delay	TD
Reactive volt-amp	VAR	Transceiver	XCVR
Receptacle	RECP	Transformer	XFMR

Transistor	Q	Very low frequency	VLF
Transmitter	XMTR	Video	VID
Tuning	TUN	Video frequency	VDF
Twisted	TW	Volt	V
Ultrahigh frequency	UHF	Voltage regulator	VR
Unfused	UNF	Voltmeter	VM
Vacuum tube	VT	Volume	VOL
Vacuum tube voltmeter	VTVM	Watt	W
Var-hour meter	VRH	Watt hour	WHR
Variable frequency		Watt-hour meter	WHM
oscillator	VFO	Wattmeter	WM
Very high frequency	VHF	Wire-wound	WW

Note: Some abbreviations are followed by a period. Most do not have a period.

These symbols are authorized for use in drawings for the Military and in specifications. A similar list has been compiled by the American National Standards Institute.

The above is not the complete Military Standard. If in doubt about abbreviation, spell out the word.

Appendix B

The Frequency Spectrum

Frequency	Designation	Abbreviation	Wavelength Meters	Centimeters
Below 3 kHz	Extremely low frequency	elf	30,000	
3–30 kHz	Very low frequency	vlf	30,000–10,000	
30–300 kHz	Low frequency	lf	10,000–1,000	
300–3000 kHz	Medium frequency	mf	1,000–100	
3–30 MHz	High frequency	hf	100–10	
30–300 MHz	Very high frequency	vhf	10–1	
300–3000 MHz	Ultrahigh frequency	uhf	1–0.1	
3000 MHz–30	Superhigh frequency	shf		10–1
30–300 GHz	Extremely high frequency	ehf		1–0.1
300–3000 GHz	As yet unnamed			0.1–0.01

NOTE: The current IEEE and international standards utilize the term *hertz* for cycles per second. Thus KC (old form) becomes kHz; MC (old form) becomes MHz; etc. kHz represents kilocycles; MHz, megacycles; and GHz, gigacycles (1 billion cps).

Width of Copper-Foil Conductors for Printed Circuits*

Width of 0.00135-in.- Thick Copper (Inches)	Width of 0.0027-in.- Thick Copper (Inches)	Current Capability (Amperes)
$\frac{1}{64}$		1.5
$\frac{1}{32}$	$\frac{1}{64}$	2.5
$\frac{1}{16}$	$\frac{1}{32}$	3.5
	$\frac{1}{16}$	5.5
$\frac{1}{8}$		6.5
	$\frac{1}{8}$	8.0
	$\frac{3}{16}$	15.0

By permission of the Aerovox Corp.
* For a temperature rise of 40°C, the maximum recommended for lucite and nylon laminates.

Example: If the desired current is 3.5 A, a conductor 0.0027 in. thick and $\frac{1}{32}$ in. wide will carry it. Also, a 0.00135-in.-thick copper foil that is $\frac{1}{16}$ in. wide will carry 3.5 A with a rise of 40°C or less.

Resistor Color Code

The colored bands around the body of a resistor indicate its value in ohms and, if there are four bands, its tolerance. As the drawing shows, the first band represents the first digit of the value and the second band the second digit. (The first band is near one edge and is often right at that edge of the device.) The third band represents the number by which the two digits are multiplied. A fourth band of gold or silver represents a tolerance of ±5 percent or ±10 percent, respectively. If no fourth band is shown, the tolerance is assumed to be ±20 percent of the indicated value of resistance.

The physical size of a composition resistor is related to its wattage rating. Size increases progressively as the wattage rating increases. The diameters of $\frac{1}{2}$-, 1-, and 2-W resistors are approximately $\frac{1}{8}$, $\frac{1}{4}$, and $\frac{5}{16}$ in., respectively.

CODE

COLOR	DIGIT 1ST	DIGIT 2ND	MULTIPLIER
BLACK	0	0	1
BROWN	1	1	10
RED	2	2	100
ORANGE	3	3	1,000
YELLOW	4	4	10,000
GREEN	5	5	100,000
BLUE	6	6	1,000,000
VIOLET	7	7	10,000,000
GRAY	8	8	100,000,000
WHITE	9	9	1,000,000,000
GOLD	−	−	.1
SILVER	−	−	.01

TOLERANCE

GOLD	±	5%
SILVER	±	10%
NO BAND	±	20%

Color Code for Chassis Wiring

Color	Abbrev.	Numerical Code	Circuit
Black	BK	0	Grounds, grounded elements and returns
Brown	BR	1	Heaters of filaments off ground
Red	R	2	Power supply B-plus
Orange	O	3	Screen grids
Yellow	Y	4	Cathodes, emitters
Green	GN	5	Control grids, base
Blue	BL	6	Plates (anodes), collectors
Violet (or Purple)	V PR	7	Power supply, minus
Gray	GY	8	AC power lines
White	W	9	Miscellaneous, returns above or below ground, AVC, etc.

Source: Mil Std 122.

Circuit-Identification Color Code for Industrial Control Wiring

Circuit	Color
Line, load, and control circuit at line voltage	Black
AC control circuit	Red
DC control circuit	Blue
Interlock panel control when energized from external force	Yellow
Equipment grounding conductor	Green
Grounded neutral conductor	White

Source: National Machine Tool Builders Association.

Transformer Color Codes

POWER TRANSFORMERS

1. Primary leads
   If tapped:
      Common — Black
      Tap — Black and yellow striped
      Finish — Black and red striped
2. High-voltage plate winding — Red
      Center tap — Red and yellow striped
3. Rectifier filament winding — Yellow
      Center tap — Green and yellow striped

4. Filament winding No. 1   Green
     Center tap   Green and yellow striped
5. Filament winding No. 2   Brown
     Center tap   Brown and yellow striped
6. Filament winding No. 3   Slate
     Center tap   Slate and yellow striped

AUDIO TRANSFORMERS

Blue	Plate (finish) lead of primary
Red	B + lead (no difference with center tap)
Brown	Plate (start) lead on center-tapped primaries.
	(Blue may be used if polarity is not important.)
Green	Grid (finish) lead to secondary.
Black	Grid return (no difference with center tap.)
Yellow	Grid (start) lead on center-tapped secondaries.
	(Green may be used if polarity is not important.)

IF TRANSFORMERS

Blue	Plate lead
Red	B + lead
Green	Grid (or diode) lead
Black	Grid (or diode) return

Control-Device Designations[1]

BR	Brake relay	MC	Magnetic clutch
CR	Control relay	MN	Manual
CRH	Control relay manual	OL	Overload relay
CRM	Control relay master	PB	Pushbutton
D	Down	R	Reverse
DISC	Disconnect switch	RH	Rheostat
ET	Electron tube	S	Switch
FLS	Flow switch	SOL	Solenoid
FS	Float switch	SS	Selector switch
IOL	Instantaneous overload	T	Transformer
LS	Limit switch	TR	Time delay relay
M	Motor starter	X	Reactor
MB	Magnetic brake		

[1] The Joint Industrial Council (JIC), "Electrical Standards for Industrial Equipment."

Approximate Radii for Aluminum Alloys for 90° Cold Bend*

Alloy and Temper	Radii for Various Thicknesses in Terms of Thickness t				
	$\frac{1}{64}$ in.	$\frac{1}{32}$ in.	$\frac{1}{16}$ in.	$\frac{1}{8}$ in.	$\frac{3}{16}$ in.
1100-0	0	0	0	0	0
1100-H12	0	0	0	0	0–1
1100-H14	0	0	0	0	0–1
1100-H16	0	0	0–1	$\frac{1}{2}$–$1\frac{1}{2}$	1–2
1100-H18	0–1	$\frac{1}{2}$–$1\frac{1}{2}$	1–2	$1\frac{1}{2}$–3	2–4
Alclad 2014-0	0	0–1	0–1	0–1	0–1
Alclad 2014-T3	1–2	$1\frac{1}{2}$–3	2–4	3–5	4–6
Alclad 2014-T4	1–2	$1\frac{1}{2}$–3	2–4	3–5	4–6
Alclad 2014-T6	2–4	3–5	3–5	4–6	5–7
2024-0	0	0–1	0–1	0–1	0–1
2024-T3	$1\frac{1}{2}$–3	2–4	3–5	4–6	4–6
2024-T36	2–4	3–5	4–6	5–7	5–7
2024-T4	$1\frac{1}{2}$–3	2–4	3–5	4–6	4–6
2024-T81	$3\frac{1}{2}$–5	$4\frac{1}{2}$–6	5–7	$6\frac{1}{2}$–8	7–9
2024-T86	4–$5\frac{1}{2}$	5–7	6–8	7–10	8–11
2219-T31			$\frac{1}{2}$–$1\frac{1}{2}$	1–$1\frac{1}{2}$	1–2
2219-T37			$\frac{1}{2}$–2	$1\frac{1}{2}$–3	2–$3\frac{1}{2}$
2219-T81			2–4	$2\frac{1}{2}$–$4\frac{1}{2}$	3–5
2219-T87			2–4	3–5	4–6
3003-0	0	0	0	0	0
3003-H12	0	0	0	0	0–1
3003-H14	0	0	0	0–1	0–1
3003-H16	0–1	0–1		1–2	$1\frac{1}{2}$–3
3003-H18	$\frac{1}{2}$–$1\frac{1}{2}$	1–2	$1\frac{1}{2}$–3	2–4	3–5

* *Alcoa Aluminum Handbook*, by permission of Aluminum Company of America.

Thickness of Wire and Metal-Sheet Gages (Inches)

Gage Number	American or Brown and Sharpe Gage 1*	United States Standard Gage 2†	Gage Number	American or Brown and Sharpe Gage 1*	United States Standard Gage 2†
0	0.3249	0.3125	16	0.0508	0.0625
1	0.2893	0.2813	17	0.0453	0.0563
2	0.2576	0.2656	18	0.0403	0.0500
3	0.2294	0.2500	19	0.0359	0.0438
4	0.2043	0.2344	20	0.0320	0.0375
5	0.1819	0.2188	21	0.0285	0.0344
6	0.1620	0.2031	22	0.0253	0.0313
7	0.1443	0.1875	23	0.0226	0.0281
8	0.1285	0.1719	24	0.0201	0.0250
9	0.1144	0.1563	25	0.0179	0.0219
10	0.1019	0.1406	26	0.0159	0.0188
11	0.0907	0.1250	27	0.0142	0.0172
12	0.0808	0.1094	28	0.0126	0.0156
13	0.0720	0.0938	29	0.0113	0.0141
14	0.0641	0.0781	30	0.0100	0.0125
15	0.0571	0.0703	31	0.0089	0.0109

* For aluminum sheet, rod, and wire. Also for copper wire, and brass, alloy and nickel silver wire and sheet.
† For steel, nickel, and Monel metal sheets.

Metric Conversion Table

Millimeters to Inches				Inches to Millimeters				Inches (Decimals) to Millimeters			
mm	in.	mm	in.	in.	mm	in.	mm	in.	mm	in.	mm
1 = 0.0394		17 = 0.6693		$\frac{1}{32}$ =	0.794	$\frac{17}{32}$ =	13.493	0.01 = 0.254		0.25 =	6.350
2 = 0.0787		18 = 0.7087		$\frac{1}{16}$ =	1.587	$\frac{9}{16}$ =	14.287	0.02 = 0.508		0.26 =	6.604
3 = 0.1181		19 = 0.7480		$\frac{3}{32}$ =	2.381	$\frac{19}{32}$ =	15.081	0.03 = 0.762		0.28 =	7.112
4 = 0.1575		20 = 0.7874		$\frac{1}{8}$ =	3.175	$\frac{5}{8}$ =	15.875	0.04 = 1.016		0.30 =	7.620
5 = 0.1969		21 = 0.8268		$\frac{5}{32}$ =	3.968	$\frac{21}{32}$ =	16.668	0.05 = 1.270		0.32 =	8.128
6 = 0.2362		22 = 0.8662		$\frac{3}{16}$ =	4.762	$\frac{11}{16}$ =	17.462	0.06 = 1.524		0.34 =	8.636
7 = 0.2756		23 = 0.9055		$\frac{7}{32}$ =	5.556	$\frac{23}{32}$ =	18.256	0.07 = 1.778		0.36 =	9.144
8 = 0.3150		24 = 0.9449		$\frac{1}{4}$ =	6.350	$\frac{3}{4}$ =	19.050	0.08 = 2.032		0.38 =	9.652
9 = 0.3543		25 = 0.9843		$\frac{9}{32}$ =	7.144	$\frac{25}{32}$ =	19.843	0.09 = 2.286		0.40 = 10.160	
10 = 0.3937		26 = 1.0236		$\frac{5}{16}$ =	7.937	$\frac{13}{16}$ =	20.637	0.10 = 2.540		0.50 = 12.699	
11 = 0.4331		27 = 1.0630		$\frac{11}{32}$ =	8.731	$\frac{27}{32}$ =	21.431	0.12 = 3.048		0.60 = 15.240	
12 = 0.4724		28 = 1.1024		$\frac{3}{8}$ =	9.525	$\frac{7}{8}$ =	22.225	0.14 = 3.556		0.70 = 17.780	
13 = 0.5118		29 = 1.1418		$\frac{13}{32}$ = 10.319		$\frac{29}{32}$ =	23.018	0.16 = 4.064		0.80 = 20.320	
14 = 0.5512		30 = 1.1811		$\frac{7}{16}$ = 11.112		$\frac{15}{16}$ =	23.812	0.18 = 4.572		0.90 = 22.860	
15 = 0.5906		50 = 1.9685		$\frac{15}{32}$ = 11.906		1 =	25.400	0.20 = 5.080		1.00 = 25.400	
16 = 0.6299		100 = 3.9370		$\frac{1}{2}$ = 12.699		12 =	304.800	0.22 = 5.588		2.00 = 50.800	

Minimum Radius of Conduit* (Inches)

Size of Conduit (Inches)	National Machine Tool Builders' Association	NEC—for Conductors without Lead Sheath	NEC—for Conductors with Lead Sheath
$\frac{1}{2}$	4	4	6
$\frac{3}{4}$	$4\frac{1}{2}$	5	8
1	$5\frac{3}{4}$	6	11
$1\frac{1}{4}$	7	8	14
$1\frac{1}{2}$	$8\frac{1}{4}$	10	16
2	$9\frac{1}{2}$	12	21
$2\frac{1}{2}$	$10\frac{1}{2}$	15	25
3	13	18	31
$3\frac{1}{2}$	15	21	36
4	16	24	40
5	24	30	50
6	30	36	60

* The radius is that of the inner edge. Fittings shall be threaded unless structural difficulties prevent assembly. A run of conduit shall not contain more than the equivalent of 4 quarter bends (360°) total.

Decimal Equivalents of Fractions*

Fraction	Equivalent	Fraction	Equivalent	Fraction	Equivalent	Fraction	Equivalent
$\frac{1}{64}$	0.0156	$\frac{17}{64}$	0.2656	$\frac{33}{64}$	0.5156	$\frac{49}{64}$	0.7656
$\frac{1}{32}$	0.0312	$\frac{9}{32}$	0.2812	$\frac{17}{32}$	0.5312	$\frac{25}{32}$	0.7812
$\frac{3}{64}$	0.0468	$\frac{19}{64}$	0.2968	$\frac{35}{64}$	0.5468	$\frac{51}{64}$	0.7968
$\frac{1}{16}$	0.0625	$\frac{5}{16}$	0.3125	$\frac{9}{16}$	0.5625	$\frac{13}{16}$	0.8125
$\frac{5}{64}$	0.0781	$\frac{21}{64}$	0.3281	$\frac{37}{64}$	0.5781	$\frac{53}{64}$	0.8281
$\frac{3}{32}$	0.0937	$\frac{11}{32}$	0.3437	$\frac{19}{32}$	0.5937	$\frac{27}{32}$	0.8437
$\frac{7}{64}$	0.1093	$\frac{23}{64}$	0.3593	$\frac{39}{64}$	0.6093	$\frac{55}{64}$	0.8593
$\frac{1}{8}$	0.1250	$\frac{3}{8}$	0.3750	$\frac{5}{8}$	0.6250	$\frac{7}{8}$	0.8750
$\frac{9}{64}$	0.1406	$\frac{25}{64}$	0.3906	$\frac{41}{64}$	0.6406	$\frac{57}{64}$	0.8906
$\frac{5}{32}$	0.1562	$\frac{13}{32}$	0.4062	$\frac{21}{32}$	0.6562	$\frac{29}{32}$	0.9062
$\frac{11}{64}$	0.1718	$\frac{27}{64}$	0.4218	$\frac{43}{64}$	0.6718	$\frac{59}{64}$	0.9218
$\frac{3}{16}$	0.1875	$\frac{7}{16}$	0.4375	$\frac{11}{16}$	0.6875	$\frac{15}{16}$	0.9375
$\frac{13}{64}$	0.2031	$\frac{29}{64}$	0.4531	$\frac{45}{64}$	0.7031	$\frac{61}{64}$	0.9531
$\frac{7}{32}$	0.2187	$\frac{15}{32}$	0.4687	$\frac{23}{32}$	0.7187	$\frac{31}{32}$	0.9687
$\frac{15}{64}$	0.2343	$\frac{31}{64}$	0.4843	$\frac{46}{64}$	0.7343	$\frac{63}{64}$	0.9843
$\frac{1}{4}$	0.2500	$\frac{1}{2}$	0.5000	$\frac{3}{4}$	0.7500	1	1.000

* See page 430 for metric equivalents.

Wire Numbers and Sizes

No. AWG* MCM	Stranding†	Diameter (in.)	Area (cmil)	Area‡ (sq in.)
40	Solid	0.0031	10	0.000008
38	Solid	0.0040	16	0.000013
36	Solid	0.0050	25	0.00002
34	Solid	0.0063	40	0.00003
32	Solid	0.0080	64	0.00005
30	Solid	0.0100	100	0.00008
28	Solid	0.0126	159	0.00012
26	Solid	0.0159	253	0.00020
24	Solid	0.0201	404	0.00032
22	Solid	0.0253	640	0.00050
20	Solid	0.0320	1,023	0.00080
18	Solid	0.0403	1,620	0.0013
16	Solid	0.0508	2,580	0.0020
14	Solid	0.0641	4,110	0.0032
12	Solid	0.0808	6,530	0.0051
10	Solid	0.1019	10,380	0.0081
8	Solid	0.1285	16,510	0.0130
6	7	0.184	26,240	0.027
4	7	0.232	41,740	0.042
3	7	0.260	52,620	0.053
2	7	0.292	66,360	0.067
1	19	0.332	83,690	0.087
0	19	0.372	105,600	0.109
00	19	0.418	133,100	0.137
000	19	0.470	167,800	0.173
0000	19	0.528	211,600	0.219
250	37	0.575	250,000	0.260
300	37	0.630	300,000	0.312
350	37	0.681	350,000	0.364
400	37	0.728	400,000	0.416
500	37	0.813	500,000	0.519
600	61	0.893	600,000	0.626
700	61	0.964	700,000	0.730
750	61	0.998	750,000	0.782
800	61	1.030	800,000	0.833
900	61	1.090	900,000	0.933
1000	61	1.150	1,000,000	1.039
1250	91	1.289	1,250,000	1.305
1500	91	1.410	1,500,000	1.561
1750	127	1.526	1,750,000	1.829
2000	127	1.630	2,000,000	2.087

* Wire numbers from 40 to 0000 are AWG, while those from 250 to 2000 are MCM. MCM ≅ 1000-circular-mil area.
† Solid conductors listed can be procured in stranded configurations.
‡ Area given is that of a circle equal to the overall diameter of a stranded conductor.

Small Drills—Metric*

Metric Drill Diameter	Diameter (in.)	Metric Drill Diameter	Diameter (in.)	Metric Drill Diameter	Diameter (in.)	Metric Drill Diameter	Diameter (in.)
0.200	0.0078	2.300	0.0905	5.70	0.2244	9.20	0.3622
0.250	0.0098	2.35	0.0925	5.75	0.2264	9.25	0.3642
0.300	0.0118	2.400	0.0945	5.80	0.2283	9.30	0.3661
0.350	0.0138	2.45	0.0964	5.90	0.2323	9.40	0.3701
0.400	0.0157	2.50	0.0984	6.00	0.2362	9.50	0.3740
0.450	0.0177	2.60	0.1024	6.10	0.2401	9.60	0.3779
0.500	0.0197	2.70	0.1063	6.20	0.2441	9.70	0.3819
0.550	0.0216	2.75	0.1083	6.25	0.2461	9.75	0.3838
0.600	0.0236	2.80	0.1102	6.30	0.2480	9.80	0.3858
0.650	0.0256	2.90	0.1142	6.40	0.2520	9.90	0.3898
0.700	0.0275	3.00	0.1181	6.50	0.2559	10.00	0.3937
0.750	0.0295	3.10	0.1220	6.60	0.2598	10.50	0.4134
0.800	0.0315	3.20	0.1260	6.70	0.2638	11.00	0.4331
0.850	0.0335	3.25	0.1279	6.75	0.2657	11.50	0.4527
0.900	0.0354	3.30	0.1299	6.80	0.2677	12.00	0.4724
0.950	0.0374	3.40	0.1338	6.90	0.2716	12.50	0.4921
1.000	0.0394	3.50	0.1378	7.00	0.2756	13.00	0.5118
1.050	0.0413	3.60	0.1417	7.10	0.2795	13.50	0.5315
1.100	0.0433	3.70	0.1457	7.20	0.2835	14.00	0.5512
1.150	0.0453	3.75	0.1476	7.25	0.2854	14.50	0.5709
1.200	0.0472	3.80	0.1496	7.30	0.2874	15.00	0.5905
1.250	0.0492	3.90	0.1535	7.40	0.2913	15.50	0.6102
1.300	0.0512	4.00	0.1575	7.50	0.2953	16.00	0.6299
1.350	0.0531	4.10	0.1614	7.60	0.2992	16.50	0.6496
1.400	0.0551	4.20	0.1653	7.70	0.3031	17.00	0.6693
1.450	0.0571	4.25	0.1673	7.75	0.3051	17.50	0.6890
1.500	0.0590	4.30	0.1693	7.80	0.3071	18.00	0.7087
1.550	0.0610	4.40	0.1732	7.90	0.3110	18.50	0.7283
1.600	0.0630	4.50	0.1772	8.00	0.3150	19.00	0.7480
1.650	0.0650	4.60	0.1811	8.10	0.3189	19.50	0.7677
1.700	0.0669	4.70	0.1850	8.20	0.3228	20.00	0.7874
1.750	0.0689	4.75	0.1870	8.25	0.3248	20.50	0.8071
1.800	0.0709	4.80	0.1890	8.30	0.3268	21.00	0.8268
1.850	0.0728	4.90	0.1929	8.40	0.3307	21.50	0.8464
1.900	0.0748	5.00	0.1968	8.50	0.3346	22.00	0.8661
1.950	0.0768	5.10	0.2008	8.60	0.3386	22.50	0.8858
2.000	0.0787	5.20	0.2047	8.70	0.3425	23.00	0.9055
2.050	0.0807	5.25	0.2067	8.75	0.3445	23.50	0.9252
2.100	0.0827	5.30	0.2087	8.80	0.3464	24.00	0.9449
2.150	0.0846	5.40	0.2126	8.90	0.3504	24.50	0.9646
2.200	0.0866	5.50	0.2165	9.00	0.3543	25.00	0.9842
2.250	0.0886	5.60	0.2205	9.10	0.3583	25.50	1.0039

* Drills beyond the range of this table increase in diameter by increments of 0.50 mm—26.00, 26.50, 27.00, etc.

Twist-Drill Sizes*

Number Sizes

No. Size	Decimal Equivalent	Metric Equivalent	Closest Metric Drill (mm)	No. Size	Decimal Equivalent	Metric Equivalent	Closest Metric Drill (mm)
1	0.2280	5.791	5.80	41	0.0960	2.438	2.45
2	0.2210	5.613	5.60	42	0.0935	2.362	2.35
3	0.2130	5.410	5.40	43	0.0890	2.261	2.25
4	0.2090	5.309	5.30	44	0.0860	2.184	2.20
5	0.2055	5.220	5.20	45	0.0820	2.083	2.10
6	0.2040	5.182	5.20	46	0.0810	2.057	2.05
7	0.2010	5.105	5.10	47	0.0785	1.994	2.00
8	0.1990	5.055	5.10	48	0.0760	1.930	1.95
9	0.1960	4.978	5.00	49	0.0730	1.854	1.85
10	0.1935	4.915	4.90	50	0.0700	1.778	1.80
11	0.1910	4.851	4.90	51	0.0670	1.702	1.70
12	0.1890	4.801	4.80	52	0.0635	1.613	1.60
13	0.1850	4.699	4.70	53	0.0595	1.511	1.50
14	0.1820	4.623	4.60	54	0.0550	1.397	1.40
15	0.1800	4.572	4.60	55	0.0520	1.321	1.30
16	0.1770	4.496	4.50	56	0.0465	1.181	1.20
17	0.1730	4.394	4.40	57	0.0430	1.092	1.10
18	0.1695	4.305	4.30	58	0.0420	1.067	1.05
19	0.1660	4.216	4.20	59	0.0410	1.041	1.05
20	0.1610	4.089	4.10	60	0.0400	1.016	1.00
21	0.1590	4.039	4.00	61	0.0390	0.991	1.00
22	0.1570	3.988	4.00	62	0.0380	0.965	0.95
23	0.1540	3.912	3.90	63	0.0370	0.940	0.95
24	0.1520	3.861	3.90	64	0.0360	0.914	0.90

Letter Sizes

Size Letter	Decimal Equivalent	Metric Equivalent	Closest Metric Drill (mm)
A	0.234	5.944	5.90
B	0.238	6.045	6.00
C	0.242	6.147	6.10
D	0.246	6.248	6.25
E	0.250	6.350	6.40
F	0.257	6.528	6.50
G	0.261	6.629	6.60
H	0.266	6.756	6.75
I	0.272	6.909	6.90
J	0.277	7.036	7.00
K	0.281	7.137	7.10
L	0.290	7.366	7.40
M	0.295	7.493	7.50
N	0.302	7.671	7.70
O	0.316	8.026	8.00
P	0.323	8.204	8.20
Q	0.332	8.433	8.40
R	0.339	8.611	8.60
S	0.348	8.839	8.80
T	0.358	9.093	9.10
U	0.368	9.347	9.30
V	0.377	9.576	9.60
W	0.386	9.804	9.80
X	0.397	10.084	10.00
Y	0.404	10.262	10.50
Z	0.413	10.491	10.50

No.	Decimal			No.	Decimal		
25	0.1495	3.797	3.80	65	0.0350	0.889	0.90
26	0.1470	3.734	3.75	66	0.0330	0.838	0.85
27	0.1440	3.658	3.70	67	0.0320	0.813	0.80
28	0.1405	3.569	3.60	68	0.0310	0.787	0.80
29	0.1360	3.454	3.50	69	0.0292	0.742	0.75
30	0.1285	3.264	3.25	70	0.0280	0.711	0.70
31	0.1200	3.048	3.00	71	0.0260	0.660	0.65
32	0.1160	2.946	2.90	72	0.0250	0.635	0.65
33	0.1130	2.870	2.90	73	0.0240	0.610	0.60
34	0.1110	2.819	2.80	74	0.0225	0.572	0.55
35	0.1100	2.794	2.80	75	0.0210	0.533	0.55
36	0.1065	2.705	2.70	76	0.0200	0.508	0.50
37	0.1040	2.642	2.60	77	0.0180	0.457	0.45
38	0.1015	2.578	2.60	78	0.0160	0.406	0.40
39	0.0995	2.527	2.50	79	0.0145	0.368	0.35
40	0.0980	2.489	2.50	80	0.0135	0.343	0.35

* Fraction-size drills range in size from one-sixteenth to 4 in. and over in diameter, by sixty-fourths.

Standard Unified Thread Series*

Present Unified Thread Nominal Size—Diameter		Coarse (NC) (UNC)		Fine (NF) (UNF)		Extra-Fine (NEF) (UNEF)	
Inch	Metric Equiv.‡	Threads per Inch	Tap Drill†	Threads per Inch	Tap Drill†	Threads per Inch	Tap Drill†
0.060	0	1.52	—	80	$\frac{3}{64}$	—	—
0.073	1	1.85	64	No. 53	72	No. 53	—
0.086	2	2.18	56	No. 50	64	No. 50	—
0.099	3	2.51	48	No. 47	56	No. 45	—
0.112	4	2.84	40	No. 43	48	No. 42	—
0.125	5	3.17	40	No. 38	44	No. 37	—
0.138	6	3.50	32	No. 36	40	No. 33	—
0.164	8	4.16	32	No. 29	36	No. 29	—
0.190	10	4.83	24	No. 25	32	No. 21	—
0.216	12	5.49	24	No. 16	28	No. 14	32	No. 13
0.250	$\frac{1}{4}$	6.35	20	No. 7	28	No. 3	32	No. 2
0.3125	$\frac{5}{16}$	7.94	18	F	24	I	32	K
0.375	$\frac{3}{8}$	9.52	16	$\frac{5}{16}$	24	Q	32	S
0.4375	$\frac{7}{16}$	11.11	14	U	20	$\frac{25}{64}$	28	Y
0.500	$\frac{1}{2}$	12.70	13	$\frac{27}{64}$	20	$\frac{29}{64}$	28	$\frac{15}{32}$
0.5625	$\frac{9}{16}$	14.29	12	$\frac{31}{64}$	18	$\frac{33}{64}$	24	$\frac{17}{32}$
0.625	$\frac{5}{8}$	15.87	11	$\frac{17}{32}$	18	$\frac{37}{64}$	24	$\frac{19}{32}$
0.6875	$\frac{11}{16}$	17.46	—	—	—	—	24	$\frac{41}{64}$
0.750	$\frac{3}{4}$	19.05	10	$\frac{21}{32}$	16	$\frac{11}{16}$	20	$\frac{45}{64}$
0.8125	$\frac{13}{16}$	20.64	—	—	—	—	20	$\frac{49}{64}$
0.875	$\frac{7}{8}$	22.22	9	$\frac{49}{64}$	14	$\frac{13}{16}$	20	$\frac{53}{64}$
0.9375	$\frac{15}{16}$	23.81	—	—	—	—	20	$\frac{57}{64}$
1.000	1	25.40	8	$\frac{7}{8}$	12	$\frac{59}{64}$	20	$\frac{61}{64}$
1.0625	$1\frac{1}{16}$	26.99	—	—	—	—	18	1
1.125	$1\frac{1}{8}$	28.57	7	$\frac{63}{64}$	12	$1\frac{3}{64}$	18	$1\frac{5}{64}$

Decimal	Fraction	mm ‡	Threads per Inch	Tap Drill †	Threads per Inch	Tap Drill †	Threads per Inch	Tap Drill †
1.1875	$1\frac{3}{16}$	30.16	—	—	—	—	18	$1\frac{9}{64}$
1.250	$1\frac{1}{4}$	31.75	7	$1\frac{7}{64}$	12	$1\frac{11}{64}$	18	$1\frac{13}{64}$
1.3125	$1\frac{5}{16}$	33.34	—	—	—	—	18	$1\frac{17}{64}$
1.375	$1\frac{3}{8}$	34.92	6	$1\frac{13}{64}$	12	$1\frac{19}{64}$	18	$1\frac{5}{16}$
1.4375	$1\frac{7}{16}$	36.51	—	—	—	—	18	$1\frac{3}{8}$
1.500	$1\frac{1}{2}$	38.10	6	$1\frac{21}{64}$	12	$1\frac{27}{64}$	18	$1\frac{29}{64}$
1.5625	$1\frac{9}{16}$	39.69	—	—	—	—	18	$1\frac{1}{2}$
1.625	$1\frac{5}{8}$	41.27	—	—	—	—	18	$1\frac{9}{16}$
1.6875	$1\frac{11}{16}$	42.86	—	—	—	—	18	$1\frac{5}{8}$
1.750	$1\frac{3}{4}$	44.45	5	$1\frac{35}{64}$	—	—	**16**	$1\frac{11}{16}$
2.000	2	50.80	$4\frac{1}{2}$	$1\frac{25}{32}$	—	—	**16**	$1\frac{15}{16}$
2.250	$2\frac{1}{4}$	57.15	$4\frac{1}{2}$	$2\frac{1}{32}$	—	—	—	—
2.500	$2\frac{1}{2}$	63.50	4	$2\frac{1}{4}$	—	—	—	—
2.750	$2\frac{3}{4}$	69.85	4	$2\frac{1}{2}$	—	—	—	—
3.000	3	76.20	4	$2\frac{3}{4}$	—	—	—	—
3.250	$3\frac{1}{4}$	82.55	4	3	—	—	—	—
3.500	$3\frac{1}{2}$	88.90	4	$3\frac{1}{4}$	—	—	—	—
3.750	$3\frac{3}{4}$	95.25	4	$3\frac{1}{2}$	—	—	—	—
4.000	4	101.60	4	$3\frac{3}{4}$	—	—	—	—

* Adapted from ANS B1.1-1960.

Bold type indicates Unified threads. To be designated UNC or UNF.

Unified Standard—Classes 1A, 2A, 3A, 1B, 2B, and 3Ba.

For recommended hole-size limits before threading, see Tables 38 and 39, ANS B1.1-1960.

† Tap drill for a 75% thread (not Unified—American Standard).

Bold-type sizes smaller than $\frac{1}{4}$ in. are accepted for limited applications by the British, but the symbols NC or NF, as applicable, are retained.

‡ The values listed as metric equivalents of decimal inch values have been given to assist user in selecting the closest metric size to be found in the following table, Metric Screw Threads. Adherence to diameter preference is recommended, if feasible. For a metric thread use the tap-drill size recommended in the following table, Metric Screw Threads.

Metric Screw Threads

Nominal Size (mm)	Series with Graded Pitches*				Thread-diameter Preference†		
	Coarse	Tap Drill‡	Fine	Tap Drill‡	1	2	3¶
1.6	0.35	1.25	—	—	1.6	—	—
1.8	0.35	1.45	—	—	—	1.8	—
2	0.4	1.60	—	—	2	—	—
2.2	0.45	1.75	—	—	—	2.2	—
2.5	0.45	2.05	—	—	2.5	—	—
3	0.5	2.50	—	—	3	—	—
3.5	0.6	2.90	—	—	—	3.5	—
4	0.7	3.30	—	—	4	—	—
4.5	0.75	3.75	—	—	—	4.5	—
5	0.8	4.20	—	—	5	—	—
5.5	—	—	—	—	—	—	5.5
6	1	5.00	—	—	6	—	—
7	1	6.00	—	—	—	—	7
8	1.25	6.75	1	7.00	8	—	—
9	1.25	7.75	—	—	—	—	9
10	1.5	8.50	1.25	8.75	10	—	—
11	1.5	9.50	—	—	—	—	11
12	1.75	10.00	1.25	10.50	12	—	—
14	2	12.00	1.5	12.50	14	—	—
15	—	—	—	—	—	—	15
16	2	14.00	1.5	14.50	16	—	—
17	—	—	—	—	—	—	17
18	2.5	15.50	1.5	16.50	—	18	—
20	2.5	17.50	1.5	18.50	20	—	—
22	2.5	19.50	1.5	20.50	—	22	—
24	3	21.00	2	22.00	24	—	—
25	—	—	—	—	—	—	25
26	—	—	—	—	—	—	26
27	3	24.00	2	25.00	—	27	—
28	—	—	—	—	—	—	28
30	3.5	26.50	2	28.00	30	—	—
32	—	—	—	—	—	—	32
33	3.5	29.50	2	31.00	—	33	—
35	—	—	—	—	—	—	35
36	4	32.00	3	33.00	36	—	—
38	—	—	—	—	—	—	38
39	4	35.00	3	36.00	—	39	—
40	—	—	—	—	—	—	40
42	4.5	37.50	3	39.00	42	—	—
45	4.5	40.50	3	42.00	—	45	—

Metric Screw Threads (*Continued*)

Nominal Size (mm)	Series with Graded Pitches*				Thread-diameter Preference†		
	Coarse	Tap Drill‡	Fine	Tap Drill‡	1	2	3¶
48	5	43.00	3	45.00	48	—	—
50	—	—	—	—	—	—	50
52	5	47.00	3	49.00	—	52	—
55	—	—	—	—	—	—	55
56	5.5	50.50	4	52.00	56	—	—
58	—	—	—	—	—	—	58
60	5.5	54.50	4	56.00	—	60	—
62	—	—	—	—	—	—	62
64	6	58.00	4	60.00	64	—	—
65	—	—	—	—	—	—	65
68	6	62.00	4	64.00	—	68	—

* The pitches shown in bold type are those which have been estimated by ISO as a selected coarse- and fine-thread series for commercial threads and fasteners.
† Select thread diameter from columns 1, 2, or 3, with preference for selection being in that order.
‡ For an approximate 75% thread, use the formula: nominal O.D. minus 0.97 × pitch.

Appendix C

Symbols for Electrical and Electronic Devices

This list is abridged from ANS Y32.2, "Graphical Symbols for Electrical and Electronics Diagrams," which is now ANSI/IEEE Y32E, "Electrical and Electronics Graphics Symbols and Reference Designations."

Certain specialized build-up applications of basic symbols are omitted. Where the American National Standard shows both single-line and complete symbol equivalents, this chart shows only the single-line symbol. Consult the basic standard for complete symbols in these cases.

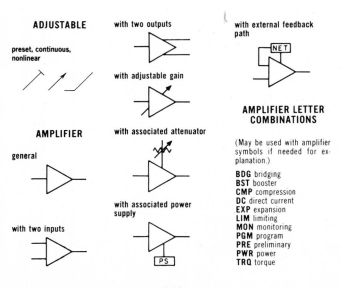

ADJUSTABLE

preset, continuous, nonlinear

AMPLIFIER

general

with two inputs

with two outputs

with adjustable gain

with associated attenuator

with associated power supply

with external feedback path

AMPLIFIER LETTER COMBINATIONS

(May be used with amplifier symbols if needed for explanation.)

BDG bridging
BST booster
CMP compression
DC direct current
EXP expansion
LIM limiting
MON monitoring
PGM program
PRE preliminary
PWR power
TRQ torque

440

ANTENNA

general

OR

dipole

loop

OR

loop antenna (alternate symbol)

counterpoise, antenna

ARRESTER, LIGHTNING

general

carbon block

electrolytic or aluminum cell

horn gap

protective gap

sphere gap

valve or film element

multigap

ATTENUATOR, FIXED

See also **PAD** (same symbols as variable attenuator without adjustment arrow.)

ATTENUATOR VARIABLE

general

balanced

unbalanced

AUDIBLE SIGNALING DEVICE

bell

buzzer

loudspeaker

LOUDSPEAKER LETTER COMBINATIONS

* **HN** horn, electrical
* **HW** howler
* **LS** loudspeaker
* **SN** siren
† **EM** electromagnetic with moving coil
† **EMN** electromagnetic, moving coil and neutralized winding
† **MG** magnetic armature
† **PM** permanent magnet

(Asterisk (*) and dagger (†) are not part of symbol.)

sounder, telegraph

BATTERY

one cell

multicell

multicell with taps

multicell with adjustable tap

CAPACITOR

general

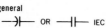   OR   IEC

polarized

OR   IEC

adjustable or variable

adjustable or variable with mechanical linkage

continuously adjustable or variable differential

phase shifter

split stator

feedthrough

CELL, PHOTOSENSITIVE

asymmetrical photoconductive transducer

symmetrical photoconductive transducer

photovoltaic transducer

CIRCUIT BREAKER

general

CIRCUIT ELEMENT

general

LETTER COMBINATIONS FOR CIRCUIT ELEMENTS

(* Asterisk not part of symbol.)
CB circuit breaker
DIAL telephone dial
EQ equalizer
FAX facsimile set
FL filter
FL-BE filter, band elimination
FL-BP filter, band pass
FL-HP filter, high pass
FL-LP filter, low-pass
NET network
PS power supply
RU reproducing unit
RG recording unit
TEL telephone station
TPR teleprinter
TTY teletypewriter

ADDITIONAL LETTER COMBINATIONS

(Specific graphical symbols preferred.)

AR amplifier
AT attenuator
C capacitor
HS handset
I indicating lamp
L inductor
LS loudspeaker
J jack
MIC microphone
OSC oscillator
PAD pad
P plug
HT receiver, headset
K relay
R resistor
S switch
T transformer
WR wall receptacle

GROUND

earth ground

chassis connection

common connections

OR

*

(Identifying marks to denote points tied together shall replace (*) asterisks.)

CLUTCH; BRAKE

clutch disengaged when
operating means
deenergized

OR

clutch engaged when
operating means
deenergized

OR

brake applied when
operating means energized

OR

brake released when
operating means energized

OR

COIL, OPERATING
(RELAY)

(Replace asterisk (*) with
device designation.)

dot shows inner end of
winding

CONNECTION,
MECHANICAL
(INTERLOCK)

with fulcrum

CONNECTOR

female contact

male contact

---

separable connectors
(engaged)

OR

separable connectors
(alternate symbol)

coaxial connector with
outside conductor carried
through

two-conductor switchboard
jack

two-conductor switchboard
plug

female contact
(convenience outlets and
mating connectors)

male contact (convenience
outlets and mating
connectors)

two-conductor nonpolarized
connector with female
contacts

two-conductor polarized
connector with male
contacts

WAVEGUIDE FLANGES

mated (symmetrical)

mated (asymmetrical)

mated (rectangular
waveguide)

CONTACT,
ELECTRICAL

fixed contact for jack, key
or relay

→ OR ⊸ OR ⟶

---

fixed contact for switch
○ OR ⟶

fixed contact for
momentary switch

sleeve
▯ OR ▯ OR ▯

moving contact,
adjustable
→ OR →

moving contact, locking
⊸

moving contact, nonlocking
⊸

segment, bridging contact
▱ OR ▱

vibrator reed
⊸▭

vibrator split reed
⊸▭

rotating contact

closed contact, break
⊁ OR ⊸⟋ OR ⊸⟋

open contact, make
⊥ OR ⊸⟋ OR ⊸⟋

transfer

make-before-break

open contact with time-
closing or time-delay-
closing
TC ⊥ OR ⊥ TDC

closed contact with time-
opening or time-delay-
opening
TO ⊁ OR ⊁ TDO

time-sequential-closing

OR

CORE

air core

NO SYMBOL

magnetic core of inductor
or transformer

core of magnet

COUNTER,
ELECTROMECHANICAL

COUPLER,
DIRECTIONAL

general

E-plane aperture coupling,
30-db loss

loop coupling, 30-db loss
 30DB

probe coupling, 30-db loss
30DB

resistance coupling, 30-db
loss
 30DB

COUPLING

(by aperture of less than
waveguide size)

(Replace asterisk (*) by E,
H or HE depending upon
type of coupling to guided
transmission path.)

DELAY FUNCTION

general

tapped delay

(Replace asterisk (*) with
value of delay.)

DIRECTION OF FLOW

one way

OR

both ways

DISCONTINUITY

equivalent series element

capacitive reactance

inductive reactance

inductance-capacitance circuit, infinite reactance at resonance

inductance-capacitance circuit, zero reactance at resonance

resistance

equivalent shunt element

capacitive susceptance

conductance

inductive susceptance

inductance-capacitance circuit with infinite susceptance at resonance

inductance-capacitance circuit with zero susceptance at resonance

ELECTRON TUBE

directly heated cathode, heater

indirectly heated cathode

cold cathode
(including ionically heated cathode)

photocathode

pool cathode

ionically heated cathode with supplementary heating

grid

deflecting electrode

ignitor

excitor

anode or plate

target or x-ray anode

dynode

composite anode-photocathode

composite anode-cold cathode

composite anode-ionically heated cathode with supplementary heating

shield, within envelope and connected to a terminal

outside envelope of x-ray tube

coupling by loop

resonator, cavity type—single-cavity envelope with grid electrodes

resonator—double cavity envelope with grid electrodes

multicavity magnetron anode and envelope

envelope

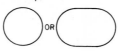

split envelope

gas-filled envelope

basing orientation, tubes with keyed bases

basing, tubes with bayonets, bosses or other reference points

base terminals

envelope terminals

triode with directly heated cathode and envelope connection to base terminal

pentode

twin triode equipotential cathode

cold-cathode voltage regulator

vacuum phototube

multiplier phototube

cathode-ray tube, electrostatic deflection

cathode-ray tube, magnetic deflection

mercury-pool tube with ignitor and control grid

mercury-pool tube with exciter, control grid and holding anode

single-anode pool-type vapor rectifier with ignitor

six-anode metal-tank pool-type rectifier with exciter

resonant magnetron with coaxial output

resonant magnetron with permanent magnet

transit-time magnetron

tunable magnetron

reflex klystron, integral cavity

double-cavity klystron, integral cavity

transmit-receive (t-r) tube

x-ray tube with directly heated cathode and focusing grid

x-ray tube with control grid

x-ray tube with grounded shield

double-focus x-ray tube with rotating anode

x-ray tube with multiple accelerating electrode

FUSE

general

OR

OR

isolating fuse switch

   IEC

OR

high-voltage fuse, oil

OR

GOVERNOR

HALL GENERATOR

HANDSET

HYBRID

general

hybrid junction

H

E

circular hybrid

(Replace asterisk (°) with E, H or HE to denote transverse field.)

INDUCTOR

general

  OR

magnetic-core inductor

tapped inductor

adjustable inductor

continuously adjustable inductor

shunt inductor

KEY, TELEGRAPH

LAMP

ballast tube

fluorescent lamp, two-terminal

fluorescent lamp, four terminal

cold-cathode glow lamp, a-c type

cold-cathode glow lamp, d-c type

incandescent lamp

MACHINE, ROTATING

generator

motor

1-phase

3-phase wye grounded

3-phase wye ungrounded

3-phase delta

MAGNET, PERMANENT

METER

METER LETTER COMBINATIONS

(Replace asterisk (*) with proper letter combination.)

A ammeter
AH ampere-hour
CMA contact-making or breaking ammeter
CMC contact-making or breaking clock
CMV contact-making or breaking voltmeter
CRO cathode-ray oscilloscope
DB decibel meter
DBM decibels referred to one milliwatt
DM demand meter
DTR demand-totalizing relay
F frequency meter
G galvanometer
GD ground detector
I indicating
INT integrating
μA or **UA** microammeter
MA milliammeter
NM noise meter
OHM ohmmeter
OP oil pressure
OSCG oscillograph, string
PH phasemeter

PI position indicator
PF power factor
RD recording demand meter
REC recording
RF reactive factor
SY synchroscope
T temperature
THC thermal converter
TLM telemeter
TT total time
V voltmeter
VA volt-ammeter
VAR varmeter
VARH varhour meter
VI volume indicating
VU standard volume indicating
W wattmeter
WH watthour meter

MICROPHONE

MODE SUPPRESSION

MODE TRANSDUCER

MOTION, MECHANICAL

translation, one direction

translation, both directions

translation, both directions

rotation, one direction

rotation, both directions

NETWORK

OSCILLATOR

OPTICAL COUPLER

PATH, TRANSMISSION

general

wire

two conductors

air or space path

dielectric path other than air

DIEL

crossing of conductors not connected

junction

junction of connected paths, conductors or wires

OR

OR ONLY IF REQUIRED BY SPACE LIMITATION

shielded single-conductor cable

coaxial cable

two-conductor cable

shielded two-conductor cable with shield grounded

twisted conductors

  OR IEC

grouping of leads

PATH, TRANSMISSION

OR

alternate or conditional wiring

associated or future wiring

- - - - -

associated or future equipment (amplifier shown)

circular waveguide

rectangular waveguide

PHASE SHIFTER

general

adjustable

PICKUP HEAD

general

recording

playback

erasing

writing, reading and erasing

stereo

PIEZOELECTRIC CRYSTAL

POLARITY

positive

+

negative

–

RECEIVER, TELEPHONE

general

headset

RECTIFIER

(Represents any method of rectification such as electron tube, solid-state device, electrochemical device, etc.)

general

controlled

bridge type

RELAY

alternating current or ringing

magnetically polarized

slow-operate

slow-release

RELAY LETTER COMBINATIONS

(Not required with specific symbol.)

AC alternating current

D differential
DB double biased
DP dashpot
EP electrically polarized
FO fast operate
FR fast release
MG marginal
NB no bias
NR nonreactive
P magnetically polarized
SA slow operate and slow release
SO slow operate
SR slow release
SW sandwich wound

RESISTOR

general

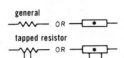 OR

tapped resistor

 OR

tapped resistor with adjustable contact

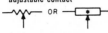 OR

adjustable or continuously adjustable

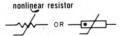 OR

instrument or relay shunt

nonlinear resistor

 OR

symmetrical varistor

 OR

OR

(Replace asterisks (*) with identification of symbol.)

RESONATOR, TUNED CAVITY

ROTARY JOINT

general (Replace asterisk (*) with transmission-path recognition symbol.)

coaxial in rectangular waveguide

circular in rectangular waveguide

SEMICONDUCTOR DEVICES

semiconductor region with one ohmic connection

semiconductor region with plurality of ohmic connections

OR   OR

rectifying junction, P on N region

OR

rectifying junction, N on P region

OR

emitter, P on N region

plurality of P emitters on N region

emitter, N on P region

plurality of N emitters on P region

collector

plurality of collectors

transition between regions of dissimilar conductivity

intrinsic region between regions of dissimilar conductivity

intrinsic region between regions of similar conductivity

intrinsic region between collector and region of dissimilar conductivity

intrinsic region between collector and region of similar conductivity

---

photosensitive

temperature dependent

t°

capacitive device

tunneling device

]

unidirectional

] OR

PNP transistor (actual device and construction of symbol)

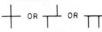

(3)

(3)

PNINIP device (actual device and construction of symbol)

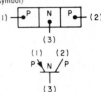

semiconductor diode
(also: rectifier)

OR

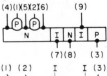

capacitive diode (also: Varicap, varactor, reactance diode, parametric diode)

 OR

breakdown diode, unidirectional
(also: backward diode, avalanche diode, voltage regulator diode, zener diode, voltage reference diode)

 OR

breakdown diode, bidirectional and backward diode
(also: bipolar voltage limiter)

 OR

tunnel diode (also esaki diode)

temperature dependent diode

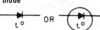

photodiode (also: solar cell)

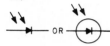

silicon controlled rectifier

PNP transistor (also: junction, point-contact, mesa, epitaxial, planar surface-barrier)

PNP transistor with one electrode connected to envelope

NPN transistor (see other names under PNP transistor)

unijunction transistor, N-type base (also: double-base diode, filamentary transistor)

unijunction transistor, P-type base (see other names above)

field-effect transistor N-type base

    OR

field-effect transistor, P-type base

    OR

semiconductor triode, PNPN switch (also: controlled rectifier)

semiconductor triode, NPNP switch (also: controlled rectifier)

NPN transistor with transverse-biased base

    OR

depletion-type MOSFET, N-channel

depletion-type MOSFET, P-channel

enhancement-type MOSFET, N-channel

enhancement-type MOSFET, P-channel

SHIELD

SQUIB

explosive

igniter

sensing link

SWITCH

single-throw

double-throw

double-pole, double-throw with terminals shown

with horn gap

knife switch

push button, circuit closing (make)

push button, circuit opening (break)

nonlocking; momentary or spring return—circuit closing (make)

OR

nonlocking; momentary or spring return—circuit opening (break)

OR

nonlocking; momentary or spring return—transfer

OR

locking—circuit closing (make)

OR

locking—circuit opening (break)

OR

locking—transfer, three-position

OFF

selector switch

OR

selector, shorting during contact transfer

OR

wafer (example shown: 3-pole, 3-circuit with 2 nonshorting and 1 shorting moving contacts)

safety interlock—circuit opening

safety interlock—circuit closing

SWITCHING FUNCTION

conducting, closed contact (break)

nonconducting, open contact (make)

transfer

    OR

SYNCHRO

general

SYNCHRO LETTER COMBINATIONS

CDX control-differential transmitter

CT control transformer

CX control transmitter

TDR torque-differential receiver

TDX torque-differential transmitter

TR torque receiver

TX torque transmitter

RS resolver

B outer winding rotable in bearings

TERMINATION

cable

open circuit

short circuit

movable short

terminating series
capacitor, path open

terminating series
capacitor, path shorted

terminating series
inductor, path open

terminating series
inductor, path shorted

terminating resistor

series resistor, path open

series resistor, path
shorted

THERMAL ELEMENT

actuating device

OR

thermal cutout

OR

thermal relay with
normally open contact

thermostat (closing on
rising temp.)

thermostat with contact
motion clarified

thermostat with integral
heater and transfer
contacts

THERMISTOR

general

OR

with integral heater

THERMOCOUPLE

general

with integral heater
internally connected

heater

with integral insulated
heater

heater

semiconductor
thermocouple, temperature
measuring

semiconductor
thermocouple, current
measuring

TRANSFORMER

general

OR

transformer with polarity
marks (instantaneous
current in to instantaneous
current out)

OR

one winding with
adjustable inductance

each winding with
adjustable inductance

adjustable mutual inductor

adjustable transformer

current transformer with
polarity marking

OR

bushing type current
transformer

OR

potential transformer

OR

TRANSFORMER
CONNECTION
WINDING

3-phase 3-wire Delta or
mesh

3-phase 3-wire Delta
grounded

3-phase open Delta
grounded at common
point

3-phase wye or star
ungrounded

TERMINAL BOARD
OR STRIP

VIBRATOR

shunt drive

separate drive

VISUAL SIGNALING
DEVICE

annunciator, general

annunciator drop or
signal, shutter type

annunciator drop or
signal, ball type

manually restored drop

electrically restored drop

switchboard-type lamp

indicating lamp

OR   OR

jeweled signal light

INDICATING LIGHT
LETTER
COMBINATIONS

(Replace asterisk (*) with
proper letter combination.)

A amber
B blue
C clear
G green
NE neon
O orange
OP opalescent
P purple
R red
W white
Y yellow

The Relationship of Basic Logic Symbology* between Various Standards

FUNCTION (TWO INPUTS SHOWN WHERE APPLICABLE)	TRUTH TABLE	ANSI Y32.14-1973 IEEE STD. 91-1973		NEMA		ANSI Y32.14-1962² IEEE 91-1962 MIL STD. 806C (NAVY)		MIL STD.² 806B
		RECTANGULAR SHAPE SYMBOLS	DISTINCTIVE SHAPE SYMBOLS	ICS 1-102 RECTANGULAR SHAPE SYMBOLS	ICS-1-103 IS5 PART 11B DISTINCTIVE SHAPE SYMBOLS	UNIFORM SHAPE	DISTINCT SHAPE	
AND	A B Y / 0 0 0 / 0 1 0 / 1 0 0 / 1 1 1	A & Y / B	(distinctive AND shape)	A	(rectangular)	A	(distinctive AND shape)	(MIL 806B AND shape)
OR	A B Y / 0 0 0 / 0 1 1 / 1 0 1 / 1 1 1	≥1	(distinctive OR shape)	OR	(circle)	OR	(distinctive OR shape)	(MIL 806B OR shape)
EXCLUSIVE OR	A B Y / 0 0 0 / 0 1 1 / 1 0 1 / 1 1 0	=1	(distinctive XOR shape)	OE	OE	OE	(distinctive XOR shape)	(MIL 806B XOR shape)
AND INVERT (NEGATED OUTPUTS)	A B Y / 0 0 1 / 0 1 1 / 1 0 1 / 1 1 0	&	(distinctive NAND shape)	A / NAND	(rectangular NAND)	A	(distinctive NAND shape)	
OR INVERT (NEGATED INPUTS)	A B Y / 0 0 1 / 0 1 1 / 1 0 1 / 1 1 0	≥1	(distinctive shape)	OR	(circle)	OR	(distinctive shape)	

* All logic symbols or combinations are not shown here; the respective standards should be consulted for additional symbology and its application.

† Superseded standards.

Function	Truth Table	Symbol	Symbol	Symbol
AND INVERT (NEGATED INPUTS)	A B Y 0 0 1 0 1 0 1 0 0 1 1 0	&	A / NOR	A
OR INVERT (NEGATED OUTPUTS)	A B Y 0 0 1 0 1 0 1 0 0 1 1 0	≥1	OR	OR
NEGATOR (NOT)	A Y 0 1 1 0	1	NOR / A / OR	N / N
ELECTRIC INVERTOR		1		
AMPLIFIER	A Y 0 0 1 1	Δ	AR	AR
OSCILLATOR		G	OSC	OSC

SINGLE SHOT		1Ω	SS	SS 1 0
SCHMITT TRIGGER		⊓	ST	ST 1 0
FLIP-FLOP LATCH		S R Z / S--R	S FF 0 / S FL C 0	FL 1 0 / S FL C 0	S FF 1 C 0
FLIP-FLOP COMPLEMENTARY		S T C / S-T-C	S T FF 0 / S T FF C	FF 1 0 / S T FF C	S T FF 1 C 0
TIME DELAY (ON/OFF- SAME TIME)		(t) / (t)	TD (t) / (t)	TD (t) / (t)	(t)
TIME DELAY (ON/OFF- SAME TIME) ADJUSTABLE		/	TD t_0-t_1 / t_0-t_1	TD t_0-t_1 / t_0-t_1	

TIME DELAY (ON)	TIME DELAY (OFF)	NEGATIVE POLARITY	POSITIVE POLARITY	GENERAL LOGIC SYMBOLS FOR FUNCTIONS NOT ELSEWHERE SPECIFIED *FUNCTION NOTED INSIDE

Symbols for Electrical and Electronic Devices for JIC Orientated Drawings*

SWITCHES

DISCONNECT — DISC

CIRCUIT INTERRUPTER — CI

CIRCUIT BREAKER — CB

LIMIT

NORMALLY OPEN	NORMALLY CLOSED	NEUTRAL POSITION	
LS	LS	LS	ACTUATED — LS
HELD CLOSED	HELD OPEN	NP	NP
LS	LS		

LIMIT (CONTINUED)

MAINTAINED POSITION — LS	PROXIMITY SWITCH	
	CLOSED — PRS	OPEN — PRS

LIQUID LEVEL

NORMALLY OPEN — FS	NORMALLY CLOSED — FS

VACUUM & PRESSURE

NORMALLY OPEN — PS	NORMALLY CLOSED — PS

TEMPERATURE

NORMALLY OPEN — TAS	NORMALLY CLOSED — TAS

FLOW (AIR, WATER ETC.)

NORMALLY OPEN — FLS	NORMALLY CLOSED — FLS

FOOT

NORMALLY OPEN — FTS	NORMALLY CLOSED — FTS

TOGGLE — TGS

CABLE OPERATED (EMERG.) SWITCH — COS

PLUGGING — PLS F, PLS F, R

NON-PLUG — PLS F, R

PLUGGING W/LOCK-OUT COIL — PLS F, LO

SELECTOR

2-POSITION — SS	3-POSITION — SS

ROTARY SELECTOR

† NON-BRIDGING CONTACTS — RSS / OR RSS	† BRIDGING CONTACTS — RSS / OR RSS

† TOTAL CONTACTS TO SUIT NEEDS

THERMOCOUPLE SWITCH — TCS (OFF, 1, 2)

PUSHBUTTONS

SINGLE CIRCUIT	DOUBLE CIRCUIT	MAINTAINED CONTACT
NORMALLY OPEN — PB	PB	PB, PB
NORMALLY CLOSED — PB	MUSHROOM HEAD — PB	

CONNECTIONS, ETC.

CONDUCTORS

NOT CONNECTED	CONNECTED

*Source: Joint Industrial Council (JIC), "Electrical Standards for Mass Production Equipment No. EMP-1–67 and General Purpose Machine Tools EPG-1–67."

CONNECTIONS, ETC. (CONT'D)			CONTACTS						
GROUND	CHASSIS OR FRAME NOT NECESSARILY GROUNDED	PLUG AND RECP.	TIME DELAY AFTER COIL				RELAY, ETC.		THERMAL OVER-LOAD
			ENERGIZED		DE-ENERGIZED		NORMALLY OPEN	NORMALLY CLOSED	
			NORMALLY OPEN	NORMALLY CLOSED	NORMALLY OPEN	NORMALLY CLOSED			
GRD	CH	PL · RECP	TR	TR	TR	TR	CR M CON	CR M CON	OL IOL

COILS							
RELAYS, TIMERS, ETC.	SOLENOIDS, BRAKES, ETC.				THERMAL OVERLOAD ELEMENT	CONTROL CIRCUIT TRANSFORMER	
	GENERAL	2-POSITION HYDRAULIC	3-POSITION PNEUMATIC	2-POSITION LUBRICATION			
CR TR M CON	SOL	SOL 2-H	SOL 3-P	SOL 2-L	OL IOL	H1 H3 H2 H4 T X1 X2	

COILS (CONTINUED)		
AUTO TRANSFORMER	LINEAR VARIABLE DIFFERENTIAL TRANSFORMER	VARIABLE AUTO-TRANSFORMER
AT	LVT	VAT

COILS (CONTINUED)			
SATURABLE TRANSFORMER	REACTORS		
	SATURABLE CORE	IRON CORE	SATURABLE CORE
ST	SX	X	SX

COILS (CONTINUED)			MOTORS	
REACTORS (CONTINUED)			3 PHASE MOTOR	DC MOTOR ARMATURE
ADJUSTABLE IRON CORE	AIR CORE	MAGNETIC AMPLIFIER WINDING		
X	X	MAX	MTR	MTR A

MOTORS (CONT'D)	RESISTORS, CAPACITORS, ETC.				
DC MOTOR FIELD	RESISTOR	HEATING ELEMENT	TAPPED RESISTOR	RHEOSTAT	POTENTIOMETER
FLD	RES	HTR	RES	RH	POT

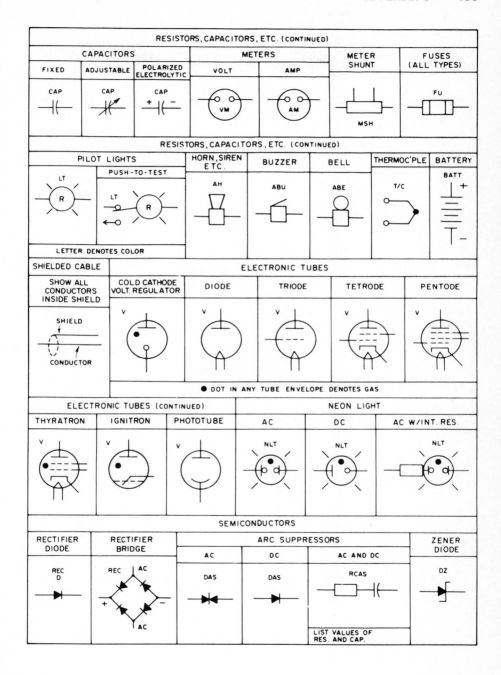

RESISTORS, CAPACITORS, ETC. (CONTINUED)						
CAPACITORS			METERS		METER SHUNT	FUSES (ALL TYPES)
FIXED	ADJUSTABLE	POLARIZED ELECTROLYTIC	VOLT	AMP		
CAP	CAP	CAP	VM	AM	MSH	FU

RESISTORS, CAPACITORS, ETC. (CONTINUED)						
PILOT LIGHTS		HORN, SIREN ETC.	BUZZER	BELL	THERMOC'PLE	BATTERY
LT R	PUSH-TO-TEST LT R	AH	ABU	ABE	T/C	BATT
LETTER DENOTES COLOR					

SHIELDED CABLE	ELECTRONIC TUBES				
SHOW ALL CONDUCTORS INSIDE SHIELD	COLD CATHODE VOLT. REGULATOR	DIODE	TRIODE	TETRODE	PENTODE
SHIELD CONDUCTOR	V	V	V	V	V
	● DOT IN ANY TUBE ENVELOPE DENOTES GAS				

ELECTRONIC TUBES (CONTINUED)			NEON LIGHT		
THYRATRON	IGNITRON	PHOTOTUBE	AC	DC	AC W/INT. RES.
V	V	V	NLT	NLT	NLT

SEMICONDUCTORS					
RECTIFIER DIODE	RECTIFIER BRIDGE	ARC SUPPRESSORS			ZENER DIODE
		AC	DC	AC AND DC	
REC D	REC AC + − AC	DAS	DAS	RCAS	DZ
				LIST VALUES OF RES. AND CAP.	

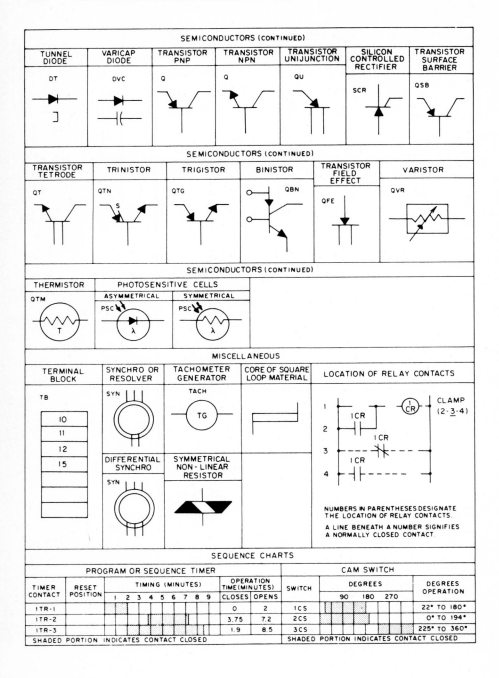

			SEMICONDUCTORS (CONTINUED)			
TUNNEL DIODE	VARICAP DIODE	TRANSISTOR PNP	TRANSISTOR NPN	TRANSISTOR UNIJUNCTION	SILICON CONTROLLED RECTIFIER	TRANSISTOR SURFACE BARRIER
DT	DVC	Q	Q	QU	SCR	QSB

			SEMICONDUCTORS (CONTINUED)		
TRANSISTOR TETRODE	TRINISTOR	TRIGISTOR	BINISTOR	TRANSISTOR FIELD EFFECT	VARISTOR
QT	QTN	QTG	QBN	QFE	QVR

SEMICONDUCTORS (CONTINUED)

THERMISTOR	PHOTOSENSITIVE CELLS	
QTM	ASYMMETRICAL PSC λ	SYMMETRICAL PSC λ

MISCELLANEOUS

TERMINAL BLOCK	SYNCHRO OR RESOLVER	TACHOMETER GENERATOR	CORE OF SQUARE LOOP MATERIAL	LOCATION OF RELAY CONTACTS
TB 10 11 12 15	SYN	TACH TG		
	DIFFERENTIAL SYNCHRO SYN	SYMMETRICAL NON-LINEAR RESISTOR		

NUMBERS IN PARENTHESES DESIGNATE THE LOCATION OF RELAY CONTACTS.

A LINE BENEATH A NUMBER SIGNIFIES A NORMALLY CLOSED CONTACT.

SEQUENCE CHARTS

	PROGRAM OR SEQUENCE TIMER					CAM SWITCH				
TIMER CONTACT	RESET POSITION	TIMING (MINUTES) 1 2 3 4 5 6 7 8 9		OPERATION TIME (MINUTES) CLOSES	OPENS	SWITCH	DEGREES 90	180	270	DEGREES OPERATION
1TR-1				0	2	1CS				22° TO 180°
1TR-2				3.75	7.2	2CS				0° TO 194°
1TR-3				1.9	8.5	3CS				225° TO 360°
SHADED PORTION INDICATES CONTACT CLOSED						SHADED PORTION INDICATES CONTACT CLOSED				

Electrical Symbols for Architectural Drawings*

1·0 Lighting outlets

	Ceiling	Wall	
1·1	◯	—◯	Surface or pendant incandescent mercury vapor or similar lamp fixture
1·2	Ⓡ	—Ⓡ	Recessed incandescent mercury vapor or similar lamp fixture
1·3	▭Ⓞ▭		Surface or pendant individual fluorescent fixture
1·4	▭Ⓞ R▭		Recessed individual fluorescent fixture
1·5	▭Ⓞ▭▭		Surface or pendant continuous-row fluorescent fixture
1·6	▭Ⓞ R▭▭		Recessed continuous-row fluorescent fixture†
1·7	├──┼──┼──┤		Bare-lamp fluorescent strip‡
1·8	Ⓧ	—Ⓧ	Surface or pendant exit light
1·9	ⓇⓍ	—ⓇⓍ	Recessed exit light
1·10	Ⓑ	—Ⓑ	Blanked outlet
1·11	Ⓙ	—Ⓙ	Junction box
1·12	Ⓛ	—Ⓛ	Outlet controlled by low-voltage switching when relay is installed in outlet box

* These symbols are taken from ANS Y32.9–1972, "Graphical Electrical Wiring Symbols for Architectural and Electrical Layout Drawings," published by the American National Standards Institute and sponsored by the Institute of Electrical and Electronics Engineers and the American Society of Mechanical Engineers.

† In the case of combination continuous-row fluorescent and incandescent spotlights, use combinations of the above standard symbols.

‡ In the case of continuous-row bare-lamp fluorescent strip above an area-wide diffusing means, show each fixture run, using the standard symbol; indicate area of diffusing means and type by light shading and/or drawing notation.

2·0 Receptacle outlets

American National Standard C1–1971, National Electric Code (NFPA 70–1978) requires that grounded receptacles be used in most installations. Therefore, when a majority of the receptacles are to be of the grounded type, the ungrounded receptacles should be identified by the notation UNG at the outlet location, and the types of receptacles required noted in the drawing list of symbols and in the specifications.

Where weatherproof, explosionproof, or other specific types of devices are to be required, use the type of uppercase subscript letters referred to under Sec. I3.2.1.2 of this standard. For example, weatherproof single or duplex receptacles would have the uppercase subscript letters noted alongside the symbol (WP, UNGWP).

	Grounded	**Ungrounded**	
2·1		UNG	Single receptacle outlet
2·2		UNG	Duplex receptacle outlet
2·3		UNG	Triplex receptacle outlet
2·4		UNG	Quadruplex receptacle outlet
2·5		UNG	Duplex receptacle outlet, split wired
2·6		UNG	Triplex receptacle outlet, split wired
2·7	*	* UNG	Single special-purpose receptacle outlet*
2·8	*	* UNG	Duplex special-purpose receptacle outlet*
2·9	R	UNG R	Range outlet

* Use numeral or letter either within the symbol or as a subscript alongside the symbol keyed to explanation in the drawing list of symbols to indicate type of receptacle or usage.

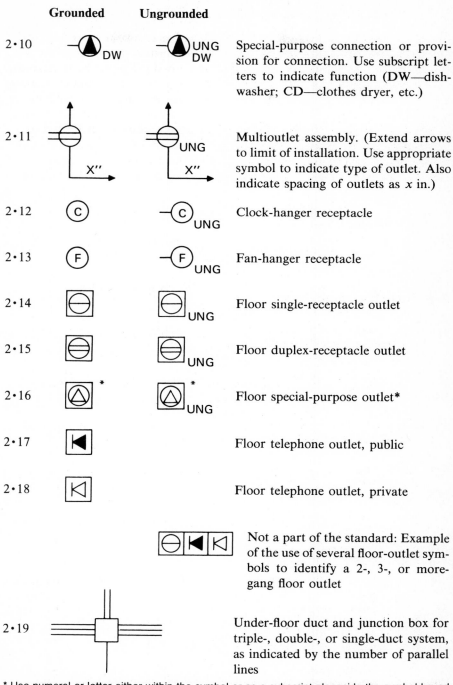

	Grounded	Ungrounded	
2·10			Special-purpose connection or provision for connection. Use subscript letters to indicate function (DW—dishwasher; CD—clothes dryer, etc.)
2·11			Multioutlet assembly. (Extend arrows to limit of installation. Use appropriate symbol to indicate type of outlet. Also indicate spacing of outlets as x in.)
2·12			Clock-hanger receptacle
2·13			Fan-hanger receptacle
2·14			Floor single-receptacle outlet
2·15			Floor duplex-receptacle outlet
2·16			Floor special-purpose outlet*
2·17			Floor telephone outlet, public
2·18			Floor telephone outlet, private
			Not a part of the standard: Example of the use of several floor-outlet symbols to identify a 2-, 3-, or more-gang floor outlet
2·19			Under-floor duct and junction box for triple-, double-, or single-duct system, as indicated by the number of parallel lines

* Use numeral or letter either within the symbol or as a subscript alongside the symbol keyed to explanation in the drawing list of symbols to indicate type of receptacle or usage.

Grounded   **Ungrounded**

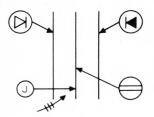

Not a part of the standard: Example of use of various symbols to identify location of different types of outlets or connections for under-floor duct or cellular floor systems

2·20   Cellular floor header duct

3·0 Switch outlets

3·1	S	Single-pole switch

3·9	⊖S	Switch and single receptacle

3·2	S_2	Double-pole switch

3·10	⊖S	Switch and double receptacle

3·3	S_3	Three-way switch

3·11	S_D	Door switch

3·4	S_4	Four-way switch

3·12	S_T	Time switch

3·5	S_K	Key-operated switch

3·13	S_{CB}	Circuit-breaker switch

3·6	S_P	Switch and pilot lamp

3·14	S_{MC}	Momentary contact switch for pushbutton for other than signaling system

3·7	S_L	Switch for low-voltage switching system

3·8	S_{LM}	Master switch for low-voltage switching system

3·15	Ⓢ	Ceiling pull switch

Signaling System Outlets

4·0 Institutional, commercial, and industrial occupancies

Basic symbol	Examples of individual item identification (not a part of the standard)	

4·1

I. Nurse-call-system devices (any type)

Nurses' annunciator (can add a number after it as 24 to indicate number of lamps)

Call station, single cord, pilot light

Call station, double cord, microphone-speaker

Corridor dome light, one lamp

Transformer

Any other item on same system: use numbers as required

4·2

II. Paging-system devices (any type)

Keyboard

Flush annunciator

Two-face annunciator

Any other item on same system: use numbers as required

Basic
symbol

Examples of
individual item
identification
(not a part of
the standard)

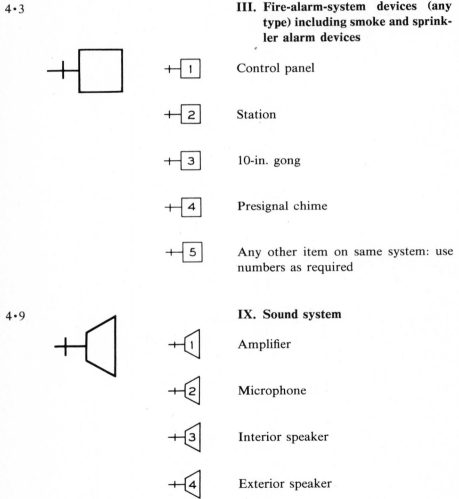

4·3

III. Fire-alarm-system devices (any type) including smoke and sprinkler alarm devices

1   Control panel

2   Station

3   10-in. gong

4   Presignal chime

5   Any other item on same system: use numbers as required

4·9

IX. Sound system

1   Amplifier

2   Microphone

3   Interior speaker

4   Exterior speaker

5   Any other item on same system: use numbers as required

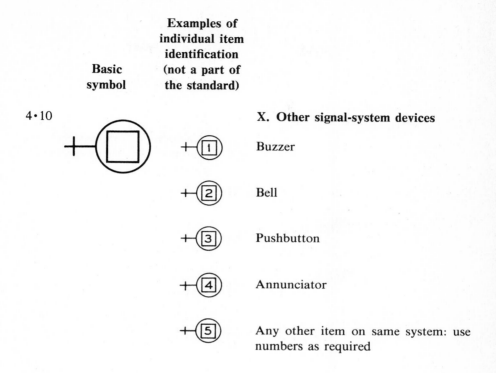

	Basic symbol	Examples of individual item identification (not a part of the standard)

4·10

X. Other signal-system devices

⊢① Buzzer

⊢② Bell

⊢③ Pushbutton

⊢④ Annunciator

⊢⑤ Any other item on same system: use numbers as required

Signaling-System Outlets

5·0 Residential occupancies

Signaling-system symbols for use in identifying standardized residential-type signal-system items on residential drawings where a descriptive symbol list is not included on the drawing. When other signal-system items are to be identified, use the above basic symbols for such items together with a descriptive symbol list.

5·1 ▪ Pushbutton

5·2 Buzzer

5·3 Bell

5·4 Combination bell-buzzer

5·5 CH Chime

5·6 ◇ Annunciator

5·7 D Electric door opener

5·8 M Maid's signal plug

5·9   ☐   Interconnection box

5·10   BT   Bell-ringing transformer

5·11   ◤▶   Outside telephone

5·12   ▷   Interconnecting telephone

5·13   R   Radio outlet

5·14   TV   Television outlet

6·0 Panelboards, switchboards, and related equipment

6·1   Flush-mounted panelboard and cabinet*

6·2   Surface-mounted panelboard and cabinet*

6·3   Switchboard, power-control center, unit substations: should be drawn to scale

6·8   Externally operated disconnection switch*

6·9   Combination controller and disconnection means*

7·0 Bus ducts and wireways

7·1 | T | T | T | Trolley duct*

7·2 | B | B | B | Busway* (service, feeder, or plug-in)*

7·3 | C | C | C | Cable trough ladder or channel*

7·4 | W | W | W | Wireway*

* Identify by notation or schedule.

9·0 Circuiting

Wiring-method identification by notation on drawing or in specifications

9·1 ——————————————— Wiring concealed in ceiling or wall

9·2 —— —— —— —— Wiring concealed in floor

9·3 — — — — — — — — — — Wiring exposed

Note: Use heavy-weight line to identify service and feeders. Indicate empty conduit by notation CO (conduit only).

9·4 ——————————————➤ Branch circuit home run to panel-board. Number of arrows indicates number of circuits. (A numeral at each arrow may be used to identify circuit number.)

Note: Any circuit without further identification indicates two-wire circuit. For a greater number of wires, indicate with cross lines, for example:
—///— three wires; —////— four wires, etc. Unless indicated otherwise, the wire size of the circuit is the minimum size required by the specification.

Identify different functions of wiring system, for example, signaling system, by notation or other means.

9·5 O——————————————— Wiring turned up

9·6 ———————————————● Wiring turned down

10·0 Electric distribution or lighting system, underground

10·1 [M] Manhole*

10·2 [H] Handhole*

* Identify by notation or schedule.

Bibliography

Electrical or Electronics Drawing

Raskhodoff, Nicholas M.: *Electronic Drafting and Design,* 3d ed., Prentice-Hall, Inc., Englewood Cliffs, N.J., 1977.

Richter, Herbert W.: *Electrical and Electronic Drafting,* John Wiley & Sons, Inc., New York, 1977.

Snow, Charles W.: *Electrical Drafting and Design,* Prentice-Hall, Inc., Englewood Cliffs, N.J., 1976.

Engineering Drawing or Graphics

French, Thomas E., Carl L. Svensen, Jay D. Helsel, and Byran Urbanick: *Mechanical Drawing,* 9th ed., McGraw-Hill Book Company, New York, 1980.

―――― and Charles J. Vierck: *Engineering Drawing and Graphic Technology,* 12th ed., McGraw-Hill Book Company, New York, 1978.

―――― and ――――: *Graphic Science and Design,* 3d ed., McGraw-Hill Book Company, New York, 1970.

Giesecke, F. E., A. Mitchell, and H. C. Spencer: *Technical Drawing,* 6th ed., The Macmillan Company, New York, 1974.

Levins, A. S.: *Graphics with an Introduction to Conceptual Design,* John Wiley & Sons, Inc., New York, 1962.

Luzadder, W. J.: *Fundamentals of Engineering Drawing,* Prentice-Hall, Inc., Englewood Cliffs, N.J., 1978.

Miscellaneous, Electronics or Electrical

Bishop Graphics, Inc.: *The Design and Drafting of Printed Circuits,* McGraw-Hill Book Company, New York, 1979.

Chute, George M., and Robert Chute: *Electronics in Industry,* 5th ed., McGraw-Hill Book Company, New York, 1979.

Deboo, G. J., and C. Burrous: *Integrated Circuits and Semiconductor Devices: Theory and Application,* 2d ed., McGraw-Hill Book Company, New York, 1977.

Grinich, Victor, and Horace Jackson: *Introduction to Integrated Circuits,* McGraw-Hill Book Company, New York, 1975.

Hall, Douglas V.: *Microprocessors and Digital Systems,* McGraw-Hill Book Company, New York, 1980.

Hamilton, Douglas, and William Howard: *Basic Integrated Circuit Engineering,* McGraw-Hill Book Company, New York, 1975.

Heumann, G. M.: *Magnetic Control of Industrial Motors,* John Wiley & Sons, Inc., New York, 1954.

Keonjian, Edward: *Microelectronics: Theory, Design, and Fabrication,* McGraw-Hill Book Company, New York, 1963.

Kiver, Milton S.: *Transistor and Integrated Electronics,* 4th ed., McGraw-Hill Book Company, New York, 1972.

Lindsey, Darryl: *The Design and Drafting of Printed Circuits,* Bishop Graphics, Inc., Westlake Village, Calif.

Malvino, A. P.: *Transistor Circuit Approximations,* 3d ed., McGraw-Hill Book Company, New York, 1980.

――― and D. P. Leach: *Digital Principles and Applications,* 2d ed., McGraw-Hill Book Company, New York, 1975.

"National Electrical Code®," NFPA, New York, 1978.

New York Institute of Technology: *A Programmed Course in Basic Pulse Circuits,* McGraw-Hill Book Company, New York, 1977.

Markus, John, and Vin Zeluff: *Handbook of Industrial Electronic Control Circuits,* McGraw-Hill Book Company, New York, 1956.

Ramirez, Edward V., and Melvyn Weiss: *Microprocessing Fundamentals: Hardware and Software,* McGraw-Hill Book Company, New York, 1980.

Slurzberg, Morris, and William Osterheld: *Essentials of Communication Electronics,* 3d ed., McGraw-Hill Book Company, New York, 1973.

Technical Manual and Catalog (for PCB layout), Bishop Graphics, Inc., Westlake Village, Calif.

Index

Abbreviations for drawings, 186, 187, 422–424
Active devices, 229, 234
Aids for drafting (*see* Drafting aids)
Aircraft electrical diagrams, 76–79
Airline diagrams, 81, 82
ALGOL (ALGOrithmic Language), 401
American National Standards, 8–10, 44, 72, 77, 82, 145, 171, 330, 348
Amplifier circuits, 166–170
  [*See also* Schematic (elementary) diagrams]
Analog-computer flow diagrams, 151–153
Appliqués, 53, 108–110, 178–180
Architectural plans, electrical drawing for, 330–357
  drawing symbols, 457–465
  (*See also* specific component)
Artwork, 108–110, 114–115, 119, 398, 399, 409
Assembler, 401
Assembly drawings, 91, 98–100
Automated machine tool, 95–97, 112, 113
Automatic drafting, 122, 123, 398
Automotive wiring diagrams, 73, 75

Balloon drawing, 282–288
Bar charts, 378–380
Baseline diagrams, 77–79, 81, 82
Battery symbol, 44, 45
Bend radii for aluminum, 429
Binary and other number systems, 146
Block diagrams, 141–144, 237–241, 270–272
  (*See also* Flow diagrams)
Branch circuit, 154, 155, 331
Bus ducts, 318, 319, 342

Cabling diagrams, 85, 91
Capacitor symbols, 45, 46, 441, 455
Cathode-ray tube (CRT), 59–61, 402, 403, 406
Cellular floor system, 345

Chassis drawings:
  holes and terminals, 95–97
  layout and construction, 93–98
  photo drawings, 100–102
Chassis manufacture, 93–97
Chassis symbol, 46, 441
Circuit breaker, function of, 253, 313, 332
  air, 305
  oil, 301
Circuit return symbol, 191
Circuits [*See* Schematic (elementary) diagrams]
COBOL (COmmon Business-Oriented Language), 401
Color codes, 73, 75, 80, 426–428
Common connection symbol, 46, 47, 171
Compiler, 401
Computer-aided design (CAD), 398–409
Computer control, 280–282
Computer flow diagrams, 153–155
Computer-output microfilm (COM), 405
Connection diagrams, 72–91
  aircraft, 76–79
  airline-type, 81, 82
  automotive, 75
  cabling, 84–90
  color coding, 73, 75, 80
  with elementary diagram, 84, 85
  harness, 86–90
  highway-type, 77–81
  interconnection diagrams, 77–79, 315–318, 323
  line spacing, 86, 87, 176
  local cabling, 86–89
  pictorial, 74
  point-to-point, 74–79
  straight-line, 83, 84
  wiring (to-and-from) lists, 89, 90
Connections and crossovers, 46, 47, 136, 171
Construction and assembly drawings, 91–100
Contactor function, 253, 255
Contactor symbols, 47, 48, 442, 454
Continuous data, 361, 363

Controls, 252–299
  circuit-breaker function, 253
  contactor function, 253, 255
  control switches, 255
  device designations, 261
  ladder diagram, 262, 276, 278
  overload protection, 253
  power-generating field, 304–312
  reduced-voltage starting, 253
  relay circuits, 286, 287
  relay function, 286
  speed controller, 292, 293
  undervoltage protection, 253
Conversion, 292
Coordination of drawings, 344, 346
Coordinatograph, 2, 113
Coupling, interstage, 167–169
Cryogenic switch, 50
Curves, 361, 362
  equations for, 374, 375
  families of, 369
  smooth, 368, 369
  straight-line, 371, 372

DC (direct coupling), 167–169
Demand-load table, 347
Detail drawings, 320, 321
Device designations, 185–187, 261
Digital-computer flow diagrams, 154, 155
Digital watch circuit, 143, 144
Digitizer, 123, 124, 402, 403, 406
Dimensioning, 95, 96
  chassis drawings, 94–97
  hole locations, 96, 97
Diode symbols, 56, 446–447, 455–456
Disconnect switch, 257, 301–303
Discrete (discontinuous) data, 362
Display console, 402, 406
Drafting aids:
  drafting machines, 3, 113
  mechanical lettering devices, 13, 179
  preprinted symbols, 53, 109, 178–180
  standard grids, 105, 176, 177
  templates, 5–7, 14, 52, 53
Drafting machines, 3, 113
Drill sizes, 95, 97, 433–435
Drilling drawing, 97, 120

Electric power installations, drawings for, 300–322

Electrical layout:
  building, 336, 337, 339
  industrial controls, 266, 267
Electromechanical controls, 260–266
Electron-tube symbols, 57–61, 443–444, 455
Elementary diagrams, 164–251
  [See also Schematic (elementary) diagrams]
Empirical equations, 373–376
Envelope symbols, 59, 61
Epitaxial growth, 214, 215
Etched circuits, 98, 99, 102–111

Feedback, AGC (automatic gain control) circuit, 142
Feeder circuits, 332, 347
Feeder lines, 77–82
Feeder-load table, 347
Film circuits, 229–234
Fixture schedule, electrical, 338
Flow diagrams, 142–163
  analog-computer programs, 152, 153
  automated machine tool, 262
  digital-computer programs, 154, 155
  microprocessors, 155, 156, 237
  (See also Block diagrams)
Foil conductor, copper, widths of, 107, 425
FORTRAN (FORmula TRANslation), 401
Fractions, decimal equivalents, 431
French curve, use of, 368, 369
Frequency spectrum, 425
Functions, logic, 146–150

Gain of amplifier, 152, 153, 165
General arrangement drawing, 313–315, 323
Glossary of electronics and electrical terms, 415–419
Graphical representation of data, 358–382
  abscissa, 361
  bar charts, 378–380
  continuous versus discrete data, 362
  conversion scale, 383
  curve fitting and drawing, 361, 368
  curve identification, 362, 363
  equations for curves, 374, 375
  independent versus dependent variable, 361
  lettering, 365–368

Graphical representation of data—
   (Cont.)
   logarithmic paper, 373–375
   origin or zero point, 359, 364
   paper, types of, 359, 360, 371, 373–377
   pictorial graphs, 382
   pie graph, 381
   plotting symbols, 363, 366
   polar graphs, 377, 378
   principles of, 358
   reproduction for copies, 370
   scales and captions, 366, 367, 369
   selection:
      of scales, 364, 366, 371
      of variables and curve fitting, 361
   semilogarithmic paper, 371, 373, 375,
      376
   steps in construction (layout), 365–369
   variables, dependent and independent,
      361
Graphs (*see* Graphical representation of
   data)
Ground symbol, 43, 46, 191, 441, 454
Grounding, 332, 333

Hardware, 400
Harness diagram, 86–90
Highway diagrams, 77–81
Hybrid circuits, 193, 194

Identification of lines in wiring diagrams,
   77, 80, 82, 83, 84
Inductor symbols, 47, 444
Industrial controls (*see* Controls)
Instrument drawing techniques, 1–6
Instrumentation, 282–291
Integrated semiconductor circuits, 209–
   228
   capacitor, resistor characteristics of,
      229–231
   diffusion steps, 213–216
   drawing sequence, 219–224
   epitaxial growth, 214, 215
   glossary, 213–214
   intraconnection pattern, 222, 226
   masking steps, 216
   packaging, 224–228
   symbols for schematics, 61–63
Interactive graphics, 402
Interconnection diagram, 77–79, 224–
   226, 264, 266
Interstage coupling, 167–169

Inversion, 292
Irregular curve and splines, 368

JIC (Joint Industrial Council) standard
   identification, 453–456

Ladder diagrams, 262, 276, 278
Layout(s):
   component, 98–100, 104, 113, 117
   sheet metal, 91–100
Legends:
   for architectural plans, 338
   for graphs, 363
Lettering, 7–14, 174–176
   alphabets, 9–10
   example of, 12
   on graphs, 362, 367, 369, 381
   guidelines for, 8–12
   mechanical lettering devices, 13, 14
   on parts list, 12, 13
Lighting-load table, 347
Lightning arrester, 302
Line spacing in wiring diagrams, 86
Line weights, typical lines and their uses,
   3
Line work, examples of, 2, 13, 23
Load centers for buildings, 342–344
Load computation, 347, 348
Local cabling in wiring diagrams, 86–90
Logarithmic paper and scales, 373–375
Logic diagrams, 145–151, 269, 270, 271,
   307
   AND, 146, 147
   control circuits, 255–257, 262, 269,
      271, 275, 279, 307
   in electric power field, 307–311
   EXCLUSIVE OR, 147, 149
   flip-flop, 147, 170, 171
   inverter (NOT), 147
   logic functions, 146–150
   NAND, 147, 149
   negative logic, 149
   NOR, 147, 149
   OR, 147, 149
Loop, crossover, 185, 186, 200, 207

Machine-tool control, 264–269
Main service entrance, 342–343
Manufacturing layout and time rate, 93,
   95

Marking drawing, 120, 121
Mechanical linkage, 188, 189, 194
Menu, 403
Metal-oxide semiconductor (MOS) technology:
  definition of, 211, 212
  process steps, 245
  schematic diagram, 222, 223
Microelectronics, 209–251
  film circuits, 229–234
  (See also Integrated semiconductor circuits; Microprocessors)
Microprocessors, 232–243
  applications of, 240, 273
  functional diagram, 236, 239
  glossary of terms, 240
  programming, 242
Monolithic (integrated) circuits, 209–228
Motor control, 252–259
Motor control center, 256–259
  layout, 257–258
  schedule, 258, 259
Motor trip, definition of, 310

National Electrical Code® (NEC), 330–334, 347, 349
  areas of coverage, 334
  definitions used in, 331–334
Negative logic, 149, 150
NEMA (National Electrical Manufacturers Association) Standard, 307, 449–452
NOR circuit, 147
Notes for schematic diagrams, 175, 182, 185–187, 191, 192
Number systems, 146
Numerical control, 93, 96, 97, 112, 113

Occupational Safety and Health Act (OSHA), 330
Office building wiring, 336–337, 339–345
One-line diagram, 301–304
Outlet symbols, 335–337, 340, 343–345
Overcurrent (overload) protection, 253, 255

Panel layout, 258, 259, 264–266, 272
  wiring diagrams, 80, 81, 257, 264–266
Panelboards for buildings, 333, 339, 341, 344

Parts-assembly drawings, 99, 113, 117
Passive devices, 229, 234
Peripheral memories, 404
Photodrawing for chassis assembly, 100–102
Photolithography, 215–219
Pictorial drawing, 29–43
  dimetric projection, 34
  graphs, 382
  isometric drawing, 30, 31, 35
  oblique projection, 32, 33, 343
  perspective drawing, 34
  sections, 35
  types of, 30–34
  wiring diagrams, 29, 74, 91
Picture tube, 60, 61
Pie graph, 381
"Pigtail" leads, 82, 83
Plotters, 399, 404–406, 408
  beltbed, 404–406, 408
  drum, 399, 404
  flatbed, 399, 404, 405
  line, 404
  photoplotters, 405
Plotting symbols, 363, 366
Point-to-point wiring diagrams, 73–76, 83, 118
Polar graphs, 377–378
Power semiconductors, 292–293
Prefixes and units, 185–187
Printed circuit boards, 102–123
  assembly drawing, 99, 101, 113, 117
  board sizes, 114, 120
  component spacing, 106
  conductor spacing, 107, 108
  double-sided layout, 115–121
  drawing sequence, 103–107, 111, 115–122
  master layout, 108, 114, 115, 119
  standard grids, 104, 105
  through connections, 105
Product directories and catalogs, 14–18
Production drawings, 72–139
  construction and assembly drawings, 91–100
  (See also Connection diagrams; Printed circuit boards)
Programmable controllers, 272–279
  block diagram, 273
  description of, 272, 273
  example of, 274–277
  program for, 279
  relays in system, 274

Raceways, 333, 344, 345
Radii for conduit bends, 431
RC (resistance-capacitance) coupling, 167
Receptacles, 74, 76, 333, 336, 343
Rectification, 292
Rectifier symbols, 56, 446, 455
Reduced voltage starting, 253
Reference designations, 175, 181, 185–188
Relay circuits, 255–257
   contact designations, 258–263
   function of, 253, 286, 287
   symbols for, 47, 48, 262, 446, 454
Resistor symbols, 48, 446, 454
Riser diagrams, 339–341, 344
Routing, harness, 89, 90

Schedule:
   lighting fixture, 338
   for panelboard, 259
Schematic (elementary) diagrams, 164–207
   amplifier, push-pull, 168, 170
   basic transistor circuits, 165–170
   biasing arrangements, 165–169
   calculator, 184
   CB receiver, 99
   common-emitter circuit, 165, 167
   coupling, interstage, 167–170
   digital clock, 197
   digital watch, 199
   electric power generation, 309–311
   flasher circuit, 174, 175
   flip-flop circuits, 171, 176, 223
   high-fidelity audio amplifier, 200
   interruption and separation, 189–191
   latch circuit, 223
   layout procedure, 173–177
   linkage (mechanical), 188–189
   NOR function, 172
   notes for drawings, 175, 182, 185–187, 191, 192
   oscillator (satellite) circuit, 207
   preamplifier circuit, 192
   prefixes and units, 186, 187
   principles for preparation, 178–181
   push-pull circuit, 168, 170
   radar display circuit, 116
   radio-receiver circuit, 204, 205
   reference designations, 186, 187
   shunt-regulator circuit, 169
   speed controller, 262, 293

Schematic (elementary) diagrams— (Cont.)
   transistor-radio circuit, 181, 182, 204
   typical patterns, 166–172
   video distribution amplifier, 181, 183
   volt-ohmmeter circuit, 203
   voltage sequence, 181
Semilogarithmic paper, 371, 373, 375, 376
Sheet metal data, 358–360
Sheet metal layouts, 93–97
Silicon controlled rectifier (SCR), 292–294
Single-line diagrams, 301–304
Size of symbols, 50, 51, 176, 177
Software, 400–402
Solid-state logic control, 266–272
Specification writing, 320–323
Speed controller, 262, 293
Stage, definition of, 141
Standards:
   abbreviations for drawings, 420–424
   alphabet, 9, 10
   for architectural drawings, 330, 337, 341, 347
   device (electrical and electronic) symbols, 440–448, 453–456
   instrument symbols and identification, 282–288
   National Electrical Code, 330–334, 347, 349
Straight-line wiring diagram, 83, 84
Straps, 82, 83
Switch symbols, 49, 447, 453
Symbols:
   for analog-computer programming, 153–154
   in block diagrams, 142, 143
   for control diagrams, 255, 269, 278, 282
   for digital-computer programming, 154
   for electrical devices, 44–71, 440–448, 453–456
     (See also specific symbols, e.g., Resistor symbols)
   for logic diagrams, 146–149
   sizes for drawing, 50, 51, 176, 177
   standard:
     for architectural wiring, 457–465
     for electronic devices, 440–448, 453–456

T square, use of, 4, 5
Television circuit, block diagram of, 143, 161

Television tubes, 59–61
Templates:
    for blocks and device symbols, 5–7, 52, 53
    for lettering, 14
    use of, 5–7
Thick-film circuits, 229–232
Thin-film circuits, 229–233
    device tolerances, 229
    materials for, 229
Threads (screw), 95–97, 436–439
Three-line diagram, 304–307, 323
Thyristor, 292, 293
To-and-from diagrams, 90
Transformer coupling, 167
Transformer symbols, 50, 301–303
Transistor symbols, 54, 55, 447, 456
Truth tables, 147–151
Tube symbols, 57–61, 443–444, 455

Under-floor systems, 343–345
Units, prefixes and, 186, 187

Vacuum-tube symbols, 57–71, 443–444, 455

Variables, dependent and independent, 361, 366
Video-tube symbols, 60, 61

Waveforms, 188
Weights of lines, 3
Wire and sheet metal gages, 430
Wiring diagrams, 72–90
    aircraft, 76–79
    airline-type, 81–82
    for architectural plans, 335–340, 342, 345
    automotive, 73, 75
    cabling diagrams, 84–90
    color coding, 75, 77, 80, 426–428
    connection and elementary, 84, 85
    harness, 86–90
    highway-type, 77, 80, 81
    interconnection, 75, 76, 78, 79, 264, 266
    line spacing, 86, 87
    local cabling, 86–90
    panel, 80, 81, 257, 264
    pictorial, 29, 41, 42, 74, 75
    point-to-point, 74–79
    straight-line, 83–84
    wiring (to-and-from) lists, 89, 90